FORTRAN 77
For Engineers

BRUCE J. TORBY

California State University
Long Beach

PRENTICE HALL
Englewood Cliffs, New Jersey 07632

Library of Congress Cataloging-in-Publication Data

Torby, Bruce J.
    FORTRAN 77 for engineers / Bruce J. Torby.
        p.   cm.
    ISBN 0-13-326745-8
    1. Engineering--Data processing. 2. FORTRAN 77 (Computer program
    language)   I. Title.
    TA345.T65   1991
    620'.00285'5133--dc20   90-31875
                                                          CIP

Editorial/production supervision
  and interior design: Bayani Mendoza de Leon
Cover design: Ben Santora
Manufacturing buyer: Lori Bulwin

© 1991 by Prentice-Hall, Inc.
A Division of Simon & Schuster
Englewood Cliffs, New Jersey 07632

The author and publisher of this book have used their best efforts in preparing
this book. These efforts include the development, research, and testing of the
theories and programs to determine their effectiveness. The author and
publisher make no warranty of any kind, expressed or implied, with regard to
these programs or the documentation contained in this book. The author and
publisher shall not be liable in any event for incidental or consequential
damages in connection with, or arising out of, the furnishing, performance, or
use of these programs.

Printed in the United States of America
10 9 8 7 6 5 4 3 2 1

ISBN 0-13-326745-8

Prentice-Hall International (UK) Limited, *London*
Prentice-Hall of Australia Pty. Limited, *Sydney*
Prentice-Hall Canada Inc., *Toronto*
Prentice-Hall Hispanoamericana, S.A., *Mexico*
Prentice-Hall of India Private Limited, *New Delhi*
Prentice-Hall of Japan, Inc., *Tokyo*
Simon & Schuster Asia Pte. Ltd., *Singapore*
Editora Prentice-Hall do Brasil, Ltda., *Rio de Janeiro*

To my wife Birgitta and daughter Linda

Contents

4 Decision Structures   86

5 The DO Loop Structure   136

Preface

Because the current practice of engineering has become so completely reliant upon the use of computers for analysis and design purposes, engineering schools have universally integrated computer usage into their curriculums. The direct intent is not only to teach the student programming skills, but to also let the student see *how* the computer can be used in his or her discipline. Ideally, this introduction to engineering practice should start as early as possible in the student's program to increase student motivation. This FORTRAN 77 textbook was written with the belief that the standard lower-division engineering FORTRAN class is the ideal place to begin this orientation. *FORTRAN 77 for Engineers,* therefore, aims to develop a complete and accurate description of the FORTRAN 77 language, and to introduce within this framework computer applications to fundamental engineering problems.

To meet the two above goals, an extensive collection of sample programs and computer exercises has been included in the text. An average of nine sample programs per chapter appears. The problems have been selected very carefully so that the student might learn by example, as well as by doing. They not only demonstrate engineering computer usage; they also illustrate thoroughly all the structures, grammar and features of the FORTRAN language. At the same time, the problems have been designed to keep pace with the text material, progressing from basic to advanced language use. They do not require an advanced engineering background, and all the needed theory and principles are self-contained within each problem description. Since a great many FORTRAN courses are taught to engineering majors by the Mathematics or Computer Science Departments, this book will prove especially helpful to non-engineering faculty in introducing engineering applications to their classes while yet meeting their rigorous requirements for a FORTRAN programming text.

To show the student that in computer programming many alternative approaches to the same problem are possible, and to provide for a thread of continuity and evolution from basic programming to advanced skills, one specific program is repeated in each chapter. The solution for the problem is updated in each chapter to demonstrate the advantages of its newly introduced language elements. Nine different complete solutions are given to the same problem within the book at various levels.

The text also emphasizes the development of problem solving skills in general. The importance of algorithm construction using a top-down design approach is repeatedly stressed. Many sample problems have their solution algorithms given in pseudocode. Several problems have flow charts included. Structured programming principles are rigorously followed throughout the text, and the advantages of using subprograms for program readability, expansions, and maintenance are clearly set forth.

In writing the book, the beginning student has been kept in mind. The completed manuscript was used in the classroom and material was consequently added to enhance the text's clarity and provide answers to questions commonly asked by students.

CHAPTER SUMMARIES

Chapter 1: Computer Fundamentals

Chapter 1 brings the reader into the engineering world of digital computers. Many new terms and concepts are defined. The parts of a computer are presented, and a description of binary code and machine language is given. Assembly and high-level languages are then discussed, and FORTRAN 77 is introduced. The chapter describes how the engineer uses FORTRAN in practice to create new software, or maintain and append existing programs. Most commercial engineering software packages are written in FORTRAN. Program development, problem solving and the importance of proper algorithm design and program documentation are stressed. Top-down design technique is introduced. A sample problem is given with its solution algorithm, flowchart, and program shown. This program is later explained extensively from a beginner's perspective. The reader is then walked through the steps of entering the source program at the keyboard, compiling the source code, and running the object program at his or her local terminal. The last part of the chapter covers time sharing, batch processing, and local system commands.

Chapter 2: FORTRAN Basics

Chapter 2 introduces the six types of FORTRAN data: INTEGER, REAL, DOUBLE PRECISION, COMPLEX, LOGICAL, and CHARACTER. How data is represented in binary form in the base-2 number system, or with ASCII or EBCDIC code, is briefly described. REAL (floating-point) and INTEGER type numeric constants are discussed. Exponential notation is explained. The concept of a FORTRAN variable, as well as the meaning of the FORTRAN equal sign, are then presented. The rules for assigning FORTRAN variable names, and implicit and explicit typing are set down. Specification statements are introduced. The proper program order for all nonexecutable and executable statements is shown in a separate table.

FORTRAN arithmetic operations, arithmetic expressions, and assignment statements are then described. The order in which arithmetic operators and expressions are processed is outlined. Mixed-mode arithmetic is introduced. Possible truncation effects brought about by division of an integer value by another fixed-point value or by assignment of a floating-point value to an integer variable address are discussed. Several commonly used intrinsic functions provided by FORTRAN are listed in this chapter. A complete set is given in the Appendix. The last topics of the chapter discuss the DATA statement, data files and the OPEN and REWIND statements.

Chapter 3: Input/Output

In Chapter 3, the proper statement syntax for control of input/output data format is investigated. Both list-directed and formatted input/output are discussed. Input/output operations described are from or to the terminal, or from or to an external memory device. The chapter's READ, PRINT, and WRITE statements list the variables to be processed, specifying the input/output unit to be used, and refer to the format edit control in effect. The FORMAT statements with its integer format identifier and FORMAT field specifiers is covered in great detail. The descriptors discussed in this chapter are: X, T, TL, TR, /, I, F, and E. Assigning the proper field widths for the I, F, and E descriptors is stressed, and right or left-justification of values entered from the field is described. Page control and vertical spacing for line printers, as well as the repeated edit descriptors, are mentioned.

Chapter 4: Decision Structures

Chapter 4 introduces FORTRAN IF-block decision and repetition structure. Relational and logical operators are listed, and logical constants are defined. Relational and simple and compound logical expressions are described. The block IF statement and the IF-THEN-END IF decision block structure are described. Nesting is discussed. IF-THEN-ELSE-END IF structure and the IF-THEN-ELSE IF-END IF construct are also covered. The chapter emphasizes structured programming forms.

The simulated WHILE loop, another IF type construct which offers the user a program module effecting repetition rather than path selection, is outlined. The chapter ends with a discussion of older FORTRAN usage. The arithmetic IF, GO TO, computed GO TO, assigned GO TO, and ASSIGN statements are briefly mentioned. Since they alter general top-down program flow, their usage is not recommended.

Chapter 5: DO Loop Structure

The work begun on WHILE repetition structure in the last chapter is continued in Chapter 5. DO loops are introduced. Nesting and transfer to and from a DO loop covered. Many sample problems are given to emphasize the DO loop's capability and its essential importance to FORTRAN programming.

Chapter 6: Arrays

Chapter 6 begins with a discussion of one-dimensional arrays. DIMENSION statements and type declaration are described. Input/output operations, and DATA statement, the short-list technique, and implied DO lists, are discussed in great detail. This material coverage is then repeated for two-dimensional arrays. Dimensioning, typing, input/output operations, the short-list technique, and implied-DO lists are all covered once more. Three-dimensional arrays are mentioned briefly. The chapter's last topic covers matrix algebra. Row, column, symmetric, diagonal, null, and identity matrices are defined. Matrix transposition, addition, subtraction, and multiplication are demonstrated.

Chapter 7: Functions and Subroutines

Chapter 7 introduces subprograms. The important role they play in proper top-down program design, program readability, and especially program maintenance is emphasized. Algorithm and program development are reiterated, and the use of stubs and drivers is described. The creation of a permanent subprogram library by the user to supplement the FORTRAN-supplied intrinsic function library is recommended. The chapter discusses fully the advantages and disadvantages of statement function, FUNCTION, and SUBROUTINE subprogram use. It describes the difference between formal or dummy arguments and actual arguments. The FUNCTION, RETURN, SUBROUTINE, and CALL statements are presented. Array dimensioning within a subprogram is treated, and adjustable dimensions are introduced. The advantages of using common blocks of storage are set forth, followed by descriptions of the blank and named COMMON statements. The use of the block data subprogram and the BLOCK DATA statement is also reviewed. The last topic of the chapter covers the EXTERNAL and INTRINSIC statements.

Chapter 8: Other Data Type

Chapter 8 treats the four remaining FORTRAN data types: CHARACTER, LOGICAL, DOUBLE PRECISION, and COMPLEX. Character constants and variables are introduced first. The CHARACTER type statement is presented. Character arrays are also included in the section's discussion. Character operations are discussed in this chapter, and substrings and character expressions are defined. Concatenation, the collating sequence, and comparison of character values are all described. The intrinsic functions INDEX, LEN, ICHAR, CHAR, LGE, LGT, LLE, and LLT are explained. Character data input/output operations are reviewed for both list-directed and formatted editing. The A descriptor is introduced. The rules for assigning field widths and for left and right-justification of data on input/output within the field are presented.

The chapter focuses next on logical constants, variables, and simple and compound logical expressions. The LOGICAL type statement is given. Since logical variables are usually employed as flags within a program, a separate chapter section is devoted solely to this topic. Logical data input/output operations are reviewed for both list-directed and formatted fields.

Double-precision constants and variables are then presented. The exponent D substitutes for E in double-precision representation. A general discussion of precision, word size, and roundoff error is provided, and the DOUBLE PRECISION type statement is introduced. Double-precision mathematical operations are reviewed and the common double-precision intrinsic functions from the FORTRAN library, e.g., DCOS, DSIN, etc., are reintroduced. The special intrinsic functions DBLE, DPROD, REAL and SNGL are also explained. Input/output operations for double-precision data are discussed for both list-directed and formatted edit control. The D descriptor is given.

Complex data is the last topic of this chapter. To introduce the subject, complex algebra is first reviewed. Sums, differences, products, quotients, and exponential forms of complex numbers are given. Complex series representation is described. FORTRAN complex constants and variables are defined. The COMPLEX type statement is listed. Complex operations are then discussed and the common complex intrinsic functions CABS, CLOG, CSQRT, CSIN, CEXP, and CCOS, are given. The special functions AIMAG, CMPLX, CONJG and REAL are also explained. Both list-directed and formatted input/output of complex data are described.

Chapter 9: File Management

Sequential and random-access files are introduced in Chapter 9. The difference between formatted and unformatted files is explained. Procedures for opening and closing a file are discussed in detail. The full syntax for the OPEN and CLOSE statements is spelled out here and the statement's parameters are explained. The INQUIRE statement and its specifiers are also discussed. The REWIND, BACK-SPACE, and ENDFILE commands are given for file positioning. The complete syntax of FORTRAN READ and WRITE statements is provided. After discussing program internal files, the chapter ends with coverage of direct-access files. Data record length and record number specifiers must be given when these files are used.

Chapter 10: Special Topics

Chapter 10 reviews the less often used commands and descriptors that are part of the FORTRAN 77 language. The IMPLICIT and EQUIVALENCE specification statements are introduced first. Then the descriptors G, Ee, P, BN, BZ, S, SP, SS, and H are discussed. Mention is made of the PAUSE control statement. New subprogram statements found in this chapter are the ENTRY and the alternative RETURN n.

ACKNOWLEDGMENTS

There are many people who assisted with this book. I would like to express my appreciation to all of them. I would like to thank my introductory FORTRAN programming classes at the California State University at Long Beach for their constructive criticisms and for testing out the review questions and computer projects. I would also like to thank the people at Microsoft Corporation who provided FORTRAN software to write the book's Solutions Manual.

Thanks and appreciation go to Ms. Patricia Baker, Mr. Edward Kraus, and to Mr. Jack Savadjan, all former students, for their valued contributions. Ms. Baker proofread and critically examined the original manuscript, and Mr. Savadjan helped with the book's production and co-authored its Solutions Manual.

Much appreciation also goes to my immediate family to whom the book is dedicated. My wife, Birgitta Torby, typed the manuscript during two summer vacations. Fortunately for me, it rained a lot both those summers. My young daughter, Linda, deserves special recognition for her tolerant acceptance of having a father who was always too busy writing. I hope to take the time now to make up for those missed moments.

I would also like to acknowledge the reviewers selected by Prentice Hall, Dr. Richard Albright, University of Delaware, and Dr. Jesa H. Kreimer, California State University at Fullerton, and those faculty who helped in the review process and who offered suggestions for the manuscript's improvement. Special recognition go to Dr. Edward Miller of CSULB for his helpful suggestions, and to the people at Prentice Hall for their excellent and professional work in the book's production. Final mention and a note of personal thanks go to Mrs. Gerry Johnson of Prentice Hall for her support and belief in the project, and for the encouragement she offered when the work appeared hopelessly stalled.

Bruce J. Torby

CHAPTER 1

Computer Fundamentals

Models of the Mirror Fusion Test Facility magnet cases. Supercomputers are required for the design of the magnets that are used in experiments to compress ionized plasmas to the fusion temperature of the sun. If the procss can be sustained and controlled, a source of energy will be made available to mankind that is safe, inexhaustible, and nonpolluting. *(Photo courtesy of Lawrence Livermore National Laboratory.)*

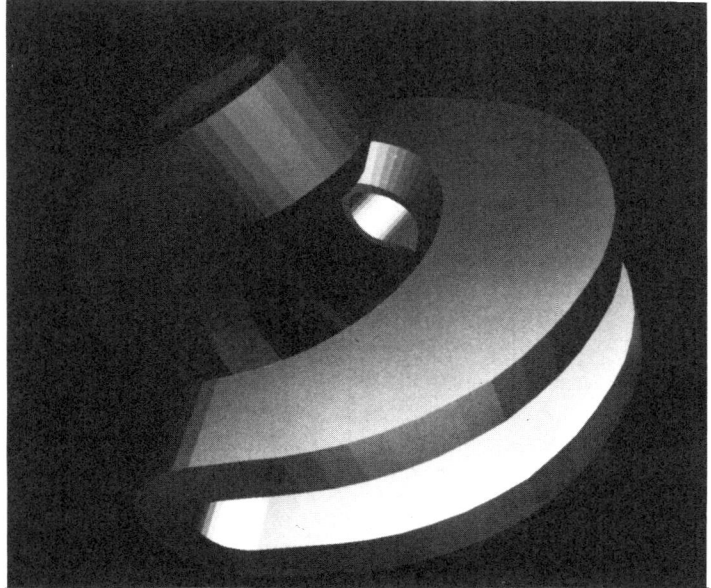

1.1 INTRODUCTION

We are in the midst of a revolution. It is a revolution so vast and significant that its ultimate effect on human evolution can barely be imagined. The industrial revolution, beginning with the invention of Watt's steam engine, led to a nearly unlimited extension of our physical selves. Humans discovered how to lift huge weights, turn great loads, dig cavernous tunnels, and erect skyscrapers. They learned to change night into day with electricity. They could even fly. Humans had transcended themselves physically by having machines perform those tasks that were burdensome, repetitive, or tedious, or those that their own fragile bodies could not endure.

The current *computer revolution,* on the other hand, now extends our mental selves. By performing repetitive arithmetic and logical operations at great speeds and with great accuracy — never tiring, always alert, and having the capacity to store (memorize) huge amounts of data — the digital computer is an electronic tool that greatly amplifies our mental powers. Its very attributes complement our thought processes, providing strengths where we are the weakest. Just as machines a century ago freed our bodies from repetitive and toilsome labor, the computer now unfetters our minds. The staggering collective release of mental energy we are just beginning to witness in this century harbors the potential for unlimited human evolvement and growth.

Today, computers are everywhere in our society. Bankers use them to help keep their financial records. Businessmen call upon them to assist in planning their financial investments. Biologists decipher genetic code, doctors track disease, and weathermen make their forecasts, all with the aid of computers. Computers help scientists discover the secrets of the stars and the basic building blocks of the atom. Engineers, one of the largest user groups, employ them in all aspects of their work. Without computers, we could not build satellites or spaceships, or control supersonic aircraft. Designs, analyses, and solutions that were impossible just three decades ago have become feasible today. For example, with the aid of finite element methods (FEM) and the computer, engineers can now design earthquake-proof structures. With the support of computer aided drawing, engineering, and manufacturing computer programs, (CAD/CAE/CAM), they can construct stronger, lighter, and vastly more efficient machines — machines that were impossible to even conceptualize before. Electrical and computer engineers now even have computers themselves design and manufacture other computers.

1.2 THE PARTS OF A COMPUTER

Digital computers may vary in size from the tiniest **microprocessor,** no larger than a fingernail, to a room-size **mainframe** computer. Depending on their speed and memory capacity, they may be classified as **microcomputers** (personal computers, or PCs), **minicomputers,** mainframes, or **supercomputers.** (see Figures 1.1 through 1.5). Any computer system must, however, incorporate the six basic elements shown in Figure 1.6: an input unit, an output unit, a processing unit, an arithmetic logic unit, and internal and external memories.

The computer "sees" through the first element, an **input unit** or device. Since the digital computer only operates on pulsed electronic information, some means must first be provided to translate real-world information into pulsed format. The most direct line of input is from a keyboard terminal, with its associated cathode-ray tube (CRT) or monitor display. The keyboard itself is a typewriterlike device that translates keystrokes of alphabetic or numeric characters (**alphanumeric** or **character data**) to electronic pulses for computer processing. Other input means carry already coded pulsed information to the computer for subsequent conversion to electronic signals at the input unit interface. This may be accomplished optically, via punched holes in punched paper tape or punched cards; magnetically, via magnetically stored information on magnetic tape or magnetic hard or floppy (soft) disks; or acoustically, via sound tones carried on telephone lines (a device called a modem is

Figure 1.1 An engineer plots a magnified schematic of a VLS1 micro-processor. The chip is 9.1×8.8 mm in actual size. *(Photo courtesy of Bull HN Information Systems, Incorporated.)*

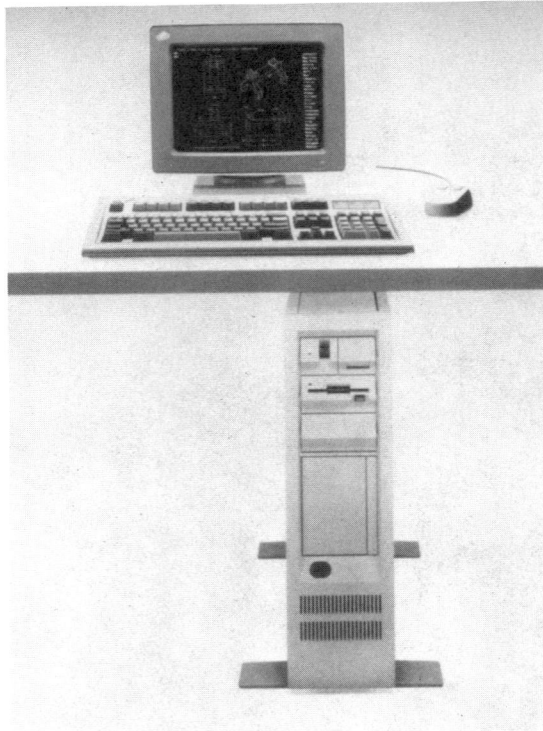

Figure 1.2 The IBM Personal System/2® Model 80 personal computer. *(Photo courtesy of IBM.)*

used to translate the information into electronic pulses). Voltage signals that are amplitude varying (for example, from measuring instruments) can first be quantified into pulsed format for the computer by using special electronic hardware called analog-to-digital converters (ADC).

The computer communicates with the outside world via an **output unit** or device. To send written messages it uses either a printer, a plotting device (plotter), a typewriter, or the CRT monitor. It can also write on magnetic media such as disks or tape, and on punched paper cards or paper tape. The digital signals it outputs may also be transmitted by telephone to another computer via a modem. Shaped as a voltage signal, digital-to-analog conversion outputs can be used directly to control a physical event, such as the motion of an airplane aileron or the cutting tool of a shop mill.

The "brain" of a computer system is the **central processing unit (CPU).** It consists of three elements (refer to Figure 1.6 again). The first part of the CPU oversees computer operation. Some of the tasks for which it is responsible are keeping track of arithmetic operations and signal timing, managing the storage and transference of intermediate results, decoding and executing commands that are taken in

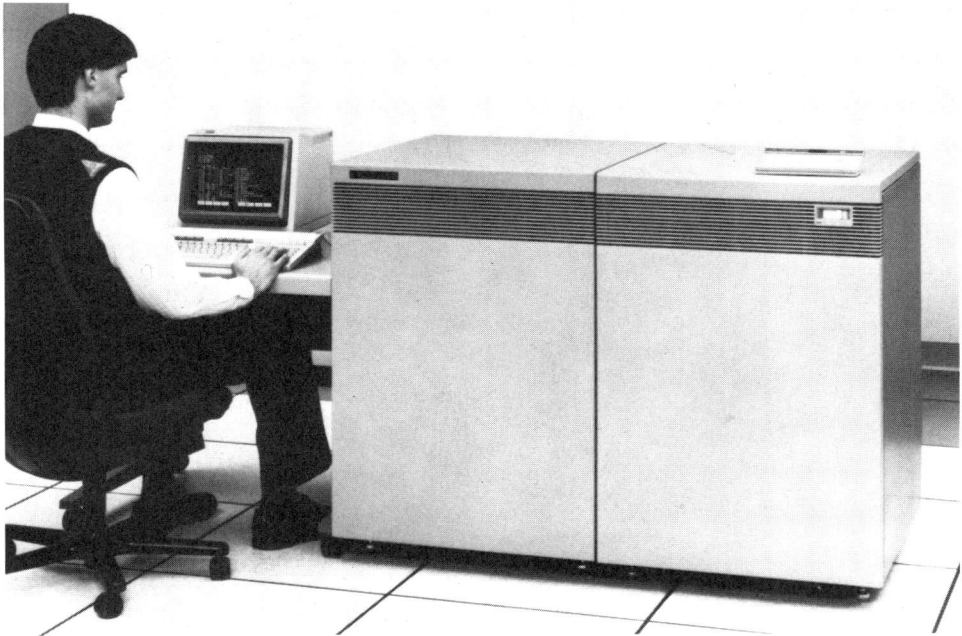

Figure 1.3   The HP9000 Model 850S superminicomputer. Model 850 can perform 7 million instructions per second. *(Photo courtesy of Hewlett-Packard Company.)*

Figure 1.4   Bull's DSP 88/82 mainframe computer. It has 128 million bytes of main memory. *(Photo courtesy of Bull HN Information Systems, Incorporated.)*

Figure 1.5 The National Magnetic Fusion Energy Computer Center at Lawrence Livermore National Laboratory houses one of the world's most powerful arrays of supercomputers. Pictured are the CRAY X-MP/22, the CRAY-1S, and the CRAY-1. *(Photo courtesy of Lawrence Livermore National Laboratory.)*

prescribed programmed order from internal computer memory, and controlling input and output operations. It should be noted that today's generation of computers execute their instructions in a **serial** manner, that is, tasks are performed in consecutive sequence. **Parallel** processors, on the other hand, perform operations simultaneously. Computers that employ parallel processing offer significantly faster processing times to the user than those obtained from serial machines. They require, however, different programming and problem-solving strategies which are more suitable to its parallel operation. Parallel machines will increasingly be in evidence in the workplace.

The **arithmetic logic unit (ALU),** also part of the CPU, performs the arithmetic and logical operations for the computer. Noteworthy here is the fact that the basic arithmetic operations of addition, subtraction, multiplication, and division are all performed in the ALU by simple addition (both negative and positive) of the variables. A supplementary counter is used to effect multiplication or division. Logical operations, such as comparison of variable numeric values (for example, greater than, equal to, or less than) are also carried out by simple addition.

The third part of the CPU is **internal memory.** Most of this memory is transient, and its contents will disappear when the user, finished with his or her session, disconnects (logs off) or turns off the power. Internal memory is used to temporarily store the program that the user wants to run and the data to be operated on. Each

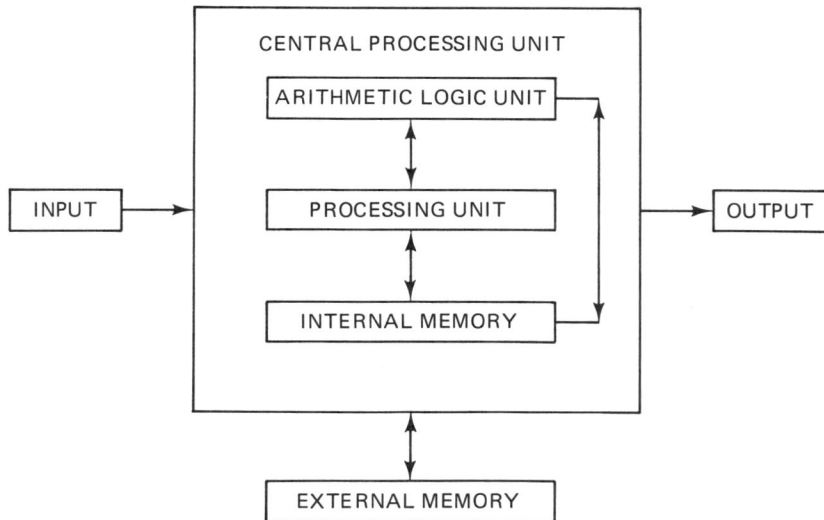

Figure 1.6

user will likely work with a different program; therefore, to control its size, memory must be erasable.

Auxiliary external memory, the sixth and last component of the computer, provides "permanent" storage for the program in a **file.** External memory usually consists of a magnetic medium, such as a tape or a disk, that is accessed by the computer's **input** or **output (I/O)** device when reading, or storing information.

1.3 LANGUAGES

1.3.1 Machine Language

One of the strongest arguments in favor of digital computers is that they are binary in nature; that is, they operate on electronic signals that are either present (the "true" or "1" state) or missing (the "false" or "0" state), similar to the controlling information presented by a solenoid or a punched card. These signals, a voltage, when present exceed a small threshold voltage, to discriminate against background noise. This means that thermal drift of electronic components or electrical signal noise levels do not have to be considered or designed around. As a consequence, the need for expensive and accurate electronic equipment to duplicate or measure voltage levels is eliminated.

In a computer these **binary signals,** or **bits** of control information, are assembled into standard-length strings called **bytes.** One byte is eight bits long. In turn, these bytes form words. Depending on the computer, these words can be eight, sixteen, or thirty-two bits long and carry data or instructions for the computer to execute. A sequence of instructions to be executed is called a **program.** If the program is written directly in binary code, using the language of the CPU, it is said to be written in **machine language.** In the early days of computing, when memory was very limited and costly (each binary bit was stored in the magnetic field of a tiny individually hand-wired, doughnut-shaped magnetic core), all programs were written in machine language. Today, machine language is mainly used to program microprocessor systems used in control applications. As one can imagine, machine language is difficult to use. Besides being tedious, machine language programming tends to introduce errors. Finding these errors in computer programs (program errors are called **bugs;** clearing errors is called **debugging**) is not very easy in binary code. All that the user sees in front of him or her is a series of 1s and 0s. (For further discussion of binary data representation, see Appendix A.)

1.3.2 High-Level Languages

Because of the complexity of programming in machine language, **assembly** and **high-level languages** were soon introduced to the user community. An assembly language program is written using alphanumeric characters. A stored program in internal memory, called an **assembler,** translates the entered program into a series of machine language instructions.

High-level languages are even more powerful and easier to use, since they are written and constructed with "English" words. The programs written in these languages are also translated by the computer into machine language for execution purposes. The common high-level languages used today are BASIC (a microcomputer language), COBOL (used for business applications), FORTRAN (used in engineering and science), and C, PL/I, ALGOL, and Pascal (powerful structured general-purpose languages). Some languages that will become important in the near future are ADA (a U.S.-government-sponsored structured language), and LISP and PROLOG (for work with machine artificial intelligence).

Of all the computer languages available today, engineers and scientists use **BASIC** and **FORTRAN** the most. BASIC is a simple language to learn, but because it translates each program statement during every execution of the program, it is much slower than other languages. FORTRAN (standing for *for*mula *tran*slation), on the other hand, first compiles the user's program into a machine language version, which is then available for the user to save on a permanent storage medium. This machine language program is then executed. Much time is saved when reruns are attempted, since the saved machine language version can always be loaded and run directly from permanent memory without having first to be translated and compiled again. (Newer and faster versions of BASIC, designed for the personal computer, also compile the program, but they will not run on a mainframe computer.)

FORTRAN, over thirty years old, is the universally accepted computer language for technical work. Developed for IBM by John Backus in 1954, it has been so extensively adopted that, rather than change to the latest languages, many users prefer to employ updated versions of FORTRAN that incorporate modern features.

The fifth version of FORTRAN, appearing in 1977, is known as FORTRAN 77 and is the version you will be studying in this book.

Engineers employ FORTRAN in various ways. They may have to develop entirely new computer code to solve specific problems. Or they may be asked to correct and amend existing computer code. When an engineer is asked to modify or append code to an existing program, the newly entered statements must delicately fit the already established program. Extreme care is taken to assure that whatever is newly entered does not cause the existing program to malfunction.

Programs that are written for computers are called **software.** Engineers also often work with commercially available application software programs that are written and serviced by software companies. These software packages are sold or leased to the user. The larger successful codes are generally written in FORTRAN. Commercial programs such as NASTRAN, SAP, EASE, and STRUDL can contain thousands of FORTRAN statements. The same is true for programs that simulate continuous systems (solve simultaneous ordinary differential equations), such as CSMP, ACSL, or ECAP. All that is required to use such programs is a specific knowledge of how to enter the pertinent data. Extensive instructions accompany each software package. When one purchases developed code, one buys not only technical expertise but also years of development, refining, and debugging.

In many instances, however, the user is operating in the dark when running such packages, having no knowledge of what the computer is doing — its assumptions, its numerical methods. This total relinquishment of control can lead to serious engineering errors. It is therefore valuable to have a general understanding of how the program works, of its practical and theoretical limitations, and a technical "feel" for what the answers should look like. To customize these packages, one must have a knowledge of FORTRAN.

1.4 PROGRAM DEVELOPMENT

Before beginning a new program, the engineer first establishes which physical principles, formulas, and equations govern his or her analysis. Once this is done, the required program input and output information can be determined, and a logical computer-oriented sequence of calculation steps can be laid out. This plan for problem solution is called an **algorithm.** The algorithm must be completed before the computer program is actually written. It is important that the reader distinguish between these two separate programming stages. The first stage concerns itself with analyzing a problem and planning its solution. The second stage deals with translating the plan into high-level computer language (FORTRAN) statements that can be processed by the computer. Both stages of program development are equally important, but it should be clear to the reader that program construction (writing) should only begin *after* the problem's algorithm has been completed.

Algorithms are written in **pseudocode.** Pseudocode is an informal blending of English words, arithmetic operators, and selected symbols representing specific computer operations. The programmer "pads" or builds upon the pseudocode when writing the actual program statements. Sometimes, in addition to writing pseudocode, a graphical representation of program tasks, called a **flowchart,** is drawn to help visualize the solution steps and the program flow. The order in which the calculations are to be executed is indicated by flow lines. Table 1.1 indicates some typical symbols used in flow charting.

Planning a problem's solution can at times be quite cumbersome. There may be so much detail present in the problem that one becomes lost among all the required solution steps. To avoid this, and to help systematically build a solution

TABLE 1.1

⬭	Start or end of program
▱	Input or output of data
▭	Assignment of values or calculation
⬭	Printed output of data
◇	Decision block
⬡	Loop structure
→	Flow lines
⊶	Program junction

algorithm, most programmers follow a **top-down design** principle. In top-down design, the major tasks that the program must perform are first laid out. These steps are then subsequently refined into subtasks or modules (see Chapter 7) that must be performed to accomplish these major tasks. This level of refinement may require further division into even smaller subtasks until, finally, a level of detail is reached sufficient to allow a direct translation to computer instructions (FORTRAN statements). A typical program structure is shown in Figure 1.7. Note that there are generally five sections or blocks to any program: a specification section in which variable characteristics are described, an input and initialization block to enter program data and constants, a main calculation block, an output section to report program results, and

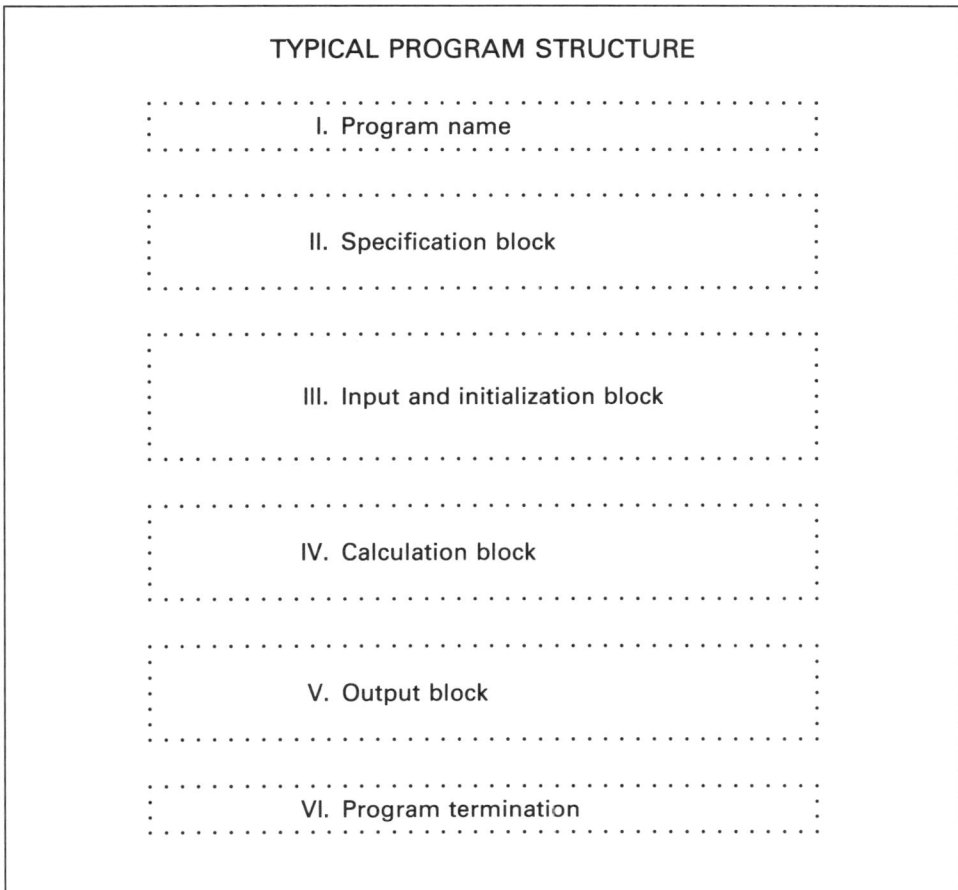

TYPICAL PROGRAM STRUCTURE

I. Program name

II. Specification block

III. Input and initialization block

IV. Calculation block

V. Output block

VI. Program termination

Figure 1.7

a termination block. These sections appear in a sequential order following the program's first statement, which assigns a name to the given program unit.

The process of writing the high-level language statements, or **source code,** for the algorithm is called coding. The completely written code is often called a source program and is entered into the computer by typing the program at the terminal's keyboard. It is later compiled and executed. Once it runs, it is tested before it is released to verify that the program's algorithm works for all sets of possible data. If the program is being used to model a physical event or system, the program's results must be **validated** as well, to ensure that they accurately portray the real system.

Sample Problem 1.1

Suppose that it is desired to simulate the motion of a fighter pilot who has ejected from his airplane. Of great concern is whether the pilot will clear the tail fin, which is H meters high. The pilot's ejection speed is V_E, and he leaves the plane at a relative angle θ with respect to the vertical. The airplane's speed is V_A when the pilot ejects; then the pilot's vertical speed upward relative to the airplane is $V_E \cos \theta$, and his horizontal speed relative to the plane is $-V_E \sin \theta$. The distance from the cockpit to the tail fin is known to be B. (See Figure 1.8.) Neglecting aerodynamic drag, the equations describing his path relative to the airplane are given by

$$x = -V_E \sin \theta\, t \qquad (1.1)$$

and

$$y = V_E \cos \theta\, t - \frac{gt^2}{2} \qquad (1.2)$$

where t is time, and $g = 9.81$ m/s^2. It is desired to write a FORTRAN program that will print out the pilot's height when he is B meters behind the cockpit.

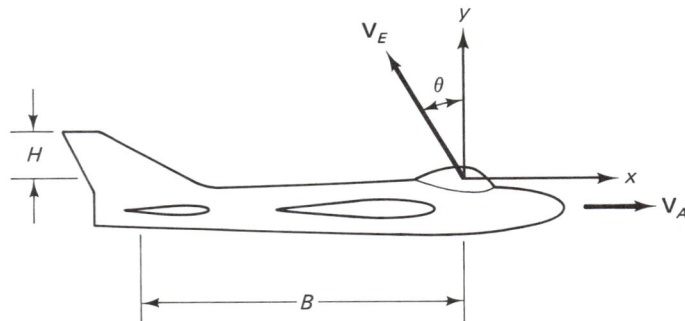

Figure 1.8

Analysis: The two equations for the path are given by equations 1.1 and 1.2. The time required to travel a distance $x = -B$ is given by

$$t = \frac{B}{V_E \sin \theta} \tag{1.3}$$

and the height is then given by

$$y = B \cot \theta - \frac{g}{2} \left(\frac{B}{V_E \sin \theta} \right)^2 \tag{1.4}$$

Given B, V_E, θ, and g, it is possible to solve for the one unknown, y.

1.4.1 Algorithm Development: Sample Problem 1.1

The three major tasks that the program must perform here are (1) input of data, (2) calculation of the unknown quantity y, and (3) output of the solution. To begin the problem, the computer must know B, V_E, θ, and g. This information must be read in as data, since it cannot be solved for. The calculation task can be further refined by noting that equation 1.4 can be used directly to solve for y, or t can first be solved for from equation 1.3 and then substituted into equation 1.2 to yield y. This latter method is elected here so that the final time can be printed out. In addition, though not required, the clearance height H_C, which equals $y - H$, is also calculated. A further refinement requires θ to be converted to radians. The final task is completed by printing the solution for y, t, and H_C at the terminal. The program algorithm appears in pseudocode as follows:

■ **Algorithm for Sample Problem 1.1**

1. Read B, VE, THETA, G
2. Calculate
   ```
   THETA=THETA/57.296
   T=B/(VE*SIN(THETA))
   Y=(-G*T/2.0+VE*COS(THETA))*T
   ```
3. Print T, Y
4. Calculate
   ```
   HC=Y-H
   ```
5. Print HC

A flowchart for the problem solution is shown in Figure 1.9. Since the clearance, H_C was not required, it was decided to calculate it separately after displaying t

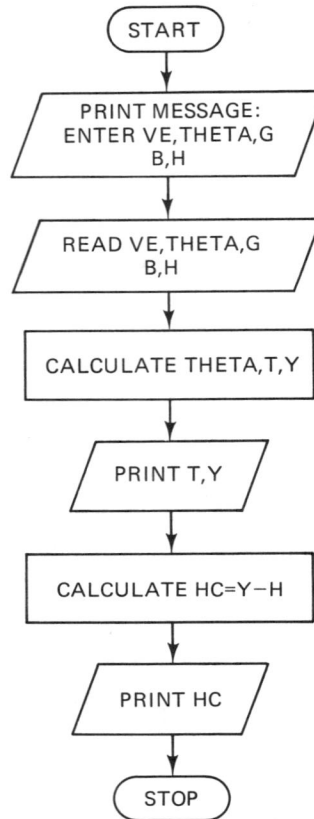

```
            ( START )
                │
                ▼
        / PRINT MESSAGE:  /
       / ENTER VE,THETA,G /
      /       B,H        /
                │
                ▼
       / READ VE,THETA,G /
      /       B,H        /
                │
                ▼
     │ CALCULATE THETA,T,Y │
                │
                ▼
         /  PRINT T,Y  /
                │
                ▼
     │ CALCULATE HC=Y−H │
                │
                ▼
         /  PRINT HC   /
                │
                ▼
            ( STOP )
```

Figure 1.9

and y. The source listing for the problem is shown in Figure 1.10. Interactive processing is assumed; that is, input and output will be performed at the terminal's keyboard when the program is actually run.

1.4.2 Brief Program Notes for the Timid Beginner

Each line of the program source code, shown in Figure 1.10, causes a specific function to be performed when the program is executed (see Section 1.7, Running a FORTRAN Program). To begin with, the first line of the listing must assign the program a symbolic name. This is accomplished with a PROGRAM **statement.** The lines that immediately follow line 1 with an asterisk in the first column are descriptive comment statements. They do not affect program operation but are kept in the listing to provide program commentary. The first executable statement tells the com-

```
                PROGRAM BAILOUT

        ****************************************************************
        *                                                              *
        *  THIS PROGRAM CALCULATES THE CLEARANCE HEIGHT, HC, ABOVE THE  *
        *  TAIL FOR A PILOT WHO HAS EJECTED FROM AN AIRCRAFT.  VE IS    *
        *  THE PILOT'S EXIT VELOCITY RELATIVE TO THE PLANE, THETA IS    *
        *  THE RELATIVE ANGLE WITH THE VERTICAL THAT THE EJECTION       *
        *  VELOCITY MAKES, G IS THE ACCELERATION OF GRAVITY, B IS THE   *
        *  DISTANCE FROM THE COCKPIT TO THE TAIL, H IS THE TAIL HEIGHT, *
        *  AND T IS TIME.                                               *
        *                                                              *
        ****************************************************************
                PRINT*,'ENTER VE, THETA, G, B, H'
                PRINT*,'ENTER THETA IN DEGREES ONLY'
                READ*,VE,THETA,G,B,H

                THETA=THETA/57.296
                T=B/(VE*SIN(THETA))
                Y=(-G*T/2.+VE*COS(THETA))*T

                PRINT*
                PRINT*,'THE TIME WHEN THE PILOT CLEARS THE TAIL IS = ',T
                PRINT*,'THE HEIGHT ABOVE THE PLANE THEN IS = ',Y
                PRINT*
                HC=Y-H
                PRINT*,'THE CLEARANCE = ',HC

                STOP
                END
```

Figure 1.10

puter, at run time, to print the message enclosed inside the apostrophes on the monitor screen as a prompt to the user:

```
PRINT *, 'ENTER VE, THETA, G, B, H'
```

The next statement will also be written on the monitor and contains important information for the user:

```
PRINT *, 'ENTER THETA IN DEGREES ONLY'
```

Programs that assist in this manner, clearly defining the responses expected of the user, are called **user friendly.** The statement that follows next requires the computer to read from the terminal:

```
READ *, VE, THETA, G, B, H
```

When running the program, the computer will wait until all the specified data is entered before proceeding further. Note that θ is divided by 57.296 to convert it to radians. Succeeding lines of the program will cause t and y to be calculated, and their values to be displayed on the monitor. The asterisk that appears in the arithmetic expressions denotes multiplication. Note that the PRINT * statement causes a blank line to appear in the output. The clearance height, H_C, is calculated next and the result is also shown.

The final END **statement,** which must be included in all FORTRAN programs, is used both to stop the program and to signal to the FORTRAN compiler that the program unit is complete. (A STOP **statement** is used when it is desired to stop program execution at a point other than at the normal end exit point. The STOP statement may appear anywhere in the program. The final END statement must still be included even when a STOP statement is employed.)

In the PRINT commands, the character part of the message, the **literals,** are always enclosed in apostrophes. They will appear exactly as typed. Note how in this program the equal sign will be handled upon printing. The variables themselves are printed in a form that the computer itself decides upon, for example, exponential notation. This is called **list-directed** format or **format-free** output and is specified by the asterisk — PRINT *,. On input, the READ *, statement will let the computer store the entered variables as it wishes, based on the type of variables that appear on the list, for example, integer, decimal, or complex. Different type variables take up different amounts of storage.

1.5 COMPILATION

High-level languages require a special program, residing in internal memory, called a **compiler** to translate them into machine-language instructions. One high-level language statement will eventually be replaced with a sequence of machine language commands that will perform the prescribed task. The programs written in high-level language are called **source programs.** The resulting machine-language program that is produced after the **compilation** process is called the **object program.** The compiler commonly also has the capability of detecting **syntax** errors (for example, misspelled language commands and missing punctuation) and reporting them to the user as **diagnostic** messages. *It is very important when using a high-level language, such as FORTRAN, to pay very close attention to the details of spelling and punctuation.* This is the most common form of error made by the beginning programmer. Expect errors and learn how to interpret the diagnostic messages the computer displays.

The object program is what is **executed** or run by the machine. If the results are what the user expected, the object program can be saved, or stored, in auxiliary memory. For the next session it can be rapidly entered, loaded, and run by the machine without again going through the compilation phase. If the program runs but the results are not right, because the algorithm is incorrect, the programmer must search for **logic errors.** They say that no large program is error free. There is always one case or situation encountered that the programmer has forgotten to include in the source code's logic. For larger programs, it may take years before this bug appears, but it will. Therefore, when learning to program, it is very important to be as thorough as possible: consider all eventual cases and their implications, and consider all possible users. It is also important to document one's program either with clarifying **comment** statements in the program, or by a separate report, so that later users of the program can understand what has been done and know what input or output the program requires. They may even be able to find that deeply hidden bug long after you, the original programmer, have moved on to greener job pastures.

1.6 TIME SHARING AND BATCH PROCESSING

The processes of program input/output, compilation, and execution we have so far discussed can be managed by a computer system in either one of two ways: by **batch processing** or by **time sharing.** In batch processing, jobs from many users are grouped together and run sequentially, after perhaps being given priority ordering, at the computer center. There is no interaction or dialogue with the user before running, or after processing. Because there is no dialogue, batch processing is more economical with computer time than time-sharing methods. It is also cheaper, since it allows jobs to be performed at night when regular office personnel are not present. To execute a program with batch processing, the computer must be given a complete file with identification information (password, identity numbers, etc.), job-control information for processing, the object program, and any necessary data that the program will want to read. (Remember, there is no chance for the user to interact and type in the data). The file, or job, is submitted to the computer center where it is processed, and the results, usually output from high-speed line printers, are then returned to the user, a process called turnaround.

Though batch processing is economical of computer time, offering faster processing, the actual turnaround time for the user can be torturously slow. It can take anywhere from several hours to a whole day to receive returned results. Any errors that are found and need correction (usually in the program-development stage) must wait till the next batch run before proceeding further. The job must simply be set aside. Because of this shortcoming, the present trend of computer processing leans toward time sharing. With time sharing, the user can constantly interact with the computer. Sitting at a terminal with keyboard and monitor, the user may create a program, correct (**edit**) it, run the program, type in necessary data that it requests, and receive the results all in one session. This immediate feedback allows for efficient computer use.

In time sharing, the mainframe computer processes many users simultaneously, switching (scanning) from terminal to terminal whenever there is a task completed and free computer processing space available for another job. While one program may be reading data, the computer can use its ALU to process another program's numerical calculations. The switching from program to program is done so rapidly that it appears to the user that he or she has the computer's constant, undivided attention.

In addition, the user of a time-sharing system is assigned a personal memory workspace from the mainframe computer memory. The illusion of working with a private personal computer is therefore complete. The user can access his or her permanent files (the file space assigned only to his or her identification number) or *temporarily* assigned internal memory workspace during any computer session. Figure 1.11 depicts a computer system that supports both batch processing and time sharing.

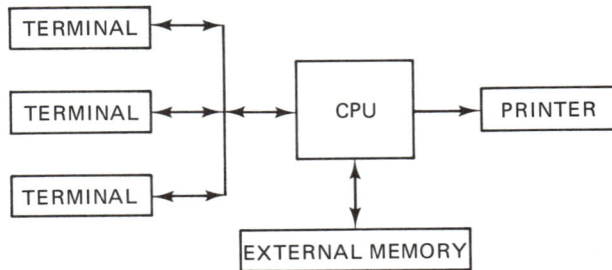

Figure 1.11

1.7 RUNNING A FORTRAN PROGRAM

To key in a program, the terminal user must first be admitted to the computer system. The procedure of getting on the computer is called **logging on.** The actual details of the log-on process vary from computer facility to computer facility. One system may only require a password and an identity number, for instance, whereas another may require several more actions from the user. After the network is entered, the user employs **system commands** to control the processing of data. System commands are instructions for the operating system. For example, operating commands could ask the computer to RUN, SAVE, EDIT, DELETE, or REPLACE a file. A system command is also needed to load the FORTRAN compiler into internal memory (see Section 1.5).

To create a new file, or correct an old one, a system command is used to call forth the **text editor.** For example, a **full-screen editor** allows the user to move freely about the monitor image of the text input to correct, insert, copy, or delete characters and line entries. (A single line of entry is called a data **record.**) A **line editor** allows the user to access and edit only one line at a time. In the edit mode, the source program is keyed in exactly as it appears on the FORTRAN coding form. The RETURN key on the keyboard serves the purpose of a carriage return on a typewriter.

As seen in Figure 1.12, there is a special format of spacing that must be followed exactly for each line of entry of a FORTRAN program. Each program line, or record, can be from one to seventy-two columns wide. Columns 1 to 5 are reserved for any FORTRAN statement address numbers that may be required. For instance, one can have a FORTRAN statement in a program that says GO TO 14. Computer execution will automatically **branch** or skip from the current statement being executed to the statement whose statement address number is 14. The computer will resume execution at this new statement and then go on to process those that sequentially follow. (Program statements reside serially in internal memory. Not all statements need be numbered.) Column 1 also has a special function. If a C or an * appears there, what is typed on the card or line is interpreted as a comment statement or explanation and will not be compiled or executed. The actual FORTRAN statements must fit into columns 7 through 72; otherwise, they will have to be con-

```
0         1         2         3         4         5         6         7         8
1234567890123456789012345678901234567890123456789012345678901234567890
```

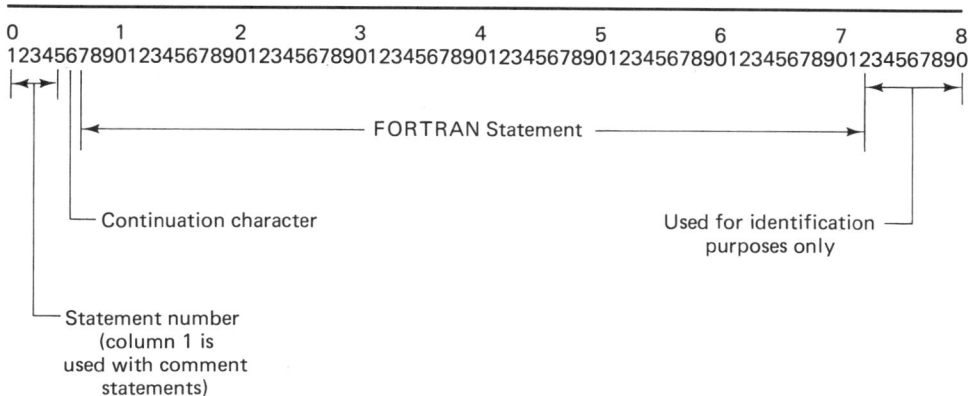

Figure 1.12

tinued on the next line or card. (Columns 73 to 80 are ignored by the computer.) Column 6, another reserved position, is intended for this purpose. By placing any nonblank character except zero there, the computer will interpret this card or line as a continuation of the previous one. It must therefore be left blank except when needed. The amount of continuation cards or lines that can follow one another is dependent on the specific computer system used.

After the program has been entered, the FORTRAN compiler is loaded using a system command. The program is then compiled. If any diagnostic errors appear, they must now be corrected. When all the error messages are remedied, both the source program and the object code (created by the compiler) should be saved (stored) for future work. The names that are chosen when storing these two files do not have to be the program name. The program name, found on line 1 of the source listing, is only there so that the compiler can recognize a program unit. When loading the program from external memory to internal memory for compilation, editing, or execution purposes, it is the **file** name that is used to retrieve the program from one's personal library. When you save a file (or replace the older version of the file with the new update), the resident file in internal memory is sent to auxiliary memory.

To see your file, you can use the system LIST command. This command will cause the local file (that which is currently in internal memory) to be displayed on the monitor. **Listing** the file at the same time that a local printer is able to receive the monitor's messages will result in a printout **hard copy** of the program. This copy is called a **source listing** (see Figure 1.10). You do not have to wait till the program is corrected before requesting a printout. In fact, it is often easier to debug a program from a source listing than from the monitor. With the programs saved and a hard copy of the source listing at hand for reference purposes, the object program is loaded into the computer's memory. Typing in your system RUN command will cause the program to be executed and the results displayed. (Some computer systems are designed to operate differently, however. For instance, with a source program as a **local file** in temporary working memory, the system command RUN will cause the computer to compile the program, load the resulting compiled object code plus any other files called out [see Chapter 3], and then execute the program—all in one step.)

Figure 1.13 shows a record of terminal input and output that was performed for Sample Problem 1.1. Note that the particular computer operating system used for this book places a question mark on the monitor to remind the user to enter information. Each variable entry should be separated by a comma, or by a blank space, or by a carriage return. Having the local printer respond to keyboard inputs as well as monitor messages will result in a written record of keyboard responses along with the problem solution. The reader is well advised to become familiar with his or her local log-on and system commands as soon as possible.

It is not at all unusual for different computer manufacturers to provide a syntax that varies ever so slightly from that of the accepted "standard" FORTRAN version. A FORTRAN program or commercial application package written for one machine may not function on, or carry to, another machine without rework. Descriptive material written in this textbook with one specific machine in mind may not exactly fit the reader's machine in all instances. The reader must check with his or her computer center when problems arise. Fortunately, the differences from machine to machine are small. Typing in the sample program of Figure 1.10 and finding out how to compile and run it on your PC, on a more powerful desktop minicomputer, or at a local terminal of your school's mainframe computer will help you get started. Good luck!

REVIEW

This introductory chapter, which contains many new terms and concepts, brings the reader into the engineering world of digital computers. A computer can be said to be an electronic device that processes information in a user-defined, preprogrammed manner. Any digital computer system must include six basic elements: It receives data or instructions from an input unit (1) and passes this information along to a central processing unit (CPU). The CPU is the collective name given to three components; one component controls computer processes (2), another element performs arithmetic (3) and logic calculations (4), and the third component internally stores

```
       RUN

       ENTER VE, THETA, G, B, H
       ENTER THETA IN DEGREES ONLY
       ? 29.8,23,9.81,15.24,3.66

       THE TIME WHEN THE PILOT CLEARS THE TAIL IS = 1.308857216225
       THE HEIGHT ABOVE THE PLANE THEN IS = 27.50055338153

       THE CLEARANCE = 23.84055338153
```

Figure 1.13

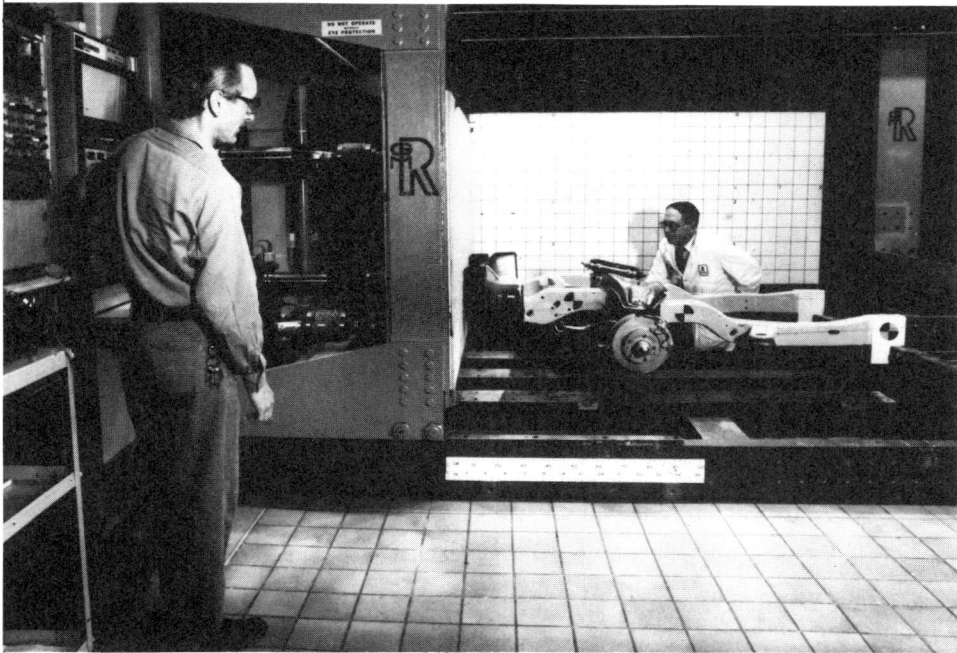

Figure 1.14   Computer simulations reduce the need for barrier crash tests and markedly speed up the development process. Crushing of vehicle components and assemblies, such as this small truck frame, validates computer simulation programs. *(Photo courtesy of General Motors Corporation.)*

instructions, data, or intermediate results (5). When processing is complete, the results are either sent to an external auxiliary memory unit or to an output device such as a printer or a monitor (6).

Internal information in the computer is carried by electronic digital pulses called bits. Eight bits are grouped together to form a byte of data. A typical computer system uses a word size of one, two, or four bytes. A program of instructions written for the computer in binary form — that is, the information is either present (represented by a 1) or missing (represented by a 0) — is said to be written in machine language, since its information is presented as symbolic electronic pulses. FORTRAN is a high-level language that allows the user to write programmed instructions for the computer using English words. An internal program called a compiler later translates these words into a series of machine-language instructions. The program written by the user in FORTRAN is called the source program. The machine-language version of the program created by the compiler is called the object program.

The engineer using FORTRAN will either write new code or modify or append code to an existing program. After analyzing a problem for which a solution is sought, a numeric algorithm is written that will accomplish the data processing. Top-down design is used. In top-down design, program tasks are broadly outlined at first and then refined into smaller and smaller subtasks. Once the program steps have been subdivided sufficiently, they are translated directly into higher-level language code and the source program is created. Program flow might be planned using a flowchart.

The user employs system operating commands to control the processing of data after logging on to the system. Typically, he or she enters the program at the keyboard, stringently following the FORTRAN rules for placement of the alphanumeric data in a line, as well as FORTRAN syntax. The compiler is then loaded and an object program file is created. When the program is run, the object program plus any required data files are loaded into the CPU for program execution. Results are sent to the output unit or auxiliary memory.

There are two ways to process data in a computer system. One method, called time sharing, is interactive and on-line. The user enters data and responds directly at the terminal during processing. The other method is called batch processing. It involves submitting to a waiting queue (as one unit) an object-program version of the source deck, any required data, and the necessary job-control information for the operating system. The job is completed at a later time and the user has no interaction with the computer during processing.

KEY TERMS

algorithm
alphanumeric (character) data
arithmetic logic unit (ALU)
assembler
assembly language
auxiliary external memory
BASIC
batch processing
binary signal
bit
branch
bug
byte
central processing unit (CPU)
comment statement
compilation, compiler
debugging
diagnostic message
edit (a program)
END statement
execute (a program)
file

listing
literals
local file
logging on, logging off
logic error
machine language
mainframe computer
microcomputer
microprocessor
minicomputer
object program
output unit
parallel processing
PRINT *,
program
PROGRAM statement
pseudocode
READ *,
record
serial processing
software
source code, listing, program

flowchart
format-free
FORTRAN
full-screen editor
hard copy
high-level language
input unit
input or output (I/O)
internal memory
line editor
list-directed format

STOP **statement**
supercomputer
syntax
system command
temporary memory
text editor
time sharing
top-down design
user friendly
validation

REVIEW QUESTIONS

 1. What are the three parts of the CPU?
 2. Define the function of the ALU.
 3. What are the six basic parts of a computer system?
 4. What is software?
 5. FORTRAN stands for an abbreviation of what two words?
 6. What is machine language?
 7. What is the function of a compiler?
 8. What is the difference between logic errors and syntax errors?
 9. What is an object program?
10. What does CRT stand for?
11. What is an algorithm?
12. Describe the differences between batch processing and time sharing.
13. What is temporary workspace?
14. What is the difference between a local file and a permanent file?
15. Define the following:
    (a) bit
    (b) byte
    (c) a computer word
16. What is meant by syntax?
17. What is meant by top-down design?
18. How is a comment line denoted in FORTRAN?
19. What is meant by list-directed input?
20. What must the last statement of any program be?
21. What is the difference between the program STOP statement and the END statement?

FORTRAN Basics

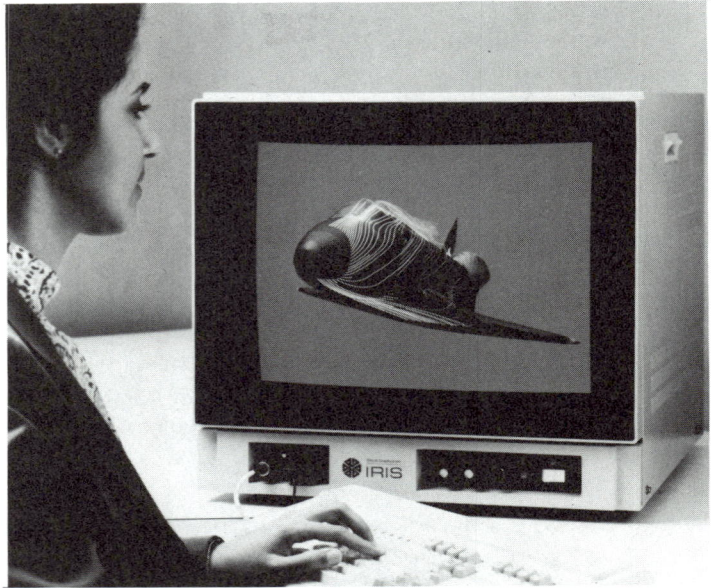

Simulation of space shuttle orbiter particle flow on IRIS graphic workstation. Calculations were performed on a CRAY-2 supercomputer. *(Photo courtesy of NASA Ames Research Center.)*

2.1 INTRODUCTION: DATA TYPES

FORTRAN provides for six types of data: *integer, real, double precision, complex, logical,* and *character*. Each type of data requires binary code for internal representation. For example, numeric data in our base-10 number system must first be converted by the computer to the binary base-2 system before machine processing can begin (see Appendix A). The number 13.5 in the base-10 system, or 13.5_{10}, is equal to 1101.1_2 in the binary base-2 system:

$$1 \times 10^1 + 3 \times 10^0 + 5 \times 10^{-1} = 1 \times 2^3 + 1 \times 2^2 + 0 \times 2^1 + 1 \times 2^0 + 1 \times 2^{-1}$$

Alphanumeric character data uses ASCII (American Standard Code for Information Interchange) or EBCDIC (Extended Binary Coded Decimal Interchange Code)

code to internally represent information (refer again to Appendix A). The letter *A* is represented by 01000001 in ASCII code and by 11000001 in EBCDIC code. The number 3_{10} is represented by 11_2 in binary code, but the character "3" is stored as 11110011 in EBCDIC and as 00110011 in ASCII code. Note that eight bits, or a one-byte memory word size, is required for each character stored. Numeric data may employ one, two, four, eight, or sixteen bytes for internal representation and processing. The actual number used depends on the size of the computer and the type of number represented, that is, whether it is integer, real (with a decimal point), double precision, or complex. A double-precision variable, for example, is stored internally with two succeeding words instead of one. This provides for greater numerical accuracy, since more digits are retained during calculations. (Problems occur when the results of a computer arithmetic operation cannot fit the assigned byte space allocated for the fractional part of a number, the part that lies to the right of the binary point. Only the number of bits allowed by the assigned word size are retained. The excess bits that are least significant are simply dropped. The total numeric error caused by this accumulated loss of information is called **roundoff error.** Refer to Appendix A for further information.) Complex variables also need two storage words: one word each for the complex constant's real and imaginary parts.

2.2 NUMERIC CONSTANTS

In FORTRAN there are four types of numeric constants, or fixed quantities: integer, real, double precision, and complex. This chapter will cover only integer and real types. A FORTRAN integer constant is a whole number or string of decimal digits represented *without a comma or decimal point*. An integer constant can be positive, negative, or zero. Some examples of integer constants are

135

−48

+237105

−0054

The range or size of the largest positive or smallest negative integer number permitted on each computer system varies; however, it is very rare for a normal integer operation to exceed a computer's capacity. Typical ranges provided exceed 2×10^9.

A real constant is a string of decimal digits with a decimal point appearing. No commas are allowed. Real numbers are also called **floating-point values.** A real constant may be positive, negative, or zero, and may carry an exponent (integer only).

Some examples of allowable real constants are

        2.6

       -48.

       135.

       135.0

     +237105.

Scientific notation is normally used in engineering and the sciences to represent very large or very small real numbers. In scientific notation, the value is expressed as a decimal number between 1.0 and 10.0, representing the mantissa or fractional part, multiplied by a power of 10. FORTRAN uses an alternative form of scientific notation called **exponential notation** to represent large or small numbers. In exponential notation, the absolute value of the number's mantissa is given greater than or equal to 0.1 but less than 1.0. The power of 10 that applies is indicated by a written exponent appearing after the mantissa. The exponent is represented by the letter E followed by an integer constant. For example,

SCIENTIFIC NOTATION	**EXPONENTIAL NOTATION**
-3.8×10^{-5}	-0.38E-4

Some examples of real constants written in exponential form are

      .516E3

     0.516E3

      -.2E4

      -.2 E-4

      .214 E+4

      .214 E4

     +0.237105 E6

The range of decimal numbers in exponential form that will fit into each computer system varies. A CDC (Control Data Corporation) CYBER 170 Series computer, for instance, provides for an exponent ranging from −293 to 322, far exceeding any normal requirement. Check with your local computer center or computer manual to find out what the numerical size limits are for your system.

2.3 VARIABLES

In mathematics a variable name is a symbol for a varying quantity. Using algebra, these symbols can be manipulated to solve for the numeric quantities they represent. A variable name in FORTRAN, however, is considered to be solely a specific address location in memory where numeric data, or constants, are stored. Different constants may enter or leave the address location at different times, thus varying the contents of the address but not the address name itself.

Each variable quantity in FORTRAN must be given a variable name, that is, an address in memory where it can receive values. The address name in FORTRAN may have some symbolic meaning to help us remember what quantity is stored in that cell, but it cannot symbolically replace the quantity that it hosts. This brings us to a very significant point. *The equal sign in FORTRAN does not represent an equality.* Instead, it means *replace* the current numeric contents (constants) at the address given to the left of the equal sign with the numeric value to the right of the equal sign. For example, the program statements

```
A=3.0
X=A
X=X+2.0
```

which are processed sequentially, would cause the numeric constant 3.0 to be stored at address A. Then the contents of address A will be sent to and stored at address X (the constant 3.0 will still remain at address A). Finally, the contents at address X will be *replaced* by the current contents plus 2.0. The contents at address X will now be 5.0. This last statement makes no algebraic sense, but its meaning is absolutely clear in FORTRAN.

The following two statements would be correct provided that Y was *initialized* beforehand, that is, some numeric value was sent to address Y.

```
X1=Y+7.0
X1=-2.0*Y+1.0
```

These FORTRAN statements do not cause the computer to solve the algebraic equations and obtain the results X1 = 5.0 and Y = −2.0. Instead, if Y were assigned a value of 3.0 beforehand, the final value at X1 would become −5.0. If Y were not initialized earlier, the results obtained would depend on what each particular computer system stores in all variable addresses when it is first turned on. For most computer systems, undefined variables are initially set to zero, and the result would then be that X1 has the value 1. However, some computer systems might print out a diagnostic message "undefined variable" at compilation time and then refuse to run the program. With the sequence of statements

```
Y=5.0
X1=Y+7.0
X1=-2.0*Y+1.0
```

the final outcome would have the numeric constant -9.0 stored at address X1. It should be clear now to the reader that the statements

```
Y=5.0
Y=X
```

which are algebraicly correct, must be written in FORTRAN as

```
Y=5.0
X=Y
```

if X is to equal 5.0.

Variable names are limited to a length of six characters on most compilers. The characters must be uppercase letters or numbers; no blanks, hyphens, and so forth, are allowed. The first character of a variable name must be a letter.

ACCEPTABLE	UNACCEPTABLE
ZERO	HOLDING
I3	3I
ALPHA1	AB, C
FLUX	R-14

The following two statements are not acceptable FORTRAN statements, since 3.0 and X+3.0 are not allowable variable names.

```
3.0=A

X+3.0=5.0
```

2.4 TYPE DECLARATIONS

Each variable name must be assigned a type so that a memory cell size and format can be assigned. Recall that the six types are integer, real, double precision, complex, logical, and character. There are two means by which typing can be performed.

The first, **explicit typing,** requires that declarative **type statements** be made in the FORTRAN program before any **executable** FORTRAN statements (that is, statements requiring an action such as reading, adding, or writing) appear in the program. All FORTRAN statements may be categorized as either executable or **nonexecutable.** The proper order that all statements must have within a FORTRAN program unit is shown in Table 2.1. Note that the **declarative statements** appear before the executable statements. Within each group, statements can be placed in any needed order. Comment and FORMAT statements may appear anywhere within a program unit. Type statements are nonexecutable statements. Some examples of type statements, such as REAL, INTEGER, COMPLEX, DOUBLE PRECISION,

TABLE 2.1 STATEMENT ORDER

I. PROGRAM, SUBROUTINE, FUNCTION, or BLOCK DATA statements

II. IMPLICIT Statement

III. Declarative Statements
    a. Type Statements: CHARACTER, COMPLEX, DOUBLE PRECISION,
       INTEGER, LOGICAL, REAL
    b. Specification Statements: COMMON, DIMENSION, EQUIVALENCE,
       EXTERNAL, INTRINSIC, SAVE
    c. PARAMETER Statements

IV. Statement Function Definitions

V. Executable Statements: BACKSPACE, CALL, CLOSE, CONTINUE, DO, ELSE,
   ELSE IF, END IF, ENDFILE, GO TO, IF, INQUIRE, OPEN, PAUSE, PRINT,
   READ, RETURN, REWIND, STOP, WRITE

VI. END Statement

Further notes:
    a. Comment and FORMAT statements may be placed anywhere in the program
       before the END statement.
    b. DATA statements may be used anywhere after the program unit's declarative
       statements.
    c. ENTRY statements may be placed anywhere within a subprogram unit except
       within the range of an IF or DO block.
    d. PARAMETER statements may appear in any order among the declarative
       statements of groups II and III.

CHARACTER, and LOGICAL, follow.

    PROGRAM ONE

    REAL INDEX

    INTEGER X,Y,Z

    COMPLEX VELOC

In the foregoing sequence of statements, after the program name is stated, the variable INDEX is declared as real, the variables, X, Y, and Z are declared to be of the integer type (whole numbers), and the variable VELOC is declared to be complex. The compiler can now prepare each named storage location to receive the correct size of numerical data.

The form the type statements take is

REAL list
INTEGER list

COMPLEX list

DOUBLE PRECISION list

LOGICAL list

where "list" is a list of variable names separated by commas. For example, the beginning code of a program might appear as

```
PROGRAM TWO
REAL INDEX,INDEX1
LOGICAL LOG
DOUBLE PRECISION TIME,DIST,VELOC
```

Character variables have the following type statement:

CHARACTER*n list

"List" is a list of variable names of the character type separated by commas, while *n, which is an optional descriptor, defines how long the character constants for each character name will be. For example, the statement

```
CHARACTER*5 CAR,NAME
```

defines two character type variables, CAR and NAME. Each variable name will be allotted storage space for five FORTRAN characters. If the descriptor is omitted, it will be assumed to have a value of 1. The statement

```
CHARACTER NUM,PLACE
```

defines two character type variables, NUM and PLACE, having a length in storage of one space (usually one byte). The statement

```
CHARACTER*10 PLANES,SHIPS,CAR*5,NUM*1,USER
```

will allot ten places in memory for the list PLANES, SHIPS, and USER, while making exceptions for CAR (length 5) and NUM (length 1). The descriptors following these variable names override the list descriptor.

The second available means of typing is called **implicit typing.** Without any explicit declarative statement to the contrary, the computer automatically assumes that any variable name that begins with the letter I, J, K, L, M, or N is of the integral type. Variable names beginning with the letters A to H or O to Z are assumed to be of the real type. If the variables used in a program are only of either the real or the integer form, it is possible to write a program without any type statement, providing that the rules that govern implicit typing are followed. For example, the FORTRAN statements

```
PROGRAM FIVE
A=3.5
INDEX=3
XI=6.2
```

will type the variable names A and XI as real (with a decimal point) and the variable INDEX as integer. But the statements

```
PROGRAM SIX
INTEGER A
REAL INDEX
A=3
INDEX=2.4
ZI=6.4
```

will type the name A as an integer, and INDEX as real explicitly, whereas the name ZI will be typed as real implicitly.

Many programmers never use implicit typing, preferring to declare the type of each variable name explicitly to avoid possible programming errors, and to provide better program documentation. In the beginning of each program, lists of all variable names that will subsequently be used appear in typed categories. This extra clarification is optional, though. What is more important is to select variable names that remind the programmer, or suggest to the user, what the variable names represent. For example,

V or VELOC	for velocity
T or TIME	for time
PRES	for pressure
FE	for iron
RI	for real electric current

In a very long program, keeping track of the meaning of each variable name can be a tedious undertaking unless the variable names themselves easily impart to the user what they stand for.

2.5 ARITHMETIC OPERATIONS

The basic building elements of a FORTRAN program are **assignment statements.** They are executable and take the form

Variable name = arithmetic expression

The arithmetic expression can be as simple as a numeric constant or as involved as a scientific formula that contains many arithmetic operations and variables. When it is a numeric constant, we usually speak of initializing a variable. The basic arithmetic operations that are provided by FORTRAN are as follows:

Operation	Symbol	Example
Addition	+	A+B
Subtraction	−	A−B
Multiplication	*	A*B
Division	/	A/B
Exponentiation	**	B**3

B**3 is the same as B^3 or B*B*B.

The order of the arithmetical operations that the computer performs in processing an assignment statement is very much like the order we would use if we were to sit down with a hand calculator and a pad of paper and set out to solve the same equation. For instance, if present, all quantities in the innermost pair of parentheses are first computed. The computer then works outward, closing all parenthetic operations. A priority ordering of operations — exponentiation, then multiplication or division, and finally addition or subtraction — is in effect within each pair of parentheses. When parentheses are cleared, all remaining exponentiations, multiplications and divisions, and additions and subtractions are carried out in the prescribed order. If operations have the same priority level, they are carried out from left to right, as they appear in the expression, except for exponentiation, which is performed from right to left. *It is apparent that two operator symbols can never appear in direct succession;* for example, A+(-B) cannot be written as A+-B.

Examples	Results
B+D*C	$B + (D \times C)$
(B+D)*C	$(B + D) \times C$
A**B**C	A^{B^C}
A/B*C	$(A/B) \times C$
A+B/C	$A + (B/C)$
A/C−B	$(A/C) - B$
A+B**C/D	$A + (B^C/D)$
(A*(B−C)**D)*E+F	$((B - C)^D \times A) \times E + F$

It is good practice to enclose complicated expressions in parentheses for purposes of clarity. In general, it is better to be faulted for having too many parenthetic operations than for having too few. When using parentheses make sure that they are placed exactly right. Also make sure that the number of left facing parentheses equals the number of those that face right. If they are not equal, a compiler diagnostic error will prevent further computer processing. Note how important the correct placement of parentheses becomes in the following arithmetic expression:

```
((C**D-E*F)**G+B)/(H*I)
12        3     4 5   6
```

Omitting parentheses 5 and 6 results in the following compiler interpretation:

```
(((C**D-E*F)**G+B)/H)*I
```

Omitting 1 and 4 produces

```
(C**D-E*F)**G+B/(H*I)
```

and with parentheses 2 and 3 missing the computer processes

```
(C**D-E*(F**G)+B)/(H*I)
```

Instead of writing long expressions—especially if they have to be continued onto the next line (using a continuation character in column 6)—it is better to break the expression up into smaller units so that they may be read more easily. As an example, instead of

```
Y=((C**D-E*F)**G+B)/(H*I)
```

one could employ the series of statements

```
A=C**D-E*F
B1=A**G+B
Y=B1/(H*I)
```

which produces the same result but is easier to understand and debug, if necessary.

When two numeric constants of the same type are involved in a single arithmetic operation, the result will have the same type. If the two operands are of different types, the operation is said to be in **mixed mode.** The value that results will depend on which type has dominance in the mix of operands present. Operand types are weighted in descending order: first complex, then double precision, then real, and then integer. For example, with one argument complex and the other argument in double-precision, real, or integer form, the operation result will be of the complex type. Similarly, if one argument is in double-precision form and the other argument is of the real or the integer type, the result will be in double precision. Some common examples of operations between real and integer values follow.

Operation	Result
N*I	integer type
F*G	real type
N/F	real type
B**2	real type

Note that in the last case, when raising a number to an integer power, we could have written either

```
B**2
```

or

    B**2.0

The former expression is preferred, since it is computed internally as a series of multiplications, that is,

    B**2

means

$$B \times B.$$

The expression B**2.0, on the other hand, requires that the ALU employ some form of logarithmic algorithm. Internally, B**2.0 means $10^{2.0 \log B}$. Depending on B and the size of the exponent, using the logarithmic algorithm might be slower. It will also involve some roundoff error. When raising a negative number to an integer power, one is forced to use an integer-type constant for the exponent, since negative logarithms are undefined (complex).

For an assignment statement having mixed mode, the final result will be sent and stored at the variable name address that appears to the left of the equal sign. Consequently, the results are ultimately changed so that they are of the same data type that fits the variable name. For example, the statement

    N=4.0

results in $N = 4$, and the statement

    N=4.0/2

results in $N = 2$. The intermediate result from the mixed-mode division 4.0/2 is real (2.0), but the final result has the integer form associated with a variable name beginning with the letter N. The result, therefore, is stored as integer (2). Study the next case.

$$N=4/2.0+6 = 2.0 + 6 = 8.0 \Rightarrow 8$$

$$F=2 \Rightarrow 2.0$$

$$F=4/2 = 2 \Rightarrow 2.0$$

$$F=7.0/4+4 \Rightarrow 1.75 + 4 \Rightarrow 5.75$$

Suppose, though, that in the last case we had

    F=7/4+4

The arithmetic division operation is not mixed; therefore, an integer result is called for. This means that the ALU truncates, or cuts off, the fractional part. Note that it does not round off the result, since it does not expect, and therefore does not keep, any fractional parts in an integer operation.

$$F=7/4+4 \Rightarrow 1 + 4 = 5 \Rightarrow 5.0$$

Quite an error indeed. The same error is present for

```
F=I/J+4.0
```

whenever I/J leaves a fractional part. This last statement, buried deep inside a program, will sometimes be correct and at other times lead to a program error — the worst of all possible situations for a programmer. Everyone, of course, consciously avoids dividing a quantity by 0, yet the assignment statement

```
R=A/(I/J)
```

will produce a program error, division by 0, whenever I is less than J, a variant form of the same problem.

The most common mistake of new programmers involves the incorrect use of mixed-mode operations. *It is strongly recommended that the use of mixed-mode expressions be avoided.*

2.6 INTRINSIC FUNCTIONS

Besides the basic arithmetic operations, FORTRAN 77 provides a library of built-in functions, called **intrinsic functions,** that the user can reference in arithmetic assignment statements. These functions, when met in the user's source program, are replaced by predefined compiler procedures at compilation time. After execution, each intrinsic function returns a single value. Table 2.2 contains a list of the most commonly used functions. Note that the trigonometric functions COS(X), SIN(X),

TABLE 2.2 COMMONLY USED INTRINSIC FUNCTIONS

Function name with REAL(X) or INTEGER(N) ARGUMENT	Resultant value
ABS(X)	absolute value of x
IABS(N)	absolute value of n
ALOG(X)	$\log_e x = \ln x$
ALOG10(X)	$\log_{10} x$
INT(X)	x truncated, integer number
REAL(N)	n as real number
SQRT(X)	square root of x
COS(X)	cosine of x (radians)
SIN(X)	sine of x (radians)
TAN(X)	tangent of x (radians)
ATAN(X)	arctangent of x
ATAN2(X1,X2)	arctangent(x_1/x_2)
EXP(X)	e^x
COSH(X)	hyperbolic cosine of x
SINH(X)	hyperbolic sine of x
TANH(X)	hyperbolic tangent of x

and TAN(X) have X expressed in *radians*. (For a complete listing of intrinsic functions, see Appendix B.)

Sample Problem 2.1

During manufacturing processes, samples and measurements are taken to ensure that the end product meets its design specifications. Two important statistical measurements for quality-control purposes are the average value, $\overline{X}$, and the standard deviation, S, which is a measure of the spread of values about the sample average. Suppose that company XYZ is manufacturing stainless steel bolts that are designed to be 3.000 centimeters long. Taking ten random samples, the lengths of the bolts chosen are found to be 2.999, 3.001, 3.010, 3.005, 2.995, 2.998, 3.003, 3.003, 2.999, and 3.001. Write and run a FORTRAN program that will calculate the sample's average and standard deviation.

Analysis: The average value of the samples is given by

$$\overline{X} = \frac{1}{n}\left(\sum_{i=1}^{n} X_i\right)$$

or the total sum of the measured values divided by the total number of samples present. The standard deviation is defined by

$$S = \left[\frac{\sum_{i=1}^{n} (X_i^2 - n\overline{X}^2)}{(n-1)}\right]^{1/2}$$

Solution: The flowchart for this problem is given in Figure 2.1. The algorithm and the problem solution are given in Figures 2.2a and 2.2b. Notice that the program's design and solution have three parts: entering of data, calculations, and output of results.

2.7 DATA AND PARAMETER STATEMENTS

If you keyed in and executed the sample program, you probably had to type the data in (X1 to X10) several times before you were finished. It is safe to assume that you will get tired of this very quickly, especially when a lot of data is presented. There are two ways to modify the program so as to avoid this problem. Both methods, however, require relinquishment of interactive dialogue with the computer.

The first method, suitable for small amounts of data, is to use the DATA **statement** to initialize variables. (The second method is covered in the next section.) A DATA statement is a nonexecutable statement that is usually placed immediately after the specification statements are given, though it can appear anywhere among the executable statements. It has the form

DATA list of variable names/list of constants/

The DATA statement causes the compiler to initialize the listed variables to the values

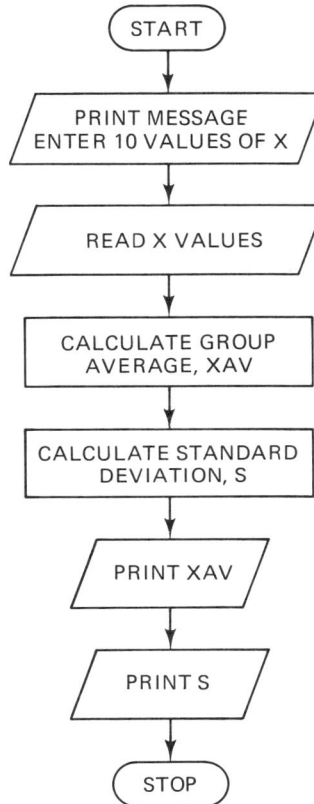

Figure 2.1

Algorithm for Sample Problem 2.1

1. Enter the ten sample values: X1, X2, . . . , X10.
2. Calculate
   ```
   XAV=(X1+X2+X3+X4+X5+X6+X7+X8+X9+X10)/10.0
   STO=(X1**2+X2**2+ ... +X10**2)-10.0*XAV**2
   S=SQRT(STO/9.0)
   ```
3. Print the average and the standard deviation: XAV,S.

Figure 2.2a

```
PROGRAM STAT

REAL X1,X2,X3,X4,X5,X6,X7,X8,X9,X10
REAL ST,ST1,STO,S,XAV

PRINT *,'ENTER 10 VALUES OF THE SAMPLES'
READ *,X1,X2,X3,X4,X5,X6,X7,X8,X9,X10

XAV=(X1+X2+X3+X4+X5+X6+X7+X8+X9+X10)/10.
ST=X1**2+X2**2+X3**2+X4**2+X5**2
ST1=X6**2+X7**2+X8**2+X9**2+X10**2
STO=ST+ST1-10.*XAV**2
S=SQRT(STO/9.)

PRINT *
PRINT *,'THE AVERAGE IS ',XAV
PRINT *,'THE STANDARD DEVIATION IS ',S

END


RUN

ENTER 10 VALUES OF THE SAMPLES
? 2.999,3.001,3.010,3.005,2.995,2.998,3.003,3.003,2.999,3.001

THE AVERAGE IS 3.0014
THE STANDARD DEVIATION IS .004168666204058
```

Figure 2.2b

specified at compilation time. Returning to the sample Problem 2.1, one can write

```
DATA N,X1,X2,X3,X4,X5,/10,2.999,3.001,3.010,3.005,2.995/
DATA X6,X7,X8,X9,X10/2.998,3.003,3.003,2.999,3.001/
```

to accommodate program input. Since some values are equal, one can plan ahead and use the DATA statement's repeat feature to reduce the required typing.

```
DATA N,X1,X9,X2,X10/10,2*2.999,2*3.001/
DATA X3,X4,X5,X6,X7,X8/3.010,3.005,2.995,2.998,2*3.003/
```

With the DATA included in the program, there is no need to read information from the terminal. The algorithm and rewritten program (Sample Problem 2.2) are shown in Figures 2.3a and 2.3b. Note that the program version is much more general, since the arithmetic statements needed to calculate the average and the standard deviation are written in terms of an unspecified amount, N, the number of samples. It is left to the DATA statement to provide the actual value.

The FORTRAN PARAMETER **statement** is similar to the DATA statement. It is used to assign a symbolic name to a constant value. The PARAMETER statement has the form

PARAMETER(name$_1$ = c_1, name$_2$ = c_2, . . .)

where c_1, c_2, and so on, are constants or expressions involving only constants. For example,

```
PARAMETER(PI=3.141593,GRAV=9.81,N=10)
```

■ **Algorithm for Sample Problem 2.2**

1. Initialize N and the N sample values: X1, X2, . . . , XN.

2. Calculate
```
XAV=(X1+X2+X3+ ... +XN)/N
STO=(X1**2+X2**2+ ... +XN**2)-N*XAV**2
S=SQRT(STO/(N-1))
```

3. Print the average and the standard deviation: XAV,S.

Figure 2.3a

```
PROGRAM STAT

REAL X1,X2,X3,X4,X5,X6,X7,X8,X9,X10,ST,ST1,STO,S,XAV
INTEGER N

DATA N,X1,X9,X2,X10/10,2*2.999,2*3.001/
DATA X3,X4,X5,X6,X7,X8/3.010,3.005,2.995,2.998,2*3.003/

XAV=(X1+X2+X3+X4+X5+X6+X7+X8+X9+X10)/N
ST=X1**2+X2**2+X3**2+X4**2+X5**2
ST1=X6**2+X7**2+X8**2+X9**2+X10**2
STO=ST+ST1-N*XAV**2
S=SQRT(STO/(N-1))

PRINT *
PRINT *,'THE AVERAGE IS ',XAV
PRINT *,'THE STANDARD DEVIATION IS ',S

END


RUN

THE AVERAGE IS 3.0014
THE STANDARD DEVIATION IS .004168666204058
```

Figure 2.3b

PARAMETER statements are declarative statements and must appear before the first executable statement of the program. If necessary, relevant type statements must precede the PARAMETER statement. Unlike the variables listed in the DATA statement, the parameters' constant values must not be changed in the program unit.

2.8 DATA FILES

The drawback with using the DATA statement of Section 2.7 is that if the data changes, you must change the source program itself and recompile. For large amounts of data, this type of input of numeric data is unacceptable. An alternative method, which is preferred and commonly used in practice, is to create a **data file** that stores the information so that it can be accessed by the computer later.

The data file is written or modified separately from the source program in the edit mode. The restrictions of FORTRAN line (column) format are not applicable to a data file. The file is saved, without compilation, as an independent file with its own file name. The source program must be altered to inform the compiler that input is not coming from the terminal but from another source, or unit (device) number. (The terminal is usually assigned the unit number 5 for input and the unit number 6 for output by most computer systems. If the user does not specify a unit number, the values that will be defaulted to are those for the terminal, namely 5 and 6.) The file name is assigned to the unit number through an OPEN **statement.** Consult your local computer system's operation manual to determine which unit numbers are allowable. (Remember, you cannot use the assigned numbers for the terminal.) The OPEN statement, abbreviated, may have the following form:

OPEN(UNIT=integer value,FILE='Data file name')

(See Chapter 9 for all the options available. Some systems may require more information.) A READ **statement** accessing this file must call out this unit number. A READ statement using list-directed format to retrieve information from the file would look like this:

READ(unit number,*) list of variable names

Another form this READ statement can take is

READ(unit number,*,END=r) list of variable names

where r is the statement number, appearing in the first five columns, of the next executable statement to be processed when the end of the data file is encountered. (See, for instance, Sample Problem 3.7.)

The data files that have been worked on should be rewound to their beginning before they are used again. On some computer systems, the OPEN statement causes the referenced file to be rewound automatically. Not all systems have this capability, though, and it is recommended that a REWIND **statement** be included in your source program in addition to the OPEN statement to ensure that the data file is actually rewound. The REWIND statement has the form

REWIND(UNIT=integer constant)

or

REWIND integer constant

where the unit number is that of the data file.

The rewritten program incorporating these changes is shown in Figure 2.4. The data file is accessed through unit number 8 and has the name TEST. This selection (number and name) was arbitrary. (*Caution:* The actual file begins with the number 10, not with the words File Test.)

```
PROGRAM STAT
REAL X1,X2,X3,X4,X5,X6,X7,X8,X9,X10,ST,ST1,STO,XAV,S
INTEGER N

OPEN(UNIT=8,FILE='TEST')
READ(8,*)N,X1,X2,X3,X4,X5,X6,X7,X8,X9,X10


XAV=(X1+X2+X3+X4+X5+X6+X7+X8+X9+X10)/N
ST=X1**2+X2**2+X3**2+X4**2+X5**2
ST1=X6**2+X7**2+X8**2+X9**2+X10**2
STO=ST+ST1-N*XAV**2
S=SQRT(STO/(N-1))

PRINT *
PRINT *,'THE AVERAGE IS ',XAV
PRINT *,'THE STANDARD DEVIATION IS ',S

REWIND 8

END


File Test

10
2.999
3.001
3.010
3.005
2.995
2.998
3.003
3.003
2.999
3.001


RUN

  THE AVERAGE IS 3.0014
  THE STANDARD DEVIATION IS .004168666204058
```

Figure 2.4

It should be noted that this procedure can be used to write entries on a data file as well. First the output file's unit number is specified, for example,

OPEN(UNIT=9,FILE='RESULT')

Then material is entered with a WRITE **statement.** (PRINT commands can only send material to the terminal.)

WRITE(9,*)XAV,S

And finally, the file is rewound.

REWIND 9

It is just as important to rewind the output file; otherwise the file could continue (on some computers) to accumulate and append storage space.

After saving the new output file, it can be listed and printed at the local terminal whenever convenient and without having to run the program. Figure 2.5 includes these last modifications. OPEN statements have been written to incorporate the output as well as the input file.

With the material contained in this chapter, you have already advanced considerably toward your goal of learning how to program in FORTRAN. Keep up the good work.

REVIEW

Chapter 2, FORTRAN Basics, introduces the six data types of FORTRAN: integer, real, double precision, complex, logical, and character. A FORTRAN integer constant is a whole number or string of decimal digits without a comma or decimal point. A

```
PROGRAM STAT

REAL X1,X2,X3,X4,X5,X6,X7,X8,X9,X10,ST,ST1,STO,S,XAV
INTEGER N

OPEN(UNIT=8,FILE='TEST')
REWIND 8
OPEN(UNIT=9,FILE='RESULT')
READ(8,*)N,X1,X2,X3,X4,X5,X6,X7,X8,X9,X10

XAV=(X1+X2+X3+X4+X5+X6+X7+X8+X9+X10)/N
ST=X1**2+X2**2+X3**2+X4**2+X5**2
ST1=X6**2+X7**2+X8**2+X9**2+X10**2
STO=ST+ST1-N*XAV**2
S=SQRT(STO/(N-1))

WRITE(9,*)XAV,S

REWIND 9

END


File Test

10
2.999
3.001
3.010
3.005
2.995
2.998
3.003
3.003
2.999
3.001


File Result

3.0014 .004168666204058
```

Figure 2.5

real constant, or floating-point value, must include the decimal point. Using exponential notation and floating-point values, both very large and very small numbers can be handled by the computer. A program variable must be assigned an internal memory location where it can receive data. The address of this location bears the variable's name instead of a number. The variable name, being just an address, cannot be symbolically manipulated as it is in mathematics. The equal sign in FORTRAN does not represent equality but symbolizes an assignment process: the quantity on the right of the equal sign is to be stored at the address (the variable name) to the left of the equal sign. Variable names must begin with a letter and can be up to six characters long.

Each variable name must be data typed so that proper memory size and format is assigned. There are two ways to assign type to a variable name. Explicit typing requires that a declarative type statement appear in the beginning of the program, after the program name but before the first executable statement. Character variables require a special declaration, since the length of the character string forming the word must be specified with the data. If not specified otherwise, the computer will assume one character per word. Implicit typing, or typing by default, lets the first letter of the variable name (the address) specify what type of variable will be stored there. Variable names beginning with letters form A to H or with letters from O to Z are considered REAL. Variables that begin with the letters I through N are considered integer.

The basic building elements of a FORTRAN program are assignment statements: a variable name is assigned a numeric value or the resultant value from a series of arithmetic operations. The user specifies the operations to be performed in a FORTRAN arithmetic expression. An assignment statement will then have the variable name to the left of the equal sign and an arithmetic expression to the right.

The computer starts to perform operations in the innermost parentheses first, and then works outward. A priority ordering of exponentiation, multiplication or division, and addition or subtraction is in effect within each pair of parentheses. When all parenthetic operations are cleared, the remaining stand-alone operations are carried out using the same given ordering. If operations have the same priority level, they are carried out from left to right.

In any expression, two operator symbols can never appear in direct succession. When two types of numeric data are mixed in an operation, the type of the resultant value is determined by the ordering: first complex, then double precision, then real, and then integer. The expression's final value must, however, assume the format of the type assigned to the statement's variable name, the variable name to the left of the equal sign. In addition to the fundamental arithmetic operations, FORTRAN provides a built-in library of intrinsic functions for the user.

One can enter data from the keyboard at execution time, or instead use a DATA statement in the program or provide a data file. A data file is not compiled, nor does it follow FORTRAN line (column) format. It also has a separate file name. The source program must be informed of the data file's name, and a unit number must be assigned to access it. This is accomplished by an OPEN statement. A REWIND statement resets the data files back to the beginning. Data files are read from or written onto by the program during execution time using READ or WRITE statements. A PARAMETER statement is a declarative statement that assigns a symbolic name to a constant value in the program.

KEY TERMS

assignment statement
character (data type)
complex (data type, variable,
    or constant)
data file
DATA statement
declarative statement
double precision (data type,
    variable, or constant)
executable (statement)
explicit typing
exponential notation
floating-point value
implicit typing
integer (data type, variable,
    or constant)

intrinsic function
logical (data type)
mixed-mode operation
nonexecutable (statement)
OPEN statement
PARAMETER statement
READ statement
real (data type, variable,
    or constant)
REWIND statement
roundoff error
scientific notation
type statement
typing (explicit and implicit)
WRITE statement

REVIEW QUESTIONS

1. What is ASCII code? What is EBCDIC code?
2. Distinguish between executable and nonexecutable statements.
3. What is a declarative statement?
4. Name the four types of numerical values available to FORTRAN.
5. How long can a variable name be on your system?
6. What is meant by implicit typing?
7. List the order in which arithmetic operations are performed in FORTRAN.
8. What is meant by truncation?
9. What is the cause of roundoff errors?
10. What units do the trigonometric arguments have in the intrinsic function FORTRAN library?
11. What is meant by mixed-mode operation?
12. What problem can arise with mixed-mode operation?
13. What is wrong with the following FORTRAN assignment statements?
    (a) 11=INDEX   (b) Y+70.3=100.0
14. Which of the following are invalid FORTRAN variable names?
    (a) APPLE   (d) X1.0T   (g) X1-3   (j) X(3)
    (b) X10T   (e) 2X1   (h) INDEX   (k) VELOCITY
    (c) X   (f) AX Y4   (i) IB$ (l) 'INDEX'

15. What is the base-10 equivalent number to the binary number 1101101.11_2?

16. If $J = 3$, $K = 4$, $R = 2.1$, and $S = 5.6$, evaluate the following arithmetic expressions. Clearly show the *type* of value that results.

(a) R*S   (d) J/K   (g) S-R**(J/K)

(b) J+K   (e) K/J+S   (h) S*R**R

(c) R/K   (f) (S-R)**J

17. If the arithmetic expressions from Problem 16 are now used to form assignment statements, indicate the resulting values that will be stored at the variable named address indicated. Clearly show type.

(a) MARY=R*S   (d) INDEX=J/K   (g) K=K+1.0

(b) TEST=J+K   (e) EVAL=J/K   (h) K=S*R**R

(c) TEMP=R/K   (f) I=K/J+S

18. For the following series of sequential statements, indicate the ultimate resulting numeric value. Clearly show the type of the stored results.

(a) K=5.6   (b) R=3.2E4   (c) A1=54.0   (d) A1=84.0   (e) X=4.0
    K=K+1   S=1.1E3   A2=67.0   A2=971.05   Y=6.0
    K=EXP(K)   S=R   A3=84.0   T=A1   T=Y**3
                     R=S   A1=A3   A1=A2   X=T*X
                                       A3=A2   A2=T   T=T*X
                                       A2=A1

19. Write valid FORTRAN assignment statements for the following engineering formulas. Assume that variable names I, K, and N will receive real values.

(a) Area moment of inertia of a rectangle

$$I = \frac{BH^3}{12}$$

(b) Radius of a circle

$$R = (X^2 + Y^2)^{1/2}$$

(c) Variation of the acceleration of gravity

$$g = g_S\left(\frac{R}{R + H}\right)^2$$

(d) Speed of sound in gas

$$C = \sqrt{Kg_C RT}$$

(e) Stagnation energy

$$h_S = \frac{P}{\rho} + \frac{V^2}{2g},$$

(f) Reynolds number

$$N = \frac{DV\rho}{\mu g},$$

(g) Manning formula for flow in a channel

$$C = \frac{1.49(r_H)^{1/6}}{N}$$

20. Write the algebraic equivalent to the following FORTRAN statements.

   (a) Equivalent resistance

```
REQ=1.0/(1.0/R1+1.0/R2+1.0/R3)
```

   (b) Damped frequency

```
WD=SQRT(W**2-RC**2)
```

   (c) Beam Deflection

```
Y=W/(120.0*E*BI*BL)*(4.0*BL**5-5.0*X*BL**4+X**5)
```

COMPUTER PROJECTS

CP1 through CP7. Write short computer programs that will read the necessary input from the computer terminal and print the desired results of the equations given in Problem 19, a through g. Label all output.

CP8. Write a FORTRAN program that will read the values of four electrical resistors acting in parallel and then compute their equivalent resistance. Label your answer.

CP9. Write a FORTRAN program that will read the lengths of the base and height of a right triangle from entries made at the terminal and then compute and print the value of the hypotenuse. Label your answer.

CP10. Repeat problem CP9 but this time calculate the triangle's area and perimeter.

CP11. Write a program that will read the values of the radius and height of a right-circular cylinder from entries made at the terminal and then compute the surface area and volume. Label your answers.

CP12. Write a program that will read the value of the radius of an aluminum sphere in meters and then compute its surface area and mass. Label your answers. The specific density of aluminium is 2.7.

CP13. Write a program that will read from the terminal the X- and Y-coordinate values of the endpoints (P_1 and P_2) of a line and then compute the line's length and slope. Label your answers.

CP14. The horizontal and vertical force components, F_x and F_y, respectively, of a planar force **F** that makes an angle θ with the x-axis is given by

$$F_x = F \cos \theta$$
$$F_y = F \sin \theta$$

Write a FORTRAN program that reads the values of F and θ (in degrees) from the terminal and then calculates F_x and F_y. Label your answer.

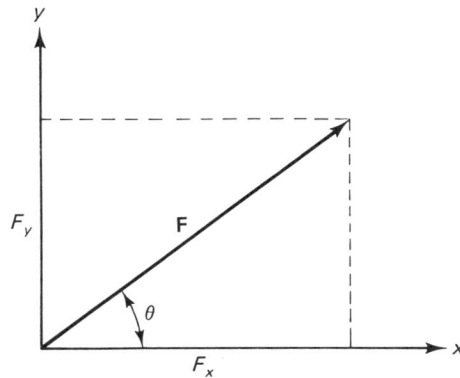

Figure CP2.14

CP15. Write a program that reads the values of F and θ (in degrees) of two forces, $\mathbf{F}_1$ and $\mathbf{F}_2$, from a data file, sums the horizontal and vertical components (see previous problem) of the forces to find the total force acting, and then computes the angle (in degrees) it makes with the x-axis. Label your answers. Why must you use the intrinsic function ATAN2 instead of ATAN?

```
FX=FX1+FX2
FY=FY1+FY2
FTOTAL=SQRT((FX)**2+(FY)**2)
THETA=ATAN2(FY,FX)
```

(The result is in radians.)

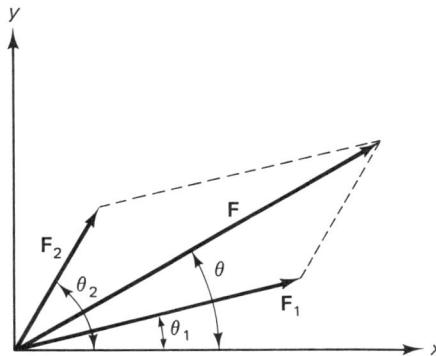

Figure CP2.15

CHAPTER 3

Input and Output

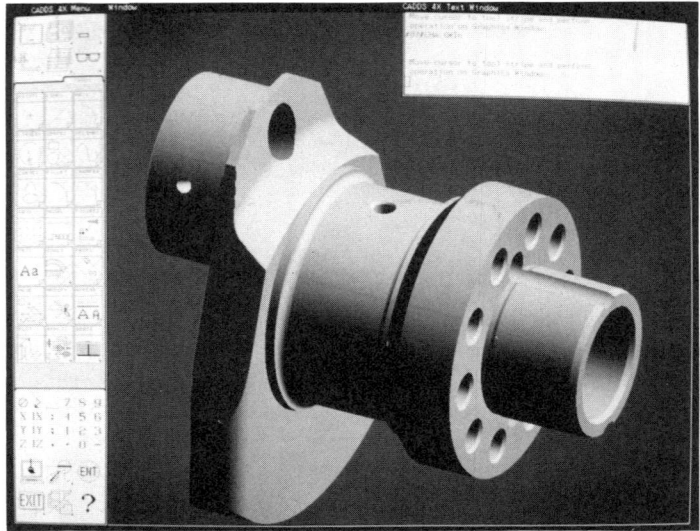

Integrated geometric modeling and shading software tools enable designers to construct accurate, unambiguous solid models of parts and shade them directly on the solid model for view at the terminal. *(Photo courtesy of Prime Computer, Incorporated.)*

3.1 INTRODUCTION

There is one aspect of engineering and scientific calculation that seldom gets the attention it warrants. No matter how clear one's programming efforts are, they will prove of little ultimate value if the end results, the numeric data, are not presented in a clear manner. How data is displayed to the reader, how it is labeled, even how tabular column output is spaced in a report, is extremely important. So far, we have discussed list-directed input and output FORTRAN commands. Though simple to use, they do not give the programmer any means by which he or she can control the format of what is presented or read. This chapter will present other means by which input and output can be handled in FORTRAN 77. We will discuss vertical and horizontal line spacing, the presentation of titles and table headings, and the various forms into which a user can put alphanumeric data. First, though, we will review list-directed input/output statements.

3.2 LIST-DIRECTED INPUT

List-directed input takes the general form

READ(*,*) variable list

The first asterisk in the parentheses represents the input unit number. Since it is not specified, the computer will by default look to the terminal or to a punched-card reader (batch processing) for the information that the READ statement requests.

The second asterisk tells the computer that the input is list-directed and that the computer can choose appropriate storage forms for the variable names. An example of list-directed input is

```
READ(*,*)A,I,CL
```

When encountering this statement in a compiled program, some computer systems will respond with a question mark on the monitor screen if it is in a time-sharing mode. A question mark will remain or reappear on the screen until sufficient data, separated by commas, blanks, or "return" keystrokes, is presented by the user. Floating-point variables, such as A and CL, can have their data entered as real numbers, as integers when they are whole numbers, or in exponential form. Integer variables, such as I, can only have data entered in integer form.

Numeric values are assigned to the variable names (addresses) *in the order in which they are listed*. Each variable name is separated from the others by a comma. For example, a possible response is

```
? 3.2,6,4.1
```

An alternative and much simpler form than the foregoing list-directed READ statement is

READ*,variable list

which we have met before. Since no specific input unit number is referred to in this statement, the CPU defaults to the terminal or punched cards. The statements

```
READ(*,*)A,I,CL
```

and

```
READ*,A,I,CL
```

will therefore cause the computer to take identical action. Similarly, if the standard input unit, the terminal, has unit number 5, one can also write

```
READ(5,*)A,I,CL
```

If, as discussed in the last chapter, a separate data file is to be read from unit number 10, the READ statement becomes

```
READ(10,*)A,I,CL
```

Like the information entered from the terminal, each of the numeric entries in the file must be separated by commas or blanks, or alternatively, they may be entered on separate lines (records). For example, the statement

```
READ(10,*)A,I,CL
```

might have a data file on unit number 10 appear as

```
3.2,6,4.7
```

or

```
3.2  6  4.7
```

or

```
3.2
6
4.7
```

The first variable name in the list, A, will receive the first value, 3.2, I will receive the second, and so forth.

3.3 LIST-DIRECTED OUTPUT

The simplest means of effecting list-directed output is through the use of the PRINT **statement.** The PRINT statement has the form

```
PRINT*,variable name list
```

Output will appear at the terminal or line printer in the order that the variable name list, separated by commas, was entered. For example,

```
PRINT*,A,I,CL
```

will print the constant that is found at address A, followed by the constant found at address I, and so on. Typical output would appear as

```
3.2  6  4.7
```

By writing

```
PRINT *
```

the PRINT command causes an output line to be skipped. It is also possible to use a PRINT statement to write literals, or *character strings,* such as

```
PRINT *,'ENTER DATA'
PRINT *,'THE ANSWER IS=',D
```

or to evaluate arithmetic expressions in the variable list. For example, using the previous data,

```
PRINT*,A+CL,I+6,A
```

will lead to an output of

    7.9 12 3.2

A WRITE **statement** can also be used for list-directed output. The WRITE statement must send output to a unit number in the system. The terminal or line printer is usually designated by the number 6, the standard unit number. If no unit number is specified, the computer will default to the standard unit number. Therefore,

    PRINT*,A
    WRITE(6,*)A

and

    WRITE(*,*)A

will each cause results to be printed at the terminal or line printer using list-directed format.

In the WRITE statement, the first value in the parentheses is the unit number (as asterisk will produce a default to the standard unit); the second value, to the right of the comma, indicates that list-directed format is called for. To direct output to a file using list-directed format, one uses

$$\text{WRITE}(i, *) \text{ A,I,CL}$$

where the file will be found at unit number i. For the variable names included on the list, currently assigned numeric values will be printed out in one record line of the file. For example,

    WRITE(11,*)A,I,CL

will lead to the following entry of data in the associated output file:

    3.2 6 4.7

Sample Problem 3.1

Write a FORTRAN program that will receive a temperature value in degrees Fahrenheit and convert it to the Celsius scale. Use list-directed input and output.

Analysis: The conversion from Fahrenheit to Celsius is given by

$$T_C = (T_F - 32.0)\frac{5}{9}$$

Solution: The algorithm and problem solution are shown in Figures 3.1a and 3.1b. An alternative solution is given in Figure 3.2. To verify the solutions, values of positive, negative, and freezing- and boiling-point temperatures should be input as data.

Sample Problem 3.2

Newton's law of gravitation states that the force of gravity, F, exerted by one body on another is given by

$$F = \frac{GMm}{r^2} \quad \text{newtons}$$

Algorithm for Sample Problem 3.1. This program converts Fahrenheit to Celsius readings.

1. Enter T.
2. Calculate
   `TI=(T-32.0)*5.0/9.0`
3. Print T,TI.

Figure 3.1a

```
       PROGRAM TEMP

*******************************************************************
*                                                                 *
*       THIS PROGRAM CONVERTS FAHRENHEIT TO CELSIUS READINGS.      *
*                                                                 *
*******************************************************************

       REAL T,TI

       PRINT*,'ENTER TEMPERATURE IN DEGREES FAHRENHEIT'
       READ *,T

       TI=(T-32.)*5./9.

       PRINT*,'TEMPERATURE IN F DEGREES = ',T
       PRINT*,'TEMPERATURE IN C DEGREES = ',TI

       END

  RUN

   ENTER TEMPERATURE IN DEGREES FAHRENHEIT
  ? 157.2
   TEMPERATURE IN F DEGREES = 157.2
   TEMPERATURE IN C DEGREES = 69.55555555556
```

Figure 3.1b

where G is the universal gravitational constant, M is the mass of the attracting body, m is the mass of the body acted upon, and r is their separation distance. Write a general program that will determine the weight of a satellite of mass m at a distance h above the earth's surface. Results should be entered on a file called TEMP and also at the terminal. Input can be from the terminal.

Analysis: Assuming a spherically shaped earth, and neglecting the effects of the sun and other bodies, the weight of an object, having mass m on the earth or above its surface, can be calculated by

$$\text{Weight} = F = \frac{GM_e m}{(R_e + h)^2}$$

In this expression, h is the altitude of the object above the earth, GM_e is given as 3.98×10^{14} m^3/s^2, and R_e is known to be 6.37×10^6 m.

```
        PROGRAM TEMP

****************************************************************
*                                                              *
*       THIS PROGRAM CONVERTS FAHRENHEIT TO CELSIUS READINGS.  *
*                                                              *
****************************************************************
        REAL T

        PRINT*,'ENTER TEMPERATURE IN DEGREES FAHRENHEIT'
        READ *,T

        PRINT*,'TEMPERATURE IN F DEGREES = ',T
        PRINT*,'TEMPERATURE IN C DEGREES = ',(T-32.)*5./9.

        END


 RUN

 ENTER TEMPERATURE IN DEGREES FAHRENHEIT
 ? -39
 TEMPERATURE IN F DEGREES = -39.
 TEMPERATURE IN C DEGREES = -39.44444444444
```

Figure 3.2

Solution: The algorithm and problem solution are shown in Figures 3.3a and 3.3b.

Sample Problem 3.3

The apparent weight of an object on the surface of the earth at a latitude angle λ is given by the formula

$$F = R - m\omega_e^2 R_e \cos^2\lambda \quad \text{newtons}$$

where ω_e is the earth's rotational rate, 7.27×10^{-5} radians/second, R_e is the earth's radius, 6.37×10^6 meters, and R is found by

$$R = \frac{GM_e m}{R_e^2} \quad \text{newtons}$$

Algorithm for Sample Problem 3.2. This program calculates the weight of a satellite in orbit.

1. Specify GRAV,RE.
2. Enter H,M.
3. Calculate
   F=GRAV*M/(RE+H)**2
4. Print F.
5. Externally store F in a file.

Figure 3.3a

```
      PROGRAM WEIGHT

***********************************************************************
*                                                                     *
*     THIS PROGRAM CALCULATES THE WEIGHT OF A SATELLITE IN ORBIT.     *
*                                                                     *
***********************************************************************
      REAL M,RE,GRAV,H,F
      PARAMETER(GRAV=3.98E14,RE=6.37E6)

      OPEN(UNIT=8,FILE='TEMP')
      PRINT*,'ENTER HEIGHT ABOVE THE EARTH IN METERS, AND MASS'
      PRINT*,'IN KILOGRAMS.'
      READ*,H,M
      PRINT*,'THE HEIGHT OF THE SATELLITE IS = ',H/1000.,' KM'
      PRINT*,'THE MASS OF THE SATELLITE IS = ',M,' KG'

      F=GRAV*M/(RE+H)**2

      PRINT*,'THE WEIGHT OF THE SATELLITE IS = ',F,' N'
      WRITE(8,*)F

      REWIND 8
      END

  RUN

   ENTER HEIGHT ABOVE THE EARTH IN METERS, AND MASS
   IN KILOGRAMS.
   ? 1.6E5
   ? 45
   THE HEIGHT OF THE SATELLITE IS = 160. KM
   THE MASS OF THE SATELLITE IS = 45. KG
   THE WEIGHT OF THE SATELLITE IS = 420.0192772667 N


  File Temp

   420.0192772667
```

Figure 3.3b

Write a FORTRAN program that will read a previously established file, TEMP, containing the value of R as input (use the program given for Sample Problem 3.2, with h set to 0 to create the required file TEMP), calculate the apparent weight, and store the final result in a file named RESULT.

Solution: The algorithm and problem solution are shown in Figures 3.4a and 3.4b. Note that in Sample Problem 3.2 the variable that had its assigned numerical value stored in file TEMP was called F. In this program, the same numerical value is being assigned to a new variable named R, not F (refer to READ(10,*) R). Keep in mind that file contents are just numerical constants having no inherent meaning until we interpret and assign them. To further emphasize this point, the variable name F is instead assigned to the apparent weight in this program. The terminal user remains unaware of this apparent conflict in meaning that the programmer introduced when assigning variable names. Of course this is not good practice, since variable names should have some association with the values they represent. What values of LAMBD1 could be used to test the program?

Algorithm for Sample Problem 3.3. This program calculates the weight of an object, allowing for the earth's rotation and the object's latitude angle. Sample Program 3.2 should be entered first with H = 0, and the mass. Answers stored in the file from that program are used as input to this program.

1. Initialize WE,RE.
2. Enter M,LAMBD1.
3. Read R from external file.
4. Calculate
   LAMBDA=LAMBD1/57.296
   F=R-M*WE**2*RE*(COS(LAMBDA))**2
5. Externally store F in a new file.
6. Print R,F.

Figure 3.4a

```
        PROGRAM WEIGHT

*******************************************************************
*                                                                 *
*    THIS PROGRAM CALCULATES THE WEIGHT OF AN OBJECT ALLOWING     *
*    FOR THE EARTH'S ROTATION AND THE OBJECT'S LATITUDE ANGLE.    *
*    THE PREVIOUS PROGRAM SHOULD BE ENTERED FIRST WITH H = 0,     *
*    AND THE MASS. ANSWERS STORED IN THE FILE FROM THAT PROGRAM   *
*    ARE USED AS INPUT TO THIS PROGRAM.                           *
*                                                                 *
*******************************************************************
        REAL M,LAMBD1,LAMBDA,WE,RE,R,F
        PARAMETER(WE=7.27E-5,RE=6.37E6)

        OPEN(UNIT=10,FILE='TEMP')
        OPEN(UNIT=11,FILE='RESULT')
        REWIND 10
        READ*,M,LAMBD1
        READ(10,*)R

        LAMBDA=LAMBD1/57.296
        F=R-M*WE**2*RE*(COS(LAMBDA))**2

        WRITE(11,*)F
        PRINT*,'THE UNCORRECTED WEIGHT IS ',R,' NEWTONS'
        PRINT*,'THE CORRECTED WEIGHT IS ',F,' NEWTONS'

        REWIND 11
        END


File Temp

 420.0192772667


RUN

? 45,30
 THE UNCORRECTED WEIGHT IS 420.0192772667 NEWTONS
 THE CORRECTED WEIGHT IS 418.8830033391 NEWTONS


File Result

 418.8830033391
```

Figure 3.4b

3.4 FORMATTED OUTPUT

To retain control (carriage control) over the vertical and horizontal spacing of printed output, or to define the type and length of a printed alphanumeric constant, requires **formatted output.** Since there are six types of data output — integer, floating point (real), double precision, complex, logical, and character — formatted output must have at least six format specifications.

The general syntax that a formatted output expression has is as follows:

PRINT format identifier,variable list

or

WRITE (unit number,format identifier) variable list

where the format identifier must be an integer. For example,

```
PRINT 8,A,I,CI

WRITE(*,8) A,I,CL

WRITE(6,8) A,I,CL
```

will all produce the same computer output at the terminal or line printer. The statement

```
WRITE(9,8) A,I,CL
```

will cause output to be entered at unit number 9.

The format identifier in the PRINT or WRITE commands refers to the statement address number found in columns 1 through 5 of an accompanying nonexecutable FORMAT **statement.** A FORMAT statement in FORTRAN has the form

k FORMAT(format specifications)

where k is the integer constant format identifier. This integer constant is entered in columns 1 through 5 of the statement line. The rest of the expression, from "FORMAT" on, is entered beginning at column 7. For example,

```
      PRINT 8,A,I,CL
    8 FORMAT(1X,F7.4,3X,I3,3X,F4.1,3X,'TESTCASE')
12345 78  . . .  (column  positions)
```

A FORMAT statement can be paired with one or more PRINT or WRITE statements. Since the FORMAT statement is nonexecutable, it may appear anywhere in the program. Most programmers prefer to have the FORMAT statement appear right after the PRINT or WRITE statement with which it is paired; others favor placing FORMAT statements in one group at the end of a program, immediately before the program's END statement.

The FORMAT specifier contains information about

1. vertical spacing: whether the printer should skip a page or a line (or lines)
2. horizontal spacing: how to control spacing between words, the length of data fields, and the placement of decimal points
3. constant type: whether the input or output is integer, floating point, complex, double precision, logical, or character data
4. literal messages for output

3.5 VERTICAL SPACING

Usually only line printers, used for batch processing, offer the capability of controlling vertical spacing. A line printer is a high-speed printer capable of printing an entire line at once at speeds in excess of 1200 lines per minute. These printers can write up to 132 characters per line on printer paper that is 37.8 centimeters wide (ask your computer center for more details). Terminal printers can also write 132 characters per line, but may use paper as small as 21.6 centimeters wide. Sixty lines of output per page is considered standard. Printer paper is continuous and fan-folded with tear perforations provided every 28 centimeters, at the folds.

Though line printers are exceedingly fast, a separate region in internal memory must be set up in the computer to form and hold printer output. A computer must have this internal memory, called **buffer memory,** because of the even greater CPU operating speed. It would be impractical for the computer to slow down to wait for the printer to finish its tasks. Each line in buffer memory is 133 characters wide. Buffer memory is filled from the left with each line's information. The leftmost position, that is, the first character in the line, provides carriage-control information for the printer. The typical control-character symbols used are as follows:

Control character	Result
blank	single-space line feed
0	double-space line feed
1	advance to next page
+	no line feed; allows overprinting

The first position in buffer memory, the carriage control position, is never printed. Actual line printer output begins at the second buffer position. It is very important to remember that the first field of output, one character wide, is reserved for carriage control, and it *must* be assigned by the user. If the user forgets to write in this field, and if the first character of the actual output fills this buffer position, this character will not appear but will be mistakenly identified as a carriage-control character. For example, if the first variable name to be printed has the value 133 and is placed in buffer memory starting at position 1 in the buffer, the printer will skip to the next page (carriage control "1") and then begin to print at the left hand margin the number 33. What happens for other numbers that mistakenly enter the first buffer position is defined differently by each local operating system. When list-directed output is used, the computer automatically places a blank in the first field position of buffer memory.

On some computer systems, carriage control may not be in effect when the terminal monitor and printer are used. The entire line, *including* the first buffer position in memory, will be printed out. Since list-directed output automatically has a blank placed in the first buffer position by the computer, a terminal printer may therefore show a blank appearing in the first column of each line's output. When writing a program that will be processed in a time-sharing mode, carriage-control characters should be left out, since they serve no useful purpose and will appear as output.

3.6 HORIZONTAL SPACING

Placement of the output data in each line's 132 print positions is controlled by the FORMAT field specifiers. Each specifier is separated by a comma and describes a unique field, or string, of printing positions along the line starting at position 1 of the printer. Remember, position 1 of the printer is position 2 in buffer memory. The fields do not have to fill up 132 positions, but they do run consecutively from the left. (As you will see, there are some specifiers that do not operate as just described, for example, tab control or slash (/) line control.) Numeric data fields are associated with the variable list of the PRINT or WRITE statements and appear in the same order that the variable names appear.

3.6.1. Literal Descriptors

Character strings forming messages to be printed when the program is executed can be entered and identified in the FORMAT specifier by enclosing it in apostrophes. Each character string occupies a field position. One can write

```
      PRINT *,'THIS IS AN EXAMPLE'
```

or

```
      PRINT 8
    8 FORMAT(' ','THIS IS AN EXAMPLE')
```

or

```
      WRITE(6,8)
    8 FORMAT(' ','THIS IS AN EXAMPLE')
```

Note that in the last two cases, two fields separated by a comma are used. The first field, one character long, is a blank and is meant to occupy buffer position 1. This blank will not appear on the line printer but instead will invoke single-space line feed. The second field, beginning immediately after the first field, in buffer position 2, will be printed next upon program execution. It will appear all the way to the left margin starting in print position 1.

3.6.2 The X Descriptor

Whenever blank spaces are desired within a field position along the line, the X **descriptor** can be used. The X descriptor has the form

nX

where n is the width of the field of blanks, or the number of blank spaces to be inserted. For example,

```
   WRITE(6,9)
9 FORMAT(1X,4X,'FIRST NAME',5X,'LAST NAME')
```

will produce single-line-spaced output owing to the first field's blank, the literal "FIRST NAME" beginning in column 5 of the printed page, then five blank spaces, then the message "LAST NAME." Note that one could instead write

```
9 FORMAT(5X,'FIRST NAME',5X,'LAST NAME')
```

or

```
9 FORMAT('_____FIRST NAME',5X,'LAST NAME')
```

3.6.3 T Descriptors

Rather than counting positions along a line, one can use the T **descriptors,** or tab controls, as field specifiers. They have the general forms

Tn, TLn, and TRn

For the specifier Tn, the line printer moves forward to buffer position n. This corresponds to print position $n - 1$. At the terminal, the print head moves forward to column position n. The next output will be placed in that print position. Using the previous example, one can instead write

```
9 FORMAT(T6,'FIRST NAME',T21,'LAST NAME')
```

The first tab control places the printer in print position 5 (buffer position 6); the second tab control places the head in print position 20.

The descriptor TLn instructs the print head to move backwards n positions (the maximum is to the first print position), and the descriptor TRn instructs the head to move forward n positions.

3.6.4 End-of-Record Slash Descriptor

When a **slash** (/) **descriptor** is encountered within a field or between fields, it signals to the printer that a new line or record should begin. Multiple slashes will cause subsequent lines to be blank. A slash at the end of a FORMAT specification will cause a line to be skipped.

```
     WRITE(6,9)
   9 FORMAT(1X,'FIRST NAME'/1X,'LAST NAME')
```

causes the following output to be printed at the leftmost margin:

```
     FIRST NAME
     LAST NAME
```

Note that no commas are required to enclose the slash. Note also that since a new line is to begin, a beginning field descriptor — carriage control — must be specified again. Writing

```
   9 FORMAT(1X,'FIRST NAME'///1X,'LAST NAME')
```

causes two lines to be skipped in the output:

```
     FIRST NAME


     LAST NAME
```

Writing

```
   9 FORMAT(1X,'NAME'//)
```

causes two lines to be skipped before the next record output begins.

3.6.5 The I Descriptor

The I **descriptor,** a field specifier for numeric data, is used to inform the computer of the output format (word size) for a given integer variable. The form the description takes is

$$Iw$$

where w is the field width. The integer constant fills the field so that it is *right-justified:* If the field width is larger than needed to fit the numeric value, leading blanks, from the left, will appear at output. If the field is too small, asterisks will be printed in the field positions to indicate to the user that there is an error present in output. Space must also be made available in the field for the placement of an eventual negative sign. Plus signs are suppressed upon output.

Some examples of Iw format with the values $J1 = 32$ and $K = -57$ follow. The statements

```
     WRITE(6,9) J1,K
   9 FORMAT(1X,I2,5X,I3,5X,'TEST')
```

will lead to the line printer output

    32_____-57_____TEST

On the other hand, had K and J1 been inadvertently reversed,

    WRITE(6,9) K,J1

the same FORMAT statement would produce

    **_____32_____TEST

The asterisks appearing in field 1 indicate that there was data overflow, that is, the data was too large for the field representation. The six blanks now appearing before the next number is a direct result of the system suppressing the + sign. Note, finally, that one cannot specify an integer field for a floating-point (real) number, or vice versa.

Sample Problem 3.4

It is desired to write a program that simulates the automatic change maker for a new type of vending machine. The total price and the amount of money received is to be entered at the terminal in dollars and cents. The computer is then to determine the correct change in dollars, half-dollars, quarters, dimes, nickels and pennies. To minimize the amount of coin needed on hand, the computer should first give back as many dollars as possible, then half dollars, then quarters, and so on. The output of the program should state how many dollars, half-dollars, quarters, and so on, make up the change. It is assumed in the program that the amount of money received is greater than or equal to the price of the items.

Solution: The algorithm and problem solution shown in Figures 3.5a and 3.5b act as an ideal vehicle to review several topics, such as mixed-mode operation and truncation effects, calls to intrinsic library routines, the use of READ and PRINT commands, and work with a data file. By displaying necessary instructions on the monitor, the program is also designed to be as user friendly as possible.

In the program, the variable name PRICE is assigned to the total price. Its value is entered from the keyboard. The total amount of money received, TOTAL, is also read from the terminal. CHANGE, the change required, is then calculated as the difference TOTAL − PRICE.

The intrinsic function NINT is used here to round the dollars-and-cents CHANGE amount to a total of cents. This step ensures that no eventual truncation error will occur. For example, if CHANGE was 101.05, the product of 100.0 times CHANGE might result in a rounded answer of 10104.9999.... Making this an integer value by assigning it to variable name ICHAN will result in a value of 10104 being stored instead of 10105. Further, note how the intrinsic function INT is used to compute the number of dollars by truncating all values to the right of the decimal point.

In a similar way, truncation effects are used to advantage in the expression NHALF=ICHAN/50 to form an integer value for the number of half-dollars. After NHALF is found and reported, a new value of ICHAN is found using ICHAN=ICHAN−NHALF*50 (remember what the FORTRAN equal sign means), and then the same reduction proc-

Algorithm for Sample Problem 3.4. This program simulates the automatic change-making procedures for a new type of vending machine. Money received should be greater than or equal to the cost of the time. ICHAN is the change remaining expressed in cents.

1. Enter PRICE,TOTAL.

2. Calculate

```
CHANGE=TOTAL-PRICE
ICHAN=NINT(100.0*CHANGE)
NDOL=INT(CHANGE)
ICHAN=ICHAN-NDOL*100
NHALF=ICHAN/50
ICHAN=ICHAN-NHALF*50
NQUART=ICHAN/25
ICHAN=ICHAN-NQUART*25
NDIME=ICHAN/10
ICHAN=ICHAN-NDIME*10
NICKEL=ICHAN/5
ICHAN=ICHAN-NICKEL*5
NPENN=ICHAN
```

3. Print NDOL,NHALF,NQUART,NDIME,NICKEL,NPENN.

Figure 3.5a

```
      PROGRAM CHANGES

******************************************************************
*                                                                *
*      THIS PROGRAM SIMULATES THE AUTOMATIC CHANGE MAKING        *
*      PROCEDURES FOR A NEW TYPE VENDING MACHINE. MONEY RECEIVED *
*      SHOULD BE GREATER THAN OR EQUAL TO THE COST OF THE ITEM.  *
*                                                                *
******************************************************************

      REAL PRICE,TOTAL,CHANGE
      INTEGER ICHAN,NDOL,NHALF,NQUART,NDIME,NICKEL,NPENN

      OPEN (UNIT=7,FILE='CHANGE')
      REWIND 7

      PRINT*,'ENTER TOTAL PRICE IN DOLLARS AND CENTS.'
      READ*,PRICE
      PRINT*
      PRINT*,'THE TOTAL PRICE IN DOLLARS AND CENTS WAS ',PRICE
      PRINT*
      PRINT*,'ENTER TOTAL AMOUNT OF MONEY RECEIVED IN DOLLARS'
      PRINT*,'AND CENTS. MONEY RECEIVED CAN NOT EXCEED $200,'
      PRINT*,'BUT MUST BE GREATER THAN OR EQUAL TO THE COST OF'
      PRINT*,'THE ITEM.'
      READ*,TOTAL
      PRINT*,'TOTAL MONEY RECEIVED WAS ',TOTAL
```

Figure 3.5b

```
      CHANGE = TOTAL-PRICE
C
C CHANGE IS THE TOTAL CHANGE TO BE MADE IN DOLLARS AND CENTS.
C
      ICHAN=NINT(100.*CHANGE)
C
C ICHAN IS THE REMAINING CHANGE TO BE MADE EXPRESSED IN CENTS.
C INTRINSIC FUNCTION NINT IS USED TO PREVENT ROUNDOFF ERROR.
C
      NDOL=INT(CHANGE)
C
C NDOL IS THE AMOUNT OF DOLLARS OF CHANGE. INTRINSIC FUNCTION INT
C TRUNCATES VARIABLE CHANGE TO DOLLARS.
C
      ICHAN=ICHAN-NDOL*100
      NHALF=ICHAN/50.
C
C NHALF IS THE AMOUNT OF HALF DOLLARS OF CHANGE. USING THE TRUNCATION
C INHERENT IN INTEGER DIVISION.
C
      ICHAN=ICHAN - NHALF*50
C
C NQUART, NDIME, NICKEL, AND NPENN ARE THE AMOUNTS OF QUARTERS, DIMES,
C NICKELS, AND PENNIES TO BE RETURNED.
C
      NQUART=ICHAN/25.
      ICHAN=ICHAN-NQUART*25

      NDIME=ICHAN/10.
      ICHAN=ICHAN-10*NDIME
      NICKEL=ICHAN/5.
      ICHAN=ICHAN-5*NICKEL
      NPENN=ICHAN

      PRINT*
      PRIN((T 1
      WRITE(7,1)
    1 FORMAT(T5,'CHANGE TO BE MADE')
      PRINT 2, NDOL
      WRITE(7,2) NDOL
    2 FORMAT('THE NUMBER OF DOLLARS OF CHANGE IS ',I3)
      PRINT 3,NHALF
      WRITE(7,3) NHALF
    3 FORMAT('THE NUMBER OF HALF DOLLARS OF CHANGE EQUALS ',I1)
      PRINT 4,NQUART
      WRITE(7,4)NQUART
    4 FORMAT('THE NUMBER OF QUARTERS OF CHANGE EQUALS ',I1)
      PRINT 5,NDIME
      WRITE(7,5)NDIME
    5 FORMAT('THE NUMBER OF DIMES OF CHANGE EQUALS ',I1)
      PRINT 6,NICKEL
      WRITE(7,6) NICKEL
    6 FORMAT('THE NUMBER OF NICKELS OF CHANGE IS ',I1)
      PRINT 7,NPENN
      WRITE(7,7) NPENN
    7 FORMAT('THE NUMBER OF PENNIES OF CHANGE IS ',I1)

      END

RUN

ENTER TOTAL PRICE IN DOLLARS AND CENTS.
? 34.53

THE TOTAL PRICE IN DOLLARS AND CENTS WAS 34.53
```

Figure 3.5b   (cont.)

```
ENTER TOTAL AMOUNT OF MONEY RECEIVED IN DOLLARS
AND CENTS. MONEY RECEIVED CAN NOT EXCEED $200,
BUT MUST BE GREATER THAN OR EQUAL TO THE COST OF
THE ITEM.
? 100.75
TOTAL MONEY RECEIVED WAS 100.75

        CHANGE TO BE MADE
THE NUMBER OF DOLLARS OF CHANGE IS   66
THE NUMBER OF HALF DOLLARS OF CHANGE EQUALS 0
THE NUMBER OF QUARTERS OF CHANGE EQUALS 0
THE NUMBER OF DIMES OF CHANGE EQUALS 2
THE NUMBER OF NICKELS OF CHANGE IS 0
THE NUMBER OF PENNIES OF CHANGE IS 2


File Change

        CHANGE TO BE MADE
THE NUMBER OF DOLLARS OF CHANGE IS   66
THE NUMBER OF HALF DOLLARS OF CHANGE EQUALS 0
THE NUMBER OF QUARTERS OF CHANGE EQUALS 0
THE NUMBER OF DIMES OF CHANGE EQUALS 2
THE NUMBER OF NICKELS OF CHANGE IS 0
THE NUMBER OF PENNIES OF CHANGE IS 2
```

Figure 3.5b   (cont.)

---

ess occurs again. The number of pennies, NPENN, is assigned the final value of ICHAN.

The reader should now test the program. How can the program be extended to include bills of five-, ten-, and twenty-dollar denominations for change?

3.6.6 The F Descriptor

The F **descriptor** is used for floating-point or real number output. It takes the form

F$w.d$

In this expression w is the field width and d is the number of decimal places to be kept to the right of the decimal point. The number is right-justified in the field: If the number of digits to the right of the decimal point for the numeric value is less than d, trailing zeros will be printed out to fill the length d. If the number of decimal positions is greater than d, the displayed number will be rounded off to the correct length, but the internal number will remain the same. If the number of digits to the left of the decimal point is smaller than that allowed, blanks will fill the field from the left. If the number is too large, asterisks will fill the field, indicating an output error. Provision must be made for the sign and decimal point.

For some compilers, when the number is negative and has an absolute value less than 1, another place must be provided for a leading zero, which will precede the decimal point. It is therefore good practice to always have the field width, w, at least three positions larger than the decimal fraction length, d. In general, specifying the correct field size may become a trial-and-error process because we do not know, or cannot estimate, the magnitude of unknown variables to be calculated.

For example, if R = 37.146 and S = −.71 (with the F format, the internal value must be real or double-precision data),

```
    WRITE(6,11) R,S
11  FORMAT(1X,F6.3,5X,F5.2)
```

will produce the output

```
    37.146_____−.71
```

on the line printer. For the format

```
11  FORMAT(1X,F7.3,5X,F6.3)
```

the result will be

```
    _37.146_____−.710
```

The FORMAT expression

```
11  FORMAT(1X,F5.2,5X,F3.2)
```

will cause the output

```
    37.15_____***
```

since no provision was made for the sign (and leading zero if required).
    If the format

```
11  FORMAT(F6.3,5X,F5.2)
```

were used, the resulting output would depend on the operating characteristics of the local computer system, since the first position in the buffer, the carriage control position for the line printer, would now contain a 3.

Sample Problem 3.5

Write a FORTRAN program that will take the position of an arbitrary point given in x- and y-coordinates and convert it to the polar coordinates r and θ. Assuming that the given point lies on the perimeter of a circle that is centered at the origin, calculate the circle's circumference and area. Results should be placed in a labeled table with X, Y, R, THETA, AREA, and CIRCUM in one row. Test your program at $x = y = 2$.

Analysis: The equations that convert x- and y-coordinates to polar coordinates are

$$r = \sqrt{x^2 + y^2}$$
$$\theta = \tan^{-1}(y/x)$$

Solution: The algorithm and problem solution are shown in Figures 3.6a and 3.6b. The value of θ is converted to degrees within the program before it is printed. Note also how the sixty-five character underscores (_) were entered in the program. By enclosing any field within parentheses and then placing an integer number, n, before it, the field is repeated n times during program execution (see Section 3.8). Finally, observe how a PARAMETER statement is employed.

Algorithm for Sample Problem 3.5

1. Define PI.

2. Enter X,Y.

3. Calculate
```
R=SQRT(X*X+Y*Y)
THETA=ATAN2(Y,X)
THETA=THETA*180.0/PI
AREA=PI*R*R/2.0
CIRCUM=2.0*PI*R
```

4. Print X,Y,R,THETA,AREA,CIRCUM.

Figure 3.6a

```
      PROGRAM CIRCLE

      REAL X,Y,R,THETA,AREA,CIRCUM,PI
      PARAMETER (PI=3.141592)

      READ*,X,Y

      R=SQRT(X*X+Y*Y)
      THETA =ATAN2(Y,X)
C
C   CHANGING THETA TO DEGREES.
C
      THETA =THETA*180./PI
      AREA=PI*R*R/2.
      CIRCUM=2.*PI*R

      WRITE(6,1)
    1 FORMAT(//T27,'CIRCLE PROPERTIES'//)
      WRITE(6,2)
C
C   NOTICE THE SHORT CUT TAKEN IN SPECIFYING A LINE OF UNDERSORES: 65'_'
C
    2 FORMAT(1X,65('_'))
      WRITE(6,3)
    3 FORMAT(T7,'X',TR8,'Y',TR8,'R',TR8,'THETA',TR8,'AREA',
     + TR8,'CIRCUM')
      WRITE(6,4)X,Y,R,THETA,AREA,CIRCUM
    4 FORMAT(/T2,F6.2,3X,F6.2,3X,F6.2,7X,F6.2,6X,F6.2,8X,F6.2)

      END


RUN

?  2,2
```

                                CIRCLE PROPERTIES

X	Y	R	THETA	AREA	CIRCUM
2.00	2.00	2.83	45.00	12.57	17.77

Figure 3.6b

3.6.7 The E Descriptor

The E **descriptor** specifies conversion of an internal floating-point (or double-precision) number to exponential notation upon output. The E field descriptor has the following format:

E*w.d*

where *w* is the field width and *d* is the number of decimal digit positions retained after the decimal point. The field width, *w*, must make provision for a sign, the decimal point, the decimal fraction, the symbol E for the exponent, the sign of the exponent, the maximum size of the exponent, and finally, the decimal fraction length itself, namely *d*. Some common compilers provide for a leading zero before the decimal point if the results are negative. Others require a three-place exponent, including sign. The recommended field width, *w*, should then be equal to *d* + 7. Details vary with the actual compiler system used, so check with your computer center.

$$-0.d\mathrm{E}-ee=d+7$$
$$1\ 23\ 4\ 5\ 67$$
$$d$$

On output, the field is right-justified: If the decimal fraction is less than *d*, trailing zeros will follow the decimal fraction to fill the field; if the internal fraction is larger than *d*, the displayed number will be rounded off to the correct size, but the internal value will not change. If insufficient places are available for output of the signs, decimal points, and exponents, asterisks will fill the entire field. If the field width, *w*, is larger than required, blanks will fill the leftmost portion of the output field. The overriding advantage of exponential output form is that one specification, for example, E14.7, can handle any size number, small or large. Since the magnitude of output variables is usually not known beforehand, E format is the preferred means of data representation.

Once the size of answers is made available, the user can switch to F format for final reports if he or she so desires. Assuming that R = 37.146, S = −0.71, and T = −0.032, then, for line printer output,

```
    WRITE(6,11) R,S,T
11  FORMAT(1X,E12.5,5X,E9.2,5X,E10.3)
```

will produce the display

```
__.37146E+02_____-.71E+00_____-.320E-01
```

The format

```
11  FORMAT(1X,E12.4,5X,E7.2,5X,E9.2)
```

will display

```
___.3715E+02_____*******_____-.32E-01
```

Sample Problem 3.6

For the alternating-current circuit shown in Figure 3.7 containing resistance, capacitance, and inductance, write a FORTRAN program to find the following quantities:

Total impedance = $Z = \sqrt{R^2 + [2\pi fL - 1/(2\pi fC)]^2}$ ohms

Maximum current = $I_{max} = V_{max}/Z$ amps

$\theta = \tan^{-1}\left[\dfrac{2\pi fL - 1/(2\pi fC)}{R}\right]$ radians

Effective current = $I_{eff} = 0.707 I_{max}$ amps

Resonant frequency = $F_{res} = \dfrac{1}{\sqrt{4\pi^2 LC}}$ hertz

Band width = $BW = R/(2\pi L)$ hertz

Take $V_{max} = 9$ volts, $R = 500$ ohms, $L = 0.06$ henrys, $C = 0.053 \times 10^{-6}$ farads, $f = 2000$ Hz, and $\pi = 3.141592$. Your results should be arranged in tabular form.

Solution: See Figures 3.8a and 3.8b for the algorithm and problem solution.

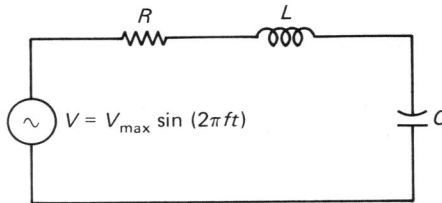

Figure 3.7

Algorithm for Sample Problem 3.6

1. Define PI.
2. Enter VMAX,R,L,C,F.
3. Calculate
   ```
   Z=SQRT(R*R+(2.0*PI*F*L-1.0/(2.0*PI*F*C))**2)
   THETA=ATAN2((2.0*PI*F*L-1.0/(2.0*PI*F*C)),R)
   THETA=THETA*180.0/PI
   IMAX=VMAX/Z
   IEFF=0.707*IMAX
   FRES=1.0/SQRT(4.0*PI*PI*L*C)
   BW=R/(2.0*PI*L)
   ```
4. Print VMAX,F,R,L,C,IMAX,IEFF,Z,THETA,BW,FRES.

Figure 3.8a

```
      PROGRAM CIRCUIT

      REAL L,IMAX,IEFF,PI,VMAX,R,C,F,Z,THETA,FRES,BW
      PARAMETER (PI=3.141592)

      PRINT*,'ENTER VMAX IN VOLTS.'
      READ*,VMAX
      PRINT*,'ENTER R IN OHMS.'
      READ*,R
      PRINT*,'ENTER L IN HENRY.'
      READ*,L
      PRINT*,'ENTER C IN FARADS.'
      READ*,C
      PRINT*,'ENTER F IN HERTZ.'
      READ*,F

      Z=SQRT(R*R+(2.*PI*F*L-1./(2.*PI*F*C))**2)
      THETA=ATAN2((2.*PI*F*L-1./(2.*PI*F*C)),R)*180./PI
      IMAX=VMAX/Z
      IEFF=0.707*IMAX
      FRES=1./SQRT(4.*PI*PI*L*C)
      BW=R/(2.*PI*L)

      PRINT 1
    1 FORMAT(///T29,'CIRCUIT PROPERTIES'/)
      PRINT 2,VMAX,F
    2 FORMAT(15X,'VMAX = ',E10.3,' VOLTS',5X,' FREQ = ',E9.2,' HZ'/)
      PRINT 3,R,L,C
    3 FORMAT(3X,'R = ',E10.3,' OHMS',5X,'L = ',E10.3,' HENRYS',
     + 5X,'C = ',E10.3,' FARADS')
      PRINT 4
    4 FORMAT(//75('_')/)
      PRINT 6
    6 FORMAT(T4,'I-AMPS',T16,'IEFF-AMPS',T29,'Z-OHMS',T41,
     + 'PHASE-DEG', T53,'BAND-HZ',T65,'RZFREQ.-HZ')
      PRINT 7
    7 FORMAT(75('_')/)
      PRINT 8,IMAX,IEFF,Z,THETA,BW,FRES
    8 FORMAT(1X,E10.3,2X,E10.3,3X,E10.3,2X,E10.3,2X,E10.3,
     + 2X,E10.3)

      END

RUN

 ENTER VMAX IN VOLTS.
 ? 9
 ENTER R IN OHMS.
 ? 500
 ENTER L IN HENRY.
 ? .06
 ENTER C IN FARADS.
 ? .53E-07
```

CIRCUIT PROPERTIES

```
        VMAX =   .900E+01 VOLTS      FREQ =   .20E+04 HZ

 R =   .500E+03 OHMS     L =   .600E-01 HENRYS    C =   .530E-07 FARADS
```

I-AMPS	IEFF-AMPS	Z-OHMS	PHASE-DEG	BAND-HZ	RZFREQ.-HZ
.100E-01	.708E-02	.899E+03	-.562E+02	.133E+04	.282E+04

Figure 3.8b

3.7 FORMATTED INPUT

Using formatted input to enter data is not as prevalent today as it was when punched-card readers were employed extensively. The exact location of data on punched cards is very critical and requires much concentration from the user. Variable names are given individual fields on the card, and the format descriptors define their lengths. Using formatted input in a time-sharing mode requires even greater care. When entering data, each decimal digit must occupy a specific column position on an unmarked monitor screen. List-directed input operations are a much simpler means for entering data. However, when reading from a preexisting data file that has a specific data field placement, the user has no other choice but to use input-format specifiers.

The input descriptors available for READ operations are the very same descriptors that were used for alphanumeric character output. There is no carriage-control field, of course, and input of data begins in the very first column position. It is also not possible to enter literal expressions, that is, character expressions that are not associated with any variable name.

3.7.1 The X Descriptor

The X descriptor, having the form

nX

results in n spaces being skipped over in a data line or record. For instance, if the current data line of unit number 8 appears as 431765.21 (nine positions), the statements

```
      READ(8,11) R
11 FORMAT(4X,F5.2)
```

will cause R to be taken from the last five entries as 65.21. The first four positions are skipped. For the data line entry 32146879.21, the statements

```
      READ(8,11) N,S
11 FORMAT(I2,3X,F6.2)
```

will cause N to be taken from the first two data positions, and S from the last six positions: N = 32 and S = 879.21.

3.7.2 T Descriptors

T descriptors for input data have the general forms

Tn, TLn, and TRn

They function almost identically to the T descriptors for output, which were discussed in Section 3.6.3, providing a means of moving rapidly over the data field. Since no carriage-control field is provided for in the reader buffer memory, the statement

Tn

will cause $n - 1$ spaces to be skipped on the data line and set the next entry position at column n. The descriptor TRn produces the same result as the use of nX.

3.7.3 The End-of-Record Slash Descriptor

The end-of-record slash descriptor, when encountered in a READ-associated FORMAT statement, will cause the current line activity to stop and the next data line to be read, starting at the first position of the new line. For example,

```
    READ(8,11) I,N
11  FORMAT(I2/I3)
```

with the assumed consecutive data lines

```
1234_42
461
```

will assign the values 12 to I and 461 to N. Commas do not have to precede or follow the slash.

3.7.4 The I Descriptor

The I descriptor for input has the general form

Iw

Except for an allowed leading positive or negative sign, no nonnumeric data may be present in an integer field. (Leading positive signs are not required.) Even a decimal-point character is not permitted. If nonnumeric data appear in the field, an error message will result. Any blanks in the field will be converted to zeros.

3.7.5 The F Descriptor

The F descriptor for input appears as

Fw.d

As before, w is the field width and d is the dimension of the decimal fraction. The field width must be large enough to accommodate any sign or decimal point that appears in the field. Blanks in the field are set to zero. If no decimal point appears with the data, the data will receive a decimal point placed in the $d + 1$ position from the right. When a decimal point appears with the data, it overrides any decimal placement defined by d. For example, with the data line

```
-461.74__641__2.64
```

the following statements

```
      READ(8,11) R,S,T
 11   FORMAT(F7.2,2X,F3.2,2X,F4.0)
```

will set R = -461.74, S = 6.41, and T = 2.64.

For the variable name S, no decimal point was given in its assigned data field. Instead, when the field was read with the format descriptor F3.2, a decimal was placed in the third position from the right. Note that to write this number now from internal memory requires the format descriptor F4.2.

The variable name T has a decimal point placed in its field position. This last means of entering data is preferred, since it reduces the risk of typing errors. The constant value for T could be entered in exponential form as .264E1 or even .264 + 1 (the plus sign replacing the E), but this would require a larger field.

For the previous given data line, the statement

```
      READ(8,11) R
 11   FORMAT(F6.2)
```

will cause the assignment

```
R=-461.7
```

Here the field for R was cut after six places ($w = 6$). The decimal place actually appearing in the field overrides the d specification.

3.7.6 The E Descriptor

The E descriptor for input has the form

$$Ew.d$$

where w is the field width and d is the dimension of the decimal fraction. The field must be large enough to include the exponent, the decimal part, and any sign or decimal point that appears. If a decimal point appears in the field, it overrides the format descriptor. If no decimal point appears, the position of the decimal point is determined by the value d. (See F Descriptor, Section 3.7.5.) Any blanks in the field are interpreted as zero. Since the E descriptor will also process data in F format, there is no difference between E and F format for input purposes; both formats are treated identically. This means that for the descriptor E11.4, all the following input data fields, which have a width of eleven data positions, will be interpreted in the same manner by the computer:

```
_.16327E+01
_0.16327E+1
```

```
_0.163270+1
__.16327 +1
__.16327+ 1
___.16327E1
_163.27E-02
_.016327+02
_____16327
1.6327_____
```

The presence of just an exponent part in an input data field is not allowed, for example, E+2 instead of 1.E+2.

3.8 REPEATED EDIT DESCRIPTORS

The numeric edit descriptors we have been discussing (I, F, and E) can be made to cover successive fields by prefixing the descriptor with an appropriate integer coefficient. The general forms then become

rIw, rF$w.d$, and rE$w.d$

For example, if N = 2, K = 30, R = 1.06, and S = 9.73 are stored in memory, the statements

```
    WRITE(6,11) N,K,R,S
11  FORMAT(1X,2I2,5X,2F4.2)
```

will cause the following output to appear on the line printer.

```
_230_____1.069.73
```

Changing the format statement to

```
11  FORMAT(1X,2(I2,5X),2(F4.2,5X))
```

will cause group repetition of the field descriptors in parentheses. The output will now appear as

```
_2_____30_____1.06_____9.73
```

If the variable list is less than the number of fields made available by the FORMAT statement, the extra fields will be ignored. If, on the other hand, the variable list is greater than the number of fields specified, the next line is begun, and the computer reinterprets the rightmost groups of fields enclosed in parentheses — that is, the last groups defined within the format statement delimiting parentheses — to obtain further specifications. When these are depleted again, the next line is begun and control goes back to the beginning of this last group, and so on. If no groups are present within the delimiting parentheses, the next line is begun and the whole FORMAT specification is repeated.

When the variable list is greater than the number of fields specified, the input/output list is said to be **overloaded.** To show how these rules function, take M = 1, N = 2, R = 3.4, S = 4.2, and T = 5.6; then the statements

```
      WRITE(6,11) M,N,R,S,T
11  FORMAT(1X,2(2X,I1),2(2X,F3.1))
```

will produce as line-printer output

```
__1__2__3.4__4.2
_5.6
```

Note that with the carry over to the next line, errors with carriage control can occur. Two more examples follow for M = 1, N = 2, M1 = 3, N1 = 4, and L = 5. If

```
      WRITE(6,9) M,N,M1,N1,L
9  FORMAT(1X,2I4)
```

is used, the line-printer output will be

```
___1___2
___3___4
___5
```

Employing the FORMAT statement

```
9  FORMAT(1X,I4)
```

results in the output

```
___1
___2
___3
___4
___5
```

Sample Problem 3.7

It is desired to evaluate the load-carrying capacity of several bars of varying materials and cross-sectional areas (see Figure 3.9). It is known that when the internal stress (load/area) existing within a material exceeds the material's stress tolerance, it will break. Knowing what this stress limit is, and providing for a factor of safety, it is possible to calculate the maximum load P that can be applied to the member. In this problem data from a file BARS is to be read, for which it is known that the first five lines of the file contain heading material only. The needed information, A (area) and S (stress

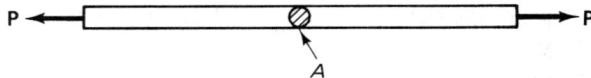

Figure 3.9

limit), are found beginning in columns 15 and 40, respectively, after line 5. Each have formats of E10.3. Taking the factor of safety $F_S = 2.0$, write a FORTRAN program that will access the data file BARS, read the necessary data, calculate the allowable loads, and then print the values of A, the allowable stress (S/F_S), and P in an output file called STRESS.

Analysis: The maximum load, P, with a factor of safety, F_S, is given by

$$P = \text{allowable stress} \times \text{area} = A(S/F_S)$$

Solution: (See Figures 3.10 and 3.11a and 3.11b for the solution's flow chart, algorithm, and source code.) This program employs two new FORTRAN tools. The first, a READ statement with an end-of-data transfer, was introduced in Section 2.8. It must be remembered that normal execution flow of a program is sequential, from top to bottom. With this particular READ statement, when the end of the file BARS is reached, control will jump — be transferred — to the statement that has statement number 99. In this case, the transfer will cause the processing to stop. To empty the file, the READ statement must be executed several times. One could write repetitive identical READ statements in the program, but since it is not known how long the data file is, and its length must be assumed to vary from day to day, the programmer cannot plan accordingly. Too little or too many READ statements will cause the program to fail.

An alternative method is to repeatedly loop back, that is, transfer control, to the given READ statement until the file is empty. A GO TO statement accomplishes this. The GO TO statement effects an unconditional jump in execution to the statement number it references. In this case, processing goes backward to the READ statement, then forward sequentially, then back to the READ statement, and so forth. A loop has been formed.

If the END "instructions" had not been included in the READ statement, providing an exit from the loop, the program would have failed after the file was emptied (the last entry read) and the next READ call was placed to it. An end-of-data-encountered error message would have appeared. In general, be very careful to provide some form of exit when you use a loop construct, otherwise the program will only stop after you run out of computer-allotted processing time, and that can be expensive. Note finally that the five end-of-record slashes included in the first FORMAT statement cause the first five records in the file to be passed over.

REVIEW

In Chapter 3 we learn how to enter and print out real- and integer-type data in a FORTRAN program. The READ, PRINT, or WRITE statements that effect input/output operations impart information to the compiler regarding the unit device number with which to communicate, a list of variables to be processed, and the formatting requirements for the data. Variable names in the list are separated by commas, and data must be entered in the order corresponding to the variable list. The simplest means of handling input/output operations is through the use of list-directed format. On input, all variable data must be separated by commas, blanks, or carriage returns (next record line). The numerical data entered should be of the same type as the variable name to which it is assigned. List-directed output is achieved either through

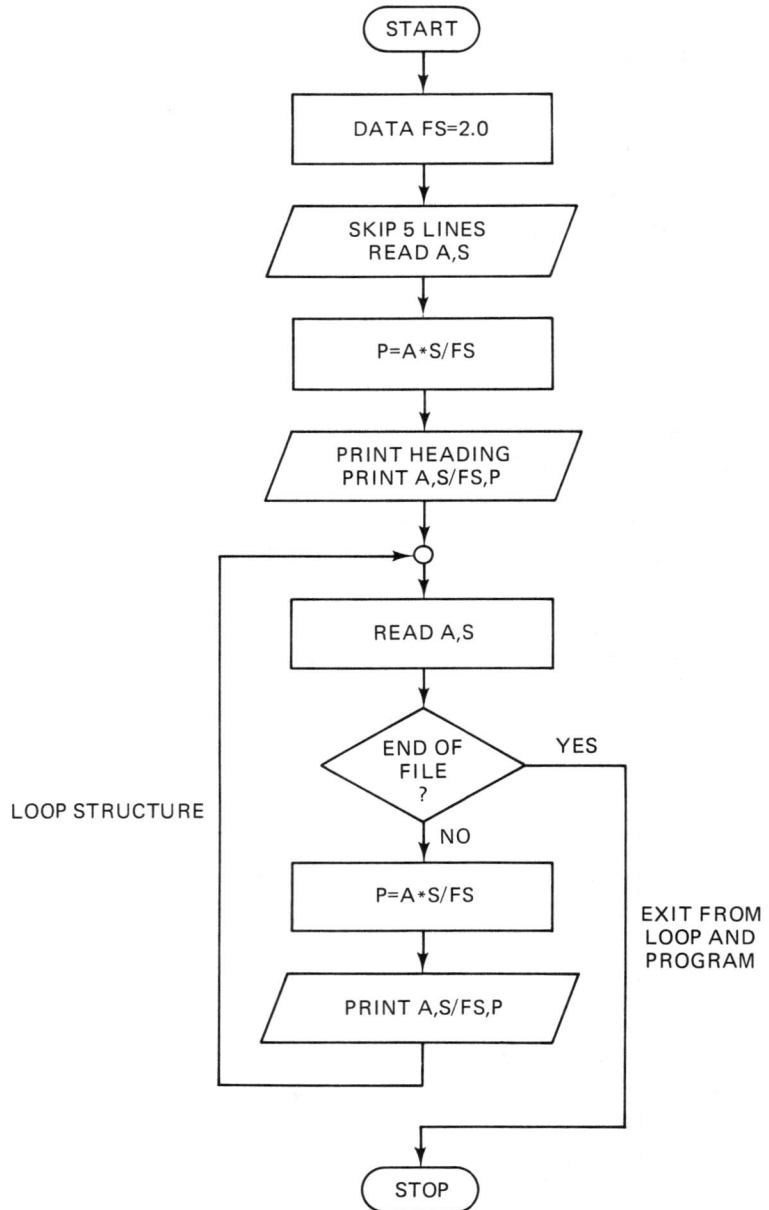

Figure 3.10

a PRINT or a WRITE statement. The PRINT statement can also be used to generate blank lines, print literal messages, or evaluate arithmetic expressions in the variable list.

Formatted input is accomplished through READ commands that work in tandem with paired FORMAT statements. The accompanying referenced FORMAT statements, which are numbered, incorporate a list of format specifiers, or field descriptors, that define data placement in an input line, or record. Field descriptors are separated by commas. The input specifiers described in this chapter are the X, T, TRn, and TLn descriptors for horizontal positioning, the end-of-record slash (/) for vertical positioning to the next line, and the Iw, F$w.d$, and E$w.d$ formats for numeric data. The field width, w, must be large enough to include sign and decimal point, decimal fraction, and any exponent that appears. The actual placement of the decimal point in the field takes precedence over the field specifier in case of disagreement. If no decimal point appears in the field, a decimal point is placed by the compiler d positions from the right. (The format size needed for output will have to be increased by one position.) An integer format cannot have a decimal point or other character in the field.

PRINT and WRITE commands are used in conjunction with FORMAT statement field specifiers to control record output. When using a line printer for output,

---

Algorithm for Sample Problem 3.7. This program calculates the allowable loads for a bar of uniform cross-sectional area. Data for the program, area and allowable stress, is read in from a data file. The first five records of this file are to be skipped over. Answers are written on an output file. A factor of safety (FS) of 2 is assumed for the part's design. The allowable load is P, A is the cross-sectional area, and S is the allowable stress. Task I is to skip five records in the input data file.

1. Initialize FS.
2. Skip 5 records and read A,S.
3. Calculate
   P=A*S/FS
4. Externally store A,S/FS,P.

Task II is to empty the input data file.

5. While the file still has data left,
   a. Read A,S
   b. Calculate
      P=A*S/FS
   c. Externally store A,S/FS,P
   d. Repeat steps 5a through 5c.

Figure 3.11a

```
      PROGRAM LOADS

*********************************************************************
*                                                                   *
*      THIS PROGRAM CALCULATES THE ALLOWABLE LOADS FOR A BAR OF     *
*      UNIFORM CROSS-SECTIONAL AREA. DATA FOR THE PROGRAM, AREA     *
*      AND ALLOWABLE STRESS, IS READ IN FROM A DATA FILE NAMED      *
*      'BARS'. ANSWERS ARE WRITTEN ON AN OUTPUT FILE NAMED          *
*      'STRESS'. A FACTOR OF SAFETY (FS) OF 2 IS ASSUMED.           *
*                                                                   *
*********************************************************************

      REAL FS,A,S,P

      DATA FS/2.0/
      OPEN(UNIT=7,FILE='BARS')
      OPEN(UNIT=8,FILE='STRESS')
      REWIND 7
      REWIND 8
      READ(7,1)A,S
    1 FORMAT(/////,T15,E10.3,T40,E10.3)

      P=A*S/FS

      WRITE(8,2)
    2 FORMAT(//T29,'LOADING ON BARS')
      WRITE(8,3)
    3 FORMAT(1X,65('_'))
      WRITE(8,4)
    4 FORMAT(/T12,'AREA',TR17,'STRESS',TR14,'ALL. LOAD')
      WRITE(8,6)
    6 FORMAT(1X,65('_')/)
      WRITE(8,5)A,S/FS,P
    5 FORMAT(5X,3(E10.3,13X))

   10 READ(7,7,END=99)A,S
C
C  NOTE THE END OF FILE INSTRUCTION. WHEN THE INPUT FILE 'BARS' IS
C  EMPTY, CONTROL PASSES TO STATEMENT NUMBER 99, A STOP STATEMENT.
C
    7 FORMAT(T15,E10.3,T40,E10.3)
      P=A*S/FS
      WRITE(8,5)A,S/FS,P
C
C  WE WANT TO CONTINUE READING THE FILE UNTIL IT IS EMPTY. THIS MEANS WE
C  MUST ALTER THE NORMAL SEQUENTIAL TOP-DOWN FLOW OF THE PROGRAM AND
C  RETURN TO THE READ STATEMENT. BY GIVING THE READ STATEMENT A
C  STATEMENT NUMBER, WE CAN TRANSFER BACK TO THE READ STATEMENT AT ANY
C  TIME VIA A "GO TO" CONTROL STATEMENT.
C
      GO TO 10

   99 STOP
      END

      File Bars

      THIS
      IS AN EXAMPLE
      OF HOW
      RECORDS
      CAN BE SKIPPED IN A FILE
                    .200E 00              .330E 05
                    .250E 00              .330E 05
                    .300E 00              .400E 05
```

Figure 3.11b

```
File Stress
```

LOADING ON BARS

AREA	STRESS	ALL. LOAD
.200E+00	.165E+05	.330E+04
.250E+00	.165E+05	.413E+04
.300E+00	.200E+05	.600E+04

Figure 3.11b (cont.)

the first field position of each line in a FORMAT statement must be reserved for carriage control information, that is, double-spacing, starting a new page, and so forth. The remaining fields are then specified by the X, T, TRn, TLn, Iw, F$w.d$, E$w.d$, and slash descriptors. Each field is separated by a comma. On output all numeric data appears right-justified in the field space allotted. The field width, w, must be large enough to include sign, decimal point, decimal fraction, and exponent when applicable. Some compilers require a leading zero to be written if the number is negative. Fractional parts will be rounded off to fit within the fraction length specified, d. Incompatibility between variable data size and maximum field width allotted will lead to the appearance of asterisks in the displayed output if the field is too small.

To repeat edit descriptors n times, the integer number n may be placed directly before the specifier, for example, nIw. To repeat a group of specifiers n times in a FORMAT statement, the integer number n may be placed before the group, with the group enclosed in parentheses. A typical error is to have an unequal number of right- and left-facing parentheses in a FORMAT statement. When there are more variables in the variable list than there are field specifiers, a new record will begin and control will pass to the first field specifier of the last enclosed group, or to the beginning of the FORMAT statement if no groups are present.

SUMMARY OF FORTRAN STATEMENTS

List-Directed Input

READ *,variable list	Read from the terminal.
READ(*, *) variable list	Read from the terminal.
READ(5, *) variable list	Read from the terminal.
READ(10, *) variable list	Read from a file at unit number 10.

List-Directed Output

PRINT *, variable list	Write to the terminal.
PRINT(*, *) variable list	Write to the terminal.

WRITE(6, *) variable list   Write to the terminal.

WRITE(10, *) variable list   Write to the file at unit number 10.

Formatted Input

READ(i, n) variable list i is the unit number and n is the FORMAT
                                 statement number. When $i = *$ or 5, entry
                                 is from the terminal.

n FORMAT(format specifiers) n is the statement number.

Formatted Output

WRITE(i, n) variable list i is the unit number and n is the FORMAT
                                 statement number. When $i = *$ or 6, out-
                                 put is to the terminal.

n FORMAT(format specifiers) n is the statement number.

Field Descriptors

nX   Skip n positions on input or output.

Tn Skip $n - 1$ positions in the input or output buffer.

TRn Skip right n positions.

TLn Backspace n positions.

/   Causes input or output to stop processing the current line and begin
           with a new line. n slashes in the middle of a FORMAT statement
           cause $n - 1$ records to be skipped or $n - 1$ blank lines to be printed.
           n slashes appearing at the end or beginning of a FORMAT state-
           ment cause n records to be skipped.

Iw Integer data with field width w, right-justified.

F$w.d$ Floating-point data with field width w and decimal fraction d,
           right-justified.

E$w.d$ Exponential notation with field width w and decimal fraction d,
           right-justified.

Repeating Fields

nF$w.d$ or n(F$w.d$, 2X) n is the number of times any field or group should
                                 be repeated.

KEY TERMS

buffer memory   **overloaded (input/output list)**
character strings   PRINT **statement**
E, F, I, X **and** T **descriptors**   **repeated edit descriptor**
formatted input or output   **right-justified**
FORMAT **statement**   **slash (/) descriptor**
literal descriptor   WRITE **statement**

REVIEW QUESTIONS

In the following problems, assume that the variables K, L, M, R, S, T, and U have the values

K = −210	L = 56	M = −195	
R = −108.1	S = 934.259	T = 7.632E4	U = 0.0008

1. For the given values of K, L, M, R, S, T, and U, determine the output that will appear using the FORMAT field specifiers shown. Assume that carriage control is in effect and that leading zeros are ignored. Start your answers in column 1 at the left side of the output field. Use dashes to indicate blank entries. All exponents bear signs.

	VARIABLE	FORMAT	OUTPUT
(a)	K	I3	* * – – – – – – – – –
(b)	L	I4	– – – – – – – – – – –
(c)	M	I2	– – – – – – – – – – –
(d)	R	F5.1	– – – – – – – – – – –
(e)	R	F7.2	– – – – – – – – – – –
(f)	R	F7.1	– – – – – – – – – – –
(g)	R	E10.3	– – – – – – – – – – –
(h)	S	F6.3	– – – – – – – – – – –
(i)	S	E9.2	– – – – – – – – – – –
(j)	S	E12.5	– – – – – – – – – – –
(k)	T	E12.3	– – – – – – – – – – –
(l)	T	F5.0	– – – – – – – – – – –
(m)	T	F7.1	– – – – – – – – – – –
(n)	U	E8.1	– – – – – – – – – – –
(o)	U	E10.2	– – – – – – – – – – –
(p)	U	F6.4	– – – – – – – – – – –
(q)	U	F6.3	– – – – – – – – – – –
(r)	U	F7.4	– – – – – – – – – – –

2. Indicate the output that will appear, in its proper page position (using a dash for a blank entry), if the following formatted PRINT statements are used.
    (a)   PRINT 11,K,L,R,S
            11 FORMAT('1',2(2X,I4),5X,F5.1,F8.2)
    (b)   PRINT 12,K,L,R,T
            12 FORMAT(1X,2I5,2E12.3)

    (c)   PRINT 13,L,S
           13 FORMAT(T5,'TRIAL',I5,T20,'X = ',E12.5)
    (d)   PRINT 14,S,T
           14 FORMAT('1',T5,F6.2/T5,F7.1)
    (e)   PRINT 15,K,S,U
           15 FORMAT(I4,3X,'THE RESULTS ARE = ',E10.3,3X,F8.5)
    (f)   PRINT 16,S,R,U
           16 FORMAT(F7.3,2(3X,E9.2))

3. In the following problems, indicate exactly — using blank entries and multiple lines where needed — how the given values for K, L, M, and so forth, should be entered into a data file (unit no. 9) so as to agree with the FORMAT statements shown. Some questions may have more than one correct answer.

    (a)   READ(9,40) K,L,R
           40 FORMAT(2(2X,I4),5X,F6.0)
    (b)   READ(9,41) R,S,U
           41 FORMAT(2F6.1)
    (c)   READ(9,42) R,T,U
           42 FORMAT(F6.2/E7.3,E8,3)
    (d)   READ(9,43) L,R,S,U
           43 FORMAT(I4,2(E12.3))
    (e)   READ(9,44) R,S
           44 FORMAT(E12.2,T20,E10.1)
    (f)   READ(9,45) R,T,U
           45 FORMAT(E8.2)

4. In the following problems indicate what the FORTRAN format specification must be to match the lines shown (taken from a data file) to the indicated READ statement. Use the given values of K, L, M, and so forth, to form your answer. Some problems may have more than one correct solution.

    (a)   READ(9,46) L,R,S

```
46 FORMAT _____
         |         |         |         |         |         |
123456789012345678901234567890123456789012345678901234567890 12345

     56   -108.1   934259
```
    (b)   READ(9,47) R,S,U

```
47 FORMAT _____
         |         |         |         |         |         |
123456789012345678901234567890123456789012345678901234567890 12345

-108.10
934.259
0.00080
```
    (c)   READ(9,48) R,T,U

```
48 FORMAT _____
         |         |         |         |         |         |
123456789012345678901234567890123456789012345678901234567890 12345

-.1081E37.632E04
0.800E-3
```

(d)   READ(9,49) M,R,S,T

    49 FORMAT _____

           | | | | | |
    12345678901234567890123456789012345678901234567890123456789012345

    -195 -.1081E+03.934259E+3
    0.7632E 05
(e)   READ(9,50) R,S

    50 FORMAT _____

           | | | | | |
    12345678901234567890123456789012345678901234567890123456789012345

    -10.81E1   .934259E3
(f)   READ(9,51) R,T,U

    51 FORMAT _____

           | | | | | |
    12345678901234567890123456789012345678901234567890123456789012345

    -108.176320.
    0.0008

COMPUTER PROJECTS

CP1. Modify the solution to Sample Problem 3.1 so that the program converts entered temperature values in degrees Celsius to Fahrenheit readings instead.

CP2. Modify the solution to Sample Problem 3.1 so that the program uses input and output files rather than the terminal for entry and display.

CP3. Modify the solution to Sample Problem 3.2 so that it treats the case where the attracting body is the moon. GM for the moon is given as 4.89×10^{12} m^3/s^2 and R is 1.738×10^6 m. Take h to be 3.5×10^5 m and the mass to be 50 kg.

CP4. Modify the solution to Sample Problem 3.5 so that a point's position can be entered in polar coordinates (r, θ) and then converted to Cartesian coordinates. Interchange the output so that R, THETA is printed first followed by X and Y.

CP5. Modify the solution to Sample Problem 3.7 so that the area and load are entered from a data file and the design stress is calculated. Change the output table headings and interchange their positions so that they appear in the order area, load, and stress. Neglect the factor of safety.

CP6. Modify Sample Problem 2.1 so that the output will appear as shown. Numbers at the top of the figure are for position reference only.

        | | | | | |
    12345678901234567890123456789012345678901234567890123456789012345

                    THE STATISTICAL VALUES ARE:

            XAVERAGE   DEVIATION
            --------   ---------
              XX.XXX   X.XXXX

In the following five problems, write FORTRAN computer programs that will duplicate exactly the given engineering tables. The numerical data shown should be read in from a data file and have the same format specification on input as is used for output when filling the table.

CP7.

```
    |           |           |           |           |           |
1234567890123456789012345678901234567890123456789012345678901 2345
```

CONSTANTS FOR THE SOLAR SYSTEM

BODY	MEAN DISTANCE TO SUN KM	ECCENTRICITY OF ORBIT E	PERIOD OF ORBIT SOLAR DAYS	MEAN DIAMETER KM	GRAVITATIONAL ACCELERATION AT SURFACE M/S*S
MERCURY	.573E+08	.2060	87.97	5000	3.47
VENUS	.108E+09	.0068	224.70	12400	8.44
EARTH	.150E+09	.0820	365.26	12742	9.82
MARS	.228E+09	.0930	686.98	6775	3.73

CP8.

```
    |           |           |           |           |           |
1234567890123456789012345678901234567890123456789012345678901 2345
```

WIRE TABLE FOR STANDARD ANNEALED COPPER

GAUGE NO.	DIAMETER MILS	CROSS SECTION CIR.MILS	OHMS PER 1000 FT	FEET PER POUND
1	289.3	.837E+05	.1239	3.947
2	257.6	.664E+05	.1563	4.977
3	229.4	.526E+05	.1970	6.276
4	204.3	.417E+05	.2485	7.914

CP9.

```
    |           |           |           |           |           |
1234567890123456789012345678901234567890123456789012345678901 2345
```

PERFORMANCE DATA FOR A 15000-KW TURBINE

EXHAUST PRESSURE IN.,HG ABS.	PER- CENT RATED LOAD	STRAIGHT CONDENS- ATING STEAM RATE LB/KWHR	EXTRACTION PERFORMANCE FOUR HEATERS		
			HEAT RATE, BTU/ KWHR	FEED- WATER TEMPER- ATURE,F	CONDENSER STEAM RATE LB/KWHR
1	25	8.63	11159	.254E+03	7.64
	50	7.87	10055	.294E+03	6.83
	75	7.68	9686	.323E+03	6.54
	100	7.73	9624	.345E+03	6.48

CP10.

```
           |          |          |          |          |          |
123456789012345678901234567890123456789012345678901234567890123456789012345
```

PROPERTIES OF AIR AT ATMOSPHERIC PRESSURE

TEMP. F	DENSITY SLUG/FT**3	DENSITY LBM/FT**3	KINEMATIC VISCOSITY FT**2/SEC	DYNAMIC VISCOSITY LBF-SEC/FT**2
0	.00268	.0862	.126E-03	.328E-06
40	.00247	.0794	.146E-03	.362E-06
80	.00228	.0735	.169E-03	.385E-06
120	.00215	.0684	.189E-03	.407E-06

CP11.

```
           |          |          |          |          |          |
123456789012345678901234567890123456789012345678901234567890123456789012345
```

PROPERTIES OF WIDE FLANGE BEAM SHAPES

NOMINAL SIZE IN.	WEIGHT FOOT PER LB.	AREA IN.**2	DEPTH IN.	FLANGE		WEB THICKNESS IN.
				WIDTH IN.	THICK- NESS IN.	
36 × 12	.194E+03	57.11	36.48	12.117	1.260	.770
	.182E+03	53.54	36.32	12.072	1.180	.725
	.170E+03	49.98	36.16	12.027	1.100	.680
	.160E+03	47.09	36.00	12.000	1.020	.653

Decision Structures

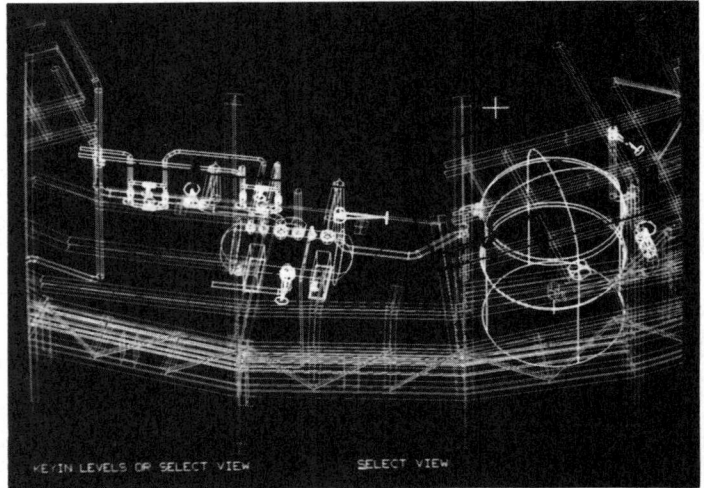

A three-dimensional view of a piping and equipment layout for a 1000-megawatt nuclear power plant. The engineer can "walk" through each room at the computer terminal. By pointing to individual pipe elements and valves using a mouse, he or she can obtain information regarding operating temperatures, pressures, materials used, pipe stresses, and even the name of the vendor who manufactured the item.

4.1 INTRODUCTION

In the previous chapters we have seen how digital computers slavishly follow the commands that we set forth. They execute our instructions accurately and rapidly, in a sequential manner, manifesting no freedom of choice or judgment. In this chapter we introduce general program structures that give the computer the capability to make decisions similarly to the way we do. The programs we write will impose our own personal logic on the machine. The machine will then be able, in a manner of speaking, to become an extension of our own mind, implementing our logical thought.

We can program the computer so that it imitates our *selection* or decision-making processes through an **IF structure.** In an IF structure we set up a condition in our program that if true requires one course of action and if false requires a different set of responses. The computer itself is left the freedom to evaluate the condition and select which of the preprogrammed alternative paths to take.

Another means we have available to impart our logical processes to a computer program is through the **WHILE repetition structure** or **WHILE loop.** This type of construct sets up a test condition that as long as it lasts, or is true, will permit repeated execution of a set of preprogrammed actions. The computer itself evaluates this test condition and acts accordingly. One of these actions must have the capability to alter the test condition so that it becomes false; otherwise, a WHILE structure establishes an unending program loop.

4.2 RELATIONAL AND LOGICAL OPERATORS

Before we discuss how we may implement IF and the WHILE loop structure in FORTRAN 77, it will be necessary to introduce the FORTRAN **relational** and **logical operators.** Relational operators are operators used to compare two arithmetic or two character expressions. The relational operators that are available are shown in Table 4.1. Note that the leading and trailing periods are part of the operator and must appear. The operators are used to form expressions as follows:

expression 1   relational operator   expression 2

The result of the comparison is a **logical constant;** that is, it is either true or false. (Logical constants and logical variables will be discussed at length in Chapter 8.) For example, if $X1 = 4.0$ and $Y1 = 3.1$ then

Relational Expression	Result
X1 . EQ . Y1	. FALSE .
X1 . GT . Y1	. TRUE .
X1 . LT . Y1	. FALSE .
X1 . LE . Y1	. FALSE .
X1 . NE . Y1	. TRUE .

Because the result of a relational expression acting alone or in combination with other relational expressions is a logical constant, it is classified as a **logical expression.** Logical expressions contain logical constants, logical variables, or relational expressions. To form **compound logical expressions** we can use the FORTRAN 77

TABLE 4.1   RELATIONAL OPERATORS

Operator	Meaning
. LT .	is less than
. LE .	is less than or equal to
. EQ .	is equal to
. NE .	is not equal to
. GT .	is greater than
. GE .	is greater than or equal to

logical operators shown in Table 4.2. The leading and trailing periods are again part of the operator and must appear. Compound logical expressions then have the following syntax:

logical expression 1   logical operator   logical expression 2

For example, if X1 = 4.0, Y1 = 3.1, and Z1 = 6.2, the results of the following expressions are as follows:

Compound Logical Expression	Result
(X1 . GT . Y1) . AND . (X1 . GT . Z1)	. FALSE .
(X1 . GT . Y1) . OR . (X1 . GT . Z1)	. TRUE .
(X1 . GT . Y1) . EQV . (X1 . GT . Z1)	. FALSE .
(X1 . GT . Y1) . NEQV . (X1 . GT . Z1)	. TRUE .
(X1 . GT . Y1) . AND . . NOT . (X1 . GT . Z1)	. TRUE .

Note that . EQV . and . NEQV . are used with logical operations, not . EQ . or . NE . Sometimes compound logical expressions can be very difficult to follow. One of the helpful visual tools that we can use to analyze logical expressions is a truth table. In Figure 4.1 the truth table for a binary half-adder is shown. (Two binary half-adders are needed to perform the addition of two binary bits.) The physical components constituting the logic circuits of the adder are represented by a black box. For the adder, if A is 1 or true and B is also 1 or true, their single-digit sum should result in 0, while the carry digit from the addition should be 1. The other three possible

TABLE 4.2 LOGICAL OPERATORS

Operator	Meaning
. NOT .	Negation. If the logical variable is true, it will be made false, that is, its complement.
. AND .	Conjunction. If two logical variables are true, the results of the conjunction will be true. If one or both are false, the conjunction result will be false.
. OR .	Disjunction. If one or both logical variables are true, the results of the disjunction will be true; otherwise, the result will be false.
. EQV .	Equivalence. If both logical variables are true or if both are false, the result will be true. If the logical variables do not have equivalent values, the result will be false.
. NEQV .	Nonequivalence. Nonequivalence is the negation of equivalence. It forms the complement of the equivalence result. If the result of equivalence is true, the result of nonequivalence will be false, and vice versa.

A (1 OR 0) → BINARY HALF-ADDER → SUM

B (1 OR 0) → BINARY HALF-ADDER → CARRY

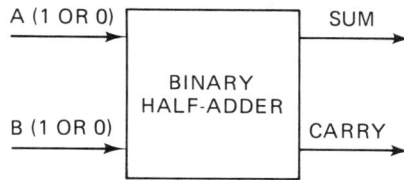

Results Desired

A	B	SUM	CARRY
1	1	0	1
1	0	1	0
0	1	1	0
0	0	0	0

Truth Table

A	B	A.NEQV.B (SUM)	A.AND.B (CARRY)
T	T	F	T
T	F	T	F
F	T	T	F
F	F	F	F

Figure 4.1

cases are also shown. FORTRAN expressions that simulate the binary half-adder's logic are

```
SUM=A.NEQV.B
CARRY=A.AND.B
```

where SUM, CARRY, A, and B must be declared as logical variables in a program type statement.

Logical operators have their own hierarchy regarding processing order. In a logical expression, the computer performs relational operations first and then logical operations. The logical operators are processed in the following order: first .NOT. operators, then .AND. conjunctions, then .OR. disjunctions, and finally .EQV. and .NEQV. expressions.

4.3 THE LOGICAL IF STATEMENT

The FORTRAN **logical IF statement** provides the most elementary form of **selection structure** to the programmer. It is limited to performing only one executable statement. It has the syntax

IF (logical expression) executable statement

When this statement is encountered in a program, the logical expression is first evaluated to see whether it is true or false. If it is true, the executable statement listed is performed. If it is not true, program control passes to the next statement appearing in the program list. It is permissible to use any executable statement in a logical IF statement except a DO, END, or another IF statement. The following are some examples of logical IF statements:

```
IF(X.GT.Y)   Y=7.1*T

IF(X.EQ.7.2) STOP

IF((X.GT.Y).AND.(Y.GT.4.1)) READ*,F

IF(ABS(Z).LT.0.001) R=5.2
```

A flowchart that diagrams the operation of the logical IF statement is shown in Figure 4.2. In the next section we will see how we can enlarge selection structure capabilities by using **IF-block structure** instead.

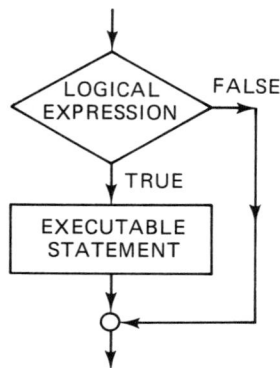

Figure 4.2

4.4 IF-BLOCK STRUCTURE

The programs we have studied thus far have been sequential in nature. The flow of the programs went from top to bottom (**top-down code**) with each statement executed in the order it appeared. A program or a part of a program that is written in this manner is said to have **sequential structure.** With the introduction of the FORTRAN IF-block structure, it will now be possible to alter a program's path or flow. As a result, two new programming structures will become fully available to us: selection or decision structure and repetition or WHILE loop structure. Good programming practice recommends that we choose only from these basic building blocks when writing a program. As we will see, each of these block structures has a single entry and a single exit point, and each incorporates top-down flow. Programs that follow these guidelines — that is, that have single entry and exit points, top-down flow, and decision and WHILE structures, are called **structured programs.** They are much easier to read and trace than programs having no clearly planned flow.

4.4.1 The IF-THEN-END IF Structure

The IF-block selection structure takes on several forms in FORTRAN. The first form is the **IF-THEN-END IF structure.** It has the following syntax:

```
IF (logical expression) THEN
    executable statement 1
    executable statement 2        Block having
         ·      ·                    n statements
         ·      ·
         ·      ·
    executable statement n
END IF
```

The IF statement that introduces the structure is called a **block IF statement.** If the logical expression enclosed in the statement's parentheses is true, the "block" of executable statements that follow are performed sequentially. If the logical expression is false, control passes to the statement immediately following the END IF **statement.** A flowchart diagram describing block operation appears in Figure 4.3. Note that the flow of control is from top to bottom. Note also that there is only one entry point and one exit point for the structure.

Sample Problem 4.1

Write a FORTRAN program that will evaluate whether three given points entered at the terminal are collinear. If they are collinear, print out this message and evaluate Y at $X = 0$.

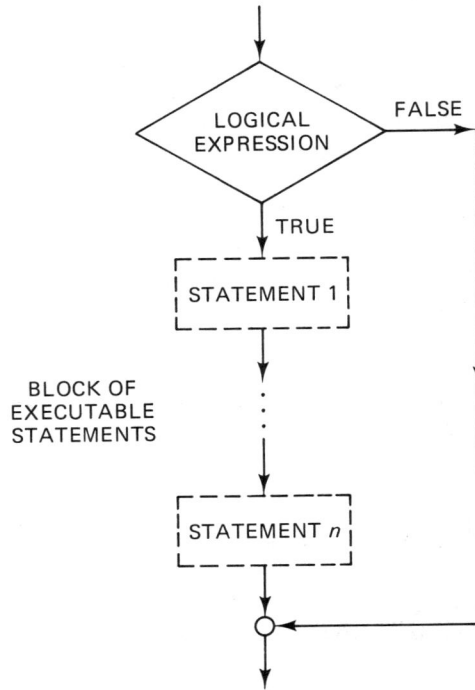

Figure 4.3

Analysis: The equation of a straight line passing through any two points can be determined by

$$Y = \frac{Y_2 - Y_1}{X_2 - X_1}(X - X_1) + Y_1$$

To test whether the third point lies on this line, we can substitute in the value X_3.

$$Y_3 = \frac{Y_2 - Y_1}{X_2 - X_1}(X_3 - X_1) + Y_1$$

If Y_3 agrees with the Y_3 value given, the point does lie on the line, and Y at $X = 0$ can then be determined.

Solution: The program's flowchart is shown in Figure 4.4, and the algorithm and problem solution are shown in Figures 4.5a and 4.5b. Several things should be noted from this figure. First, for readability, all statements in the block are indented five spaces. Secondly, as discussed in Chapter 2 and Appendix A, there will always be roundoff error introduced during ALU calculations. This means that even if the points are collinear, the entered value of Y3 may never precisely equal the calculated value of Y3. If just the last bit of the storage word is different for each of the variables, incorrect results wil be obtained. To avoid problems with rounding, the logical expression

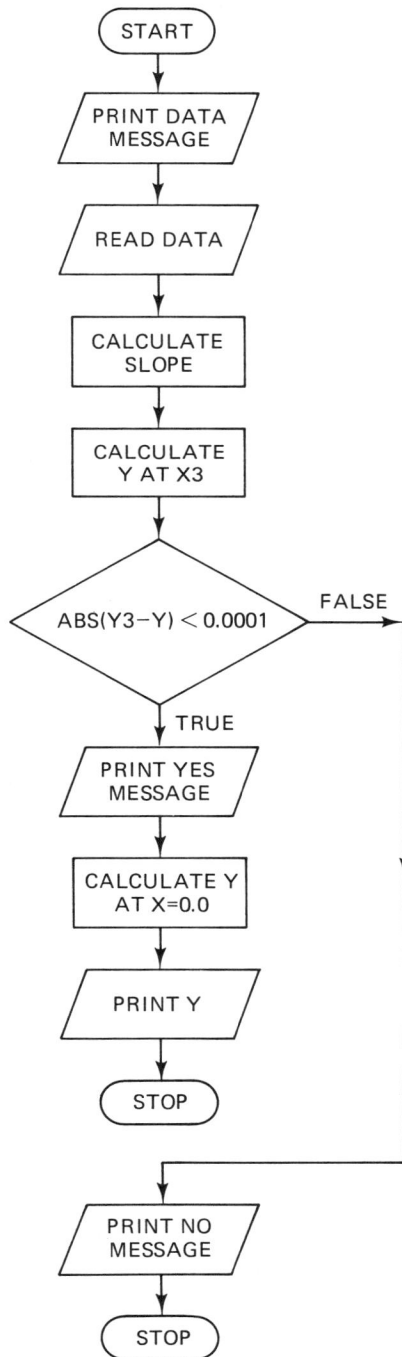

Figure 4.4

■ **Algorithm for Sample Problem 4.1.** This program determines if three given points are collinear.

1. Enter X1,Y1,X2,Y2,X3,Y3.
2. Calculate
   ```
   SLOPE=(Y2-Y1)/(X2-X1)
   Y=SLOPE*(X3-X1)+Y1
   ```
3. If $|Y3-Y|<0.0001$ then
   a. Print 'Points are collinear.'
   b. Calculate
      ```
      Y=SLOPE*(0.0-X1)+Y1
      ```
   c. Print Y
   d. Stop
4. Print 'Points are not collinear.'

Figure 4.5a

```
        PROGRAM LINES

****************************************************************
*                                                              *
*       THIS PROGRAM DETERMINES IF THREE GIVEN POINTS ARE      *
*       COLLINEAR.                                             *
*                                                              *
****************************************************************
        REAL X1,Y1,X2,Y2,X3,Y3,SLOPE,Y

        PRINT*,'ENTER X1,Y1,X2,Y2,X3,Y3...IN THAT ORDER.'
        READ*,X1,Y1,X2,Y2,X3,Y3

        SLOPE=(Y2-Y1)/(X2-X1)
        Y=SLOPE*(X3-X1)+Y1

        IF(ABS(Y3-Y).LT.0.0001) THEN

            PRINT*,'THE POINTS ARE COLLINEAR'

            Y=SLOPE*(0.-X1)+Y1
            PRINT*,'THE VALUE OF Y AT X=0 IS ',Y
            STOP

        END IF

        PRINT*,'THE POINTS ARE NOT COLLINEAR'

        END

    RUN

    ENTER X1,Y1,X2,Y2,X3,Y3...IN THAT ORDER.
    ? 1,5.5,2,11.0,3,16.5
     THE POINTS ARE COLLINEAR
     THE VALUE OF Y AT X=0 IS 0.
```

Figure 4.5b

for the IF-THEN statement has been written so that it compares the absolute difference of the two quantities with a very small user-defined number. In effect, we are asking if the difference between these two values is acceptably small enough so that we can assume collinearity. As a general rule, when you use relational operators, include enough built-in tolerance for eventual roundoff error. Also, make sure that you have planned for all possible numeric results; for example, do not use .GT. when you need .GE..

It can be seen from the solution that when using a purely IF-THEN-END IF structure, there is no way to provide for an additional printout message for the case when the points are not collinear. This IF structure provides for one set of actions to be taken when the logical expression is true, but no action when the expression is false. For example, the altered program version

```
PROGRAM LINE
      .
      .
      .
IF (ABS(Y3-Y).LT.0.0001) THEN
    PRINT *,'THE POINTS ARE COLLINEAR.'
    Y0=SLOPE*(0.0-X1)+Y1
    PRINT *,'THE VALUE OF Y0 AT X0 IS = ',Y0
END IF
PRINT *,'THE POINTS ARE NOT COLLINEAR.'
END
```

would print a correct message if the points were not collinear, skipping from the IF-THEN statement to the PRINT *,'THE POINTS ARE NOT COLLINEAR' message. However, it would print conflicting messages if the points were collinear. The first message would be printed while the program flowed through the block structure, indicating collinearity; the second (incorrect) message would appear as the program flow exited the block IF structure.

It would be possible to avoid this error by using a GO TO **statement** inside the block IF structure that transfers control to an external STOP statement. The use of GO TO statements should generally be kept to a minimum, however, to preserve programming clarity. GO TO statements are not considered to be an element of acceptable programming style. Their use leads to disjointed and hard-to-follow programs that do not maintain the top-down flow of structured programming. We will learn how to program for a set of actions initiated by a false response without recourse to GO TO statements in Section 4.4.2.

For the moment though, let us look at a similar type of error that is commonly made while using the logical IF statement. Suppose that it is desired to simulate the voltage pulse shown in Figure 4.6. The proper order of statements that would describe the pulse is given by

```
      .
      .
      .
V=0.0
IF((T.GE.0.).AND.(T.LT.T0))V=V0
```

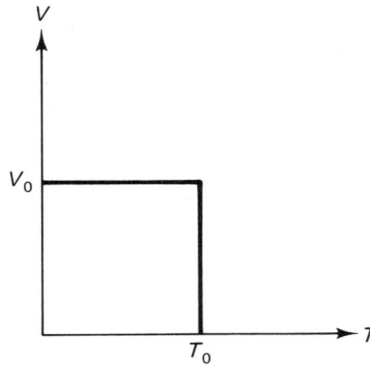

Figure 4.6

Observe what happens when this sequence is changed to

```
     .
     .
     .
IF((T.GE.0.).AND.(T.LT.T0))V=V0
V=0.0
     .
     .
     .
```

There is no way for V to keep the value V0; it changes value immediately.

IF-THEN-IF structures can also be *nested* together. The syntax required to do this is shown in Figure 4.7. Note that each block IF-THEN statement in the figure must have a matching END IF statement to close the structure. If the logical expression pertaining to the block IF-THEN statement is false, control passes to the statement immediately following its paired END IF statement. Also note how indenting the blocks, though not mandatory, greatly aids the reader in following the program's logic. To see how nesting works, refer to Sample Problems 4.2 and 4.3.

Sample Problem 4.2

A beam is a structural element designed to carry transverse loads and reactions. The simply supported beam shown in Figure 4.8 is loaded at a distance $A = 2$ m from the left-hand side by a load $P = 100{,}000$ N. The beam's length is 10 m ($B = 8$ m). Write a program that will determine the internal bending moment acting at an arbitrary cross section a distance X from point 1.

Analysis: From statics it is known that

$$R_1 = B\frac{P}{A + B}$$

$$R_2 = P - R_1$$

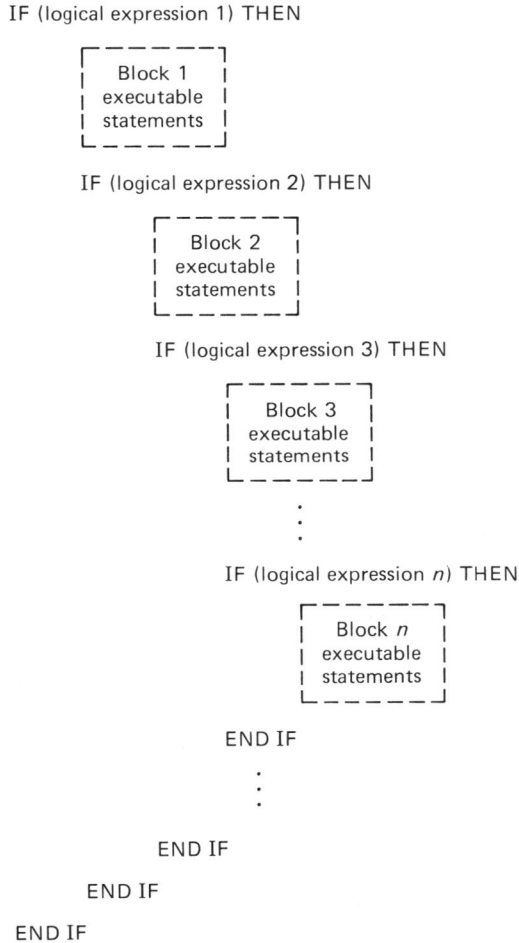

IF (logical expression 1) THEN

```
┌ ─ ─ ─ ─ ─ ┐
│  Block 1  │
│ executable │
│ statements │
└ ─ ─ ─ ─ ─ ┘
```

IF (logical expression 2) THEN

```
┌ ─ ─ ─ ─ ─ ┐
│  Block 2  │
│ executable │
│ statements │
└ ─ ─ ─ ─ ─ ┘
```

IF (logical expression 3) THEN

```
┌ ─ ─ ─ ─ ─ ┐
│  Block 3  │
│ executable │
│ statements │
└ ─ ─ ─ ─ ─ ┘
```

        .
        .
        .

IF (logical expression n) THEN

```
┌ ─ ─ ─ ─ ─ ┐
│  Block n  │
│ executable │
│ statements │
└ ─ ─ ─ ─ ─ ┘
```

END IF

    .
    .
    .

END IF

END IF

END IF

Figure 4.7

The internal bending moment is defined by

$$M = R_1 X \qquad (0 \le X \le A)$$

and

$$M = R_1 X - P(X - A) \qquad (A < X \le (A + B))$$

Solution: The flow chart, the algorithm and problem solution are shown in Figures 4.9, 4.10a, and 4.10b. It can be seen that by nesting IF structures, one additional set of actions is taken for each additional true response encountered; that is, a false reply causes the applicable subblock to be ignored. Test the program at $X < 0.0$,

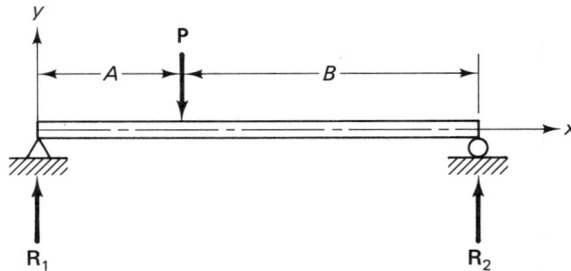

Figure 4.8

$0.0 \leq X < A$, $X \geq A$, and $X \geq 10.0$. How would you modify the program so that it runs for all values of X?

Sample Problem 4.3

It is desired to find the internal stress acting at any point X along a uniform bar, axially loaded by the forces P_1 and P_2 (see Figure 4.11). The forces P and P_3 support the bar, preventing it from changing its total length. Write a program to solve for internal stresses using selection structure and nested IF-block elements. Remember that stress = load/area. The cross-sectional area of the bar is given as $4.0 \times 10^{-4} \, m^2$, and $P = 7000 \, N$, $P_1 = 8000 \, N$, $P_2 = 5000 \, N$, and $P_3 = 6000 \, N$.

Analysis: The formulas to determine stresses in the different region are given by

$$\text{Stress}_I = P/A \qquad (0 < X < 0.1)$$

$$\text{Stress}_{II} = (P - P_1)/A \qquad (0.1 \leq X < 0.2)$$

$$\text{Stress}_{III} = (P - P_1 - P_2)/A = P_3/A \qquad (0.2 \leq X < 0.3)$$

Solution: (See Figures 4.12a and 4.12b). Note that in the solution, the stresses are established before entering each IF-block structure. If the decision of the block IF statement is false, the stress will remain unchanged and an answer can be printed out. If the decision value is true, a new stress level is specified and the next IF block is entered. Since the bar has three regions where different stress formulas apply, three IF-block structures are required. Each positive response leads to a separate course of action that negates previous actions. The value of X and its concomitant stress value are fitted into the different regions along the bar from right to left. The innermost element, if entered, changes the stress value for the third and final time. What values of X should be chosen to verify the program?

4.4.2 IF-THEN-ELSE-END IF Structures

In the preceding section we saw that the IF-THEN-END IF structure made no provision for a course of action to be taken if the response to the controlling logical expression was false. In Sample Problem 4.1, the inclusion of a GO TO statement was discussed as one possible way to introduce a false branch (so that a negative message

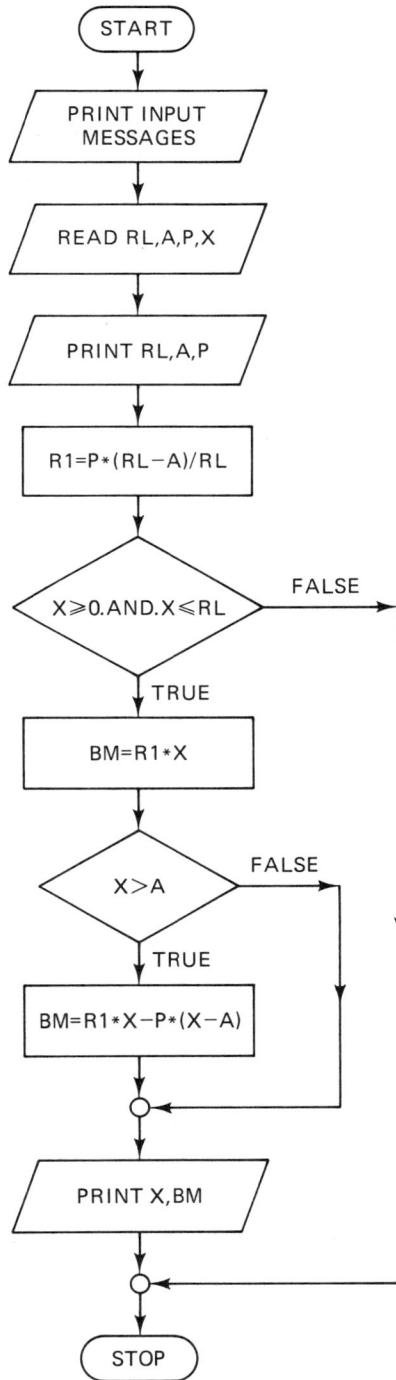

Figure 4.9

■ **Algorithm for Sample Problem 4.2.** This program finds the internal bending moment, BM, acting on a cross section located at a distance X from the left side of a simply supported beam. The beam is acted on by a single concentrated load P applied at a distance A from the left. The beam's length is RL.

1. Enter and then print RL.
2. Enter and then print A.
3. Enter and print P.
4. Enter X.
5. Calculate the beam's reaction at the left.
   ```
   R1=P*(RL-A)/RL
   ```
6. If $0.0 \leq X \leq RL$ then
   a. Calculate the bending moment BM
      ```
      BM=R1*X
      ```
   b. If $X > A$ then
      i. Calculate different BM
         ```
         BM=R1*X-P*(X-A)
         ```
7. Print X,BM.

Figure 4.10a

```
            PROGRAM BEAM

      *******************************************************************
      *                                                                 *
      *    THIS PROGRAM FINDS THE INTERNAL BENDING MOMENT BM ACTING     *
      *    UPON A CROSS SECTION, LOCATED AT A DISTANCE X FROM THE LEFT  *
      *    SIDE OF A SIMPLY SUPPORTED BEAM. THE BEAM IS ACTED UPON BY   *
      *    A SINGLE CONCENTRATED LOAD P APPLIED AT A DISTANCE A FROM    *
      *    THE LEFT. THE BEAM'S LENGTH IS RL.                           *
      *                                                                 *
      *******************************************************************

            REAL RL,A,P,X,R1,BM

            PRINT*,'ENTER THE BEAMS LENGTH'
            READ*,RL
            PRINT 1,RL
          1 FORMAT(' THE BEAMS LENGTH IS ',E10.3,' METERS')

            PRINT*,'ENTER THE LOCATION OF THE CONCENTRATED SINGLE LOAD'
            READ*,A
            PRINT 2,A
          2 FORMAT(' THE LOAD IS AT ', E10.3,' METERS FROM THE LEFT')

            PRINT*,'ENTER THE MAGNITUDE OF THE LOAD'
            READ*,P
            PRINT 3,P
          3 FORMAT(' THE CONCENTRATED LOAD IS ', E10.3,' NEWTONS')

            PRINT*,'ENTER THE DISTANCE X FOR THE CROSS SECTION'
            READ*,X
```

Figure 4.10b

```
R1=P*(RL-A)/RL

IF((X.GE.0.).AND.(X.LE.RL))THEN

    BM=R1*X

    IF(X.GT.A)THEN
        BM=R1*X-P*(X-A)
    END IF

END IF

PRINT 4,X,BM
4 FORMAT('THE INTERNAL BENDING MOMENT ACTING AT A CROSS ',
+ 'SECTION A DISTANCE ',/E10.3,' METERS FROM THE LEFT IS ',
+ E10.3,' NEWTON-METERS.')

END
```

```
RUN

ENTER THE BEAMS LENGTH

? 9
 THE BEAMS LENGTH IS   .900E+01 METERS
 ENTER THE LOCATION OF THE CONCENTRATED SINGLE LOAD
? 3
 THE LOAD IS AT   .300E+01 METERS FROM THE LEFT
 ENTER THE MAGNITUDE OF THE LOAD
? 60000
 THE CONCENTRATED LOAD IS   .600E+05 NEWTONS
 ENTER THE DISTANCE X FOR THE CROSS SECTION
? 1.5
THE INTERNAL BENDING MOMENT ACTING AT A CROSS SECTION A DISTANCE
 .150E+01 METERS FROM THE LEFT IS   .600E+05 NEWTON-METERS.
```

Figure 4.10b   (cont.)

could be printed if needed). In this section we will see that the **IF-THEN-ELSE-END IF structure** offers us a much better means to provide for a set of responses to a false result.

The IF-THEN-ELSE-END IF structure has the general form diagrammed in Figure 4.13. It can be seen from the figure that a true response to the logical expression would cause the block of statements between the block IF-THEN statement and the ELSE statement to be performed. After they are executed, the structure would be exited and control passed directly to the statement that follows the END IF statement. If the response to the logical expression were false, control would be transferred to

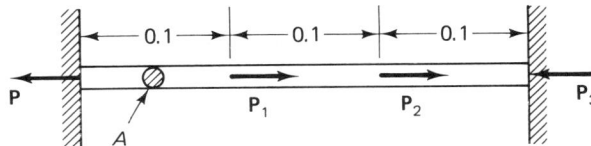

Figure 4.11

■ **Algorithm for Sample Problem 4.3.** This program finds the stresses acting on different cross sections along a bar of length L that is held between rigid supports. Loads P1 and P2 act at the points $L/3$ and $2*L/3$, respectively. The area is given as $4.0E-2$; the length is 0.3. The cross section is at a distance X from the left.

1. Initialize A,L,P,P1,P2,P3.
2. Enter X.
3. Set S=0.
4. If $0.0<X<L$ then
   a. Calculate stress
      S=P3/A
   b. If $X<2.0*L/3.0$ then
      i. Calculate new stress
         S=-(P-P1)/A
      ii. If $X<L/3.0$ then
          (a) Calculate new stress
              S=-P/A
5. Print A,L,P,P1,P2,P3,S,X.

Figure 4.12a

```
        PROGRAM STRESS
*******************************************************************
*                                                                 *
*    THIS PROGRAM FINDS THE STRESSES ACTING ON DIFFERENT CROSS    *
*    SECTIONS ALONG A BAR OF LENGTH L WHICH IS HELD BETWEEN       *
*    RIGID SUPPORTS. LOADS P1 AND P2 ACT AT THE POINTS L/3 AND    *
*    2*L/3 , RESPECTIVELY. THE AREA IS GIVEN AS 4.E-2; THE        *
*    LENGTH IS 0.3. THE CROSS SECTION IS AT A DISTANCE X FROM     *
*    THE LEFT.                                                    *
*                                                                 *
*******************************************************************

        REAL L,A,P,P1,P2,P3,X,S

        DATA A,L/4.E-2,0.3/
        DATA P,P1,P2,P3/7000.,8000.,5000.,6000./

        PRINT 8
      8 FORMAT(' ENTER X FROM THE TERMINAL. X MUST BE FROM O.',
       + ' AND 0.3.')
        READ*,X

        S=0.

        IF((X.GT.0.).AND.(X.LT.L))THEN

            S=P3/A

            IF(X.LT.(2.*L/3.))THEN

                S=-(P-P1)/A
```

Figure 4.12b

```
              IF(X.LT.L/3.)THEN

                  S=-P/A

              END IF

           END IF

        END IF

        PRINT 1
      1 FORMAT(//T22,'STRESSES FOR THE FOLLOWING BAR LOADING:')
        PRINT 2
      2 FORMAT(/T10,'AREA',TR7,'LENGTH',TR12,'P',TR11,'P1',TR11,
      + 'P2',TR11,'P3')
        PRINT 3
      3 FORMAT(T10,'M**2',TR9,'M',TR15,'N',TR12,'N',2(12X,'N'))
        PRINT 7
      7 FORMAT(5X,73('-'))
        PRINT 4,A,L,P,P1,P2,P3
      4 FORMAT(6(3X,E10.3))
        PRINT 5,S,X
```

```
RUN

 ENTER X FROM THE TERMINAL. X MUST BE FROM O. AND 0.3.
 ? .15


              STRESSES FOR THE FOLLOWING BAR LOADING:

        AREA        LENGTH          P            P1          P2          P3
        M**2          M             N            N           N           N
 ---------------------------------------------------------------------------
 .400E-01    .300E+00      .700E+04      .800E+04    .500E+04    .600E+04

     THE STRESS IS EQUAL TO    .250E+05 N/M**2 AT ADISTANCE    .150E+00 M
 FROM THE LEFT.
```

Figure 4.12b   (cont.)

IF (logical expression 1) THEN

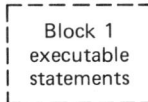

ELSE

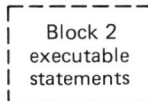

END IF

Figure 4.13

the first statement following the ELSE statement, and all statements that lie between the block IF-THEN and ELSE statements would be ignored. The block of statements that lie between the ELSE statement and the END IF statement would be processed, a normal exit would occur, and control would once again pass to the first statement occurring after the END IF statement.

Sample Problem 4.4

Write a FORTRAN program that when given an arbitrary X value will linearly interpolate between the three known (X, Y) values shown in the table to find Y. The value X is assumed to lie between 2.014 and 7.134.

Point	X	Y
1	2.014	10.01
2	5.046	15.24
3	7.134	19.36

Analysis: Between points 1 and 2 the equation for a straight line is

$$Y = \frac{Y_2 - Y_1}{X_2 - X_1}(X - X_1) + Y_1$$

Between points 2 and 3

$$Y = \frac{Y_3 - Y_2}{X_3 - X_2}(X - X_2) + Y_2$$

Solution: IF-THEN-ELSE-END IF structures are suitably employed here. The algorithm and source code for this problem are shown in Figures 4.14a and 4.14b. To verify the program, values of X less than 2.014 and greater than 7.134 should be tried. Similarly, arbitrary values of X lying between 2.014 and 5.046, or between 5.046 and 7.134, should also be entered, and the results from the program should then be compared with calculations performed "by hand." Even calculations for X values of 2.014, 5.046, and 7.134 should be attempted.

Algorithm for Sample Problem 4.4. This program linearly interpolates between three given paired values in a table.

1. Initialize X1,Y1,X2,Y2,X3,Y3.
2. Print limits X1 and X3.
3. Enter X.
4. If X1≤X≤X2 then
   a. Calculate
      Y=((Y2-Y1)/(X2-X1))*(X-X1)+Y1
   Else
   a. Calculate
      Y=((Y3-Y2)/(X3-X2))*(X-X2)+Y2
5. Print Y.

Figure 4.14a

```
          PROGRAM TABLE

   *********************************************************************
   *                                                                   *
   *      THIS PROGRAM LINEARLY INTERPOLATES BETWEEN THREE GIVEN       *
   *      PAIRED VALUES IN A TABLE.                                    *
   *                                                                   *
   *********************************************************************

          REAL X1,Y1,X2,Y2,X3,Y3,X,Y

          DATA X1,Y1,X2,Y2,X3,Y3/2.014,10.01,5.046,15.24,7.134,19.36/

          PRINT 1, X1,X3
        1 FORMAT(' ENTER VALUE OF B BETWEEN ',F7.3,' AND ',F7.3,' .')
          READ*,X

          IF((X.GE.X1) .AND. (X.LE.X2))THEN

              Y=((Y2-Y1)/(X2-X1))*(X-X1) + Y1

          ELSE

              Y=((Y3-Y2)/(X3-X2))*(X-X2) + Y2

          END IF

          PRINT 2,Y
        2 FORMAT(' THE VALUE OF Y IS ',F7.3)

          END


   RUN

    ENTER VALUE OF B BETWEEN    2.014 AND    7.134 .
    ? 3.1
     THE VALUE OF Y IS   11.883
```

Figure 4.14b

IF-THEN-ELSE-END IF structures can also be nested together as shown in Figure 4.15. In this figure we note that each block IF-THEN statement must be paired with a corresponding closing END IF statement. Also, note that if n IF-THEN-ELSE-END IF structures are included in the nesting, $n + 1$ possible alternatives can be considered.

Sample Problem 4.5

Suppose that the grading results given for the last exam you took in your computer class were divided as follows: A = 90–100, B = 80–89, C = 70–79, D = 60–69, and F = less than 60. Write a program that will accept a student's numerical score entered at the terminal and then reply with his or her grade on the monitor.

Analysis: With five possible conditions to be considered (unfortunately, even receiving an F), four IF-THEN-ELSE–END IF structures should be nested together.

Solution: See Figures 4.16a and 4.16b for the algorithm and problem solution. To test a program like this, it is wise not only to check values within each region but also to check values on the boundaries of the regions to see that no logical condition has been omitted.

```
IF (logical expression 1) THEN

        ┌─ ─ ─ ─ ─ ┐
        │  Block 1  │
        │ executable │
        │ statements │
        └─ ─ ─ ─ ─ ┘

ELSE

    IF (logical expression 2) THEN

            ┌─ ─ ─ ─ ─ ┐
            │  Block 2  │
            │ executable │
            │ statements │
            └─ ─ ─ ─ ─ ┘

    ELSE

        IF (logical expression 3) THEN

                ┌─ ─ ─ ─ ─ ┐
                │  Block 3  │
                │ executable │
                │ statements │
                └─ ─ ─ ─ ─ ┘

                        .
                        .

            ELSE

                IF (logical expression n) THEN

                        ┌─ ─ ─ ─ ─ ┐
                        │  Block n  │
                        │ executable │
                        │ statements │
                        └─ ─ ─ ─ ─ ┘

                ELSE

                        ┌─ ─ ─ ─ ─ ┐
                        │ Block n + 1 │
                        │ executable │
                        │ statements │
                        └─ ─ ─ ─ ─ ┘

                END IF

            END IF
                .
                .

        END IF

    END IF

END IF
```

Figure 4.15

■ **Algorithm for Sample Problem 4.5.** This program will compare a student's test score entered at the terminal to grade categories established by the instructor for an exam. The student's grade will be printed at the terminal.

1. Initialize A,B,C,D.
2. Enter SCORE.
3. If A≤SCORE then
   a. Print 'Grade was A.'
   Else
   a. If B≤SCORE then
      i. Print 'Grade was B.'
      Else
      i. If C≤SCORE then
         (a) Print 'Grade was C.'
         Else
         (a) If D≤SCORE then
            (1) Print 'Grade was D.'
            Else
            (1) Print 'Grade was F.'

Figure 4.16a

```
           PROGRAM GRADE

           *******************************************************************
           *                                                                 *
           *    THIS PROGRAM WILL COMPARE A STUDENT'S TEST SCORE ENTERED AT  *
           *    THE TERMINAL TO GRADE CATEGORIES ESTABLISHED BY THE          *
           *    INSTRUCTOR FOR AN EXAM. THE STUDENT'S GRADE WILL BE PRINTED  *
           *    AT THE TERMINAL.                                             *
           *                                                                 *
           *******************************************************************


           REAL A,B,C,D

           DATA A,B,C,D/90.,80.,70.,60./
           PRINT*,'ENTER YOUR TEST SCORE'
           READ*,SCORE

           IF(SCORE.GE.A)THEN
               PRINT*,'YOUR GRADE WAS AN A, GREAT!'
               PRINT*,'A GRADES RANGED FROM ',A,' TO A 100.'
           ELSE

               IF(SCORE.GE.B)THEN
                   PRINT*,'YOUR GRADE WAS A B, VERY GOOD.'
                   PRINT*,'B GRADES RANGED FROM ',B,' TO ',A-1.
               ELSE
```

Figure 4.16b

```
IF(SCORE.GE.C)THEN
    PRINT*,'YOUR GRADE WAS A C, O.K.'
    PRINT*,'C GRADES RANGED FROM ',C,' TO ',B-1.
ELSE

    IF(SCORE.GE.D)THEN
        PRINT*,'YOUR GRADE WAS A D. STUDY HARDER!'
        PRINT*,'D GRADES RANGED FROM ',D,' TO ',C-1
    ELSE

        PRINT*,'SORRY YOU FAILED THIS UNIT.'
        PRINT*,'YOU RECEIVED AN F. PLEASE REVIEW.'
        PRINT*,'F GRADES WERE BELOW ',D

    END IF

    END IF

    END IF

    END IF

    END


RUN


ENTER YOUR TEST SCORE

? 74.2
  YOUR GRADE WAS A C, O.K.
  C GRADES RANGED FROM 70. TO 79.
```

Figure 4.16b   (cont.)

4.4.3 The IF-THEN-ELSE IF-END IF Structure

The **IF-THEN-ELSE IF-END IF structure** is not really different from the IF-THEN-ELSE-END IF construct described in Section 4.4.2. Instead, it represents a simplification of the syntax that makes it easier for the reader to follow the logic branches. The syntax for IF-THEN-ELSE IF-END IF is shown in the flowchart of Figure 4.17. Here, only one END IF statement is required and indentation, if used, would only be required once. Note that if n logical expressions are presented, n or $n + 1$ alternative actions may be taken. The final ELSE block, block $n + 1$, is optional with this syntax. In the structure the only block that is executed is the one whose logical expression is first evaluated as true. After this block is processed, in which processing proceeds from top-to-bottom, control passes out of the structure to the next statement after END IF. If none are true, the ELSE block, the last block, will be evaluated. If it is missing, an "empty-handed" normal exit will occur.

Sample Problem 4.6

Write a program that will solve for the roots of a quadratic equation having the form $AX^2 + BX + C = 0$. A, B, and C are to be entered at the terminal, and answers are to

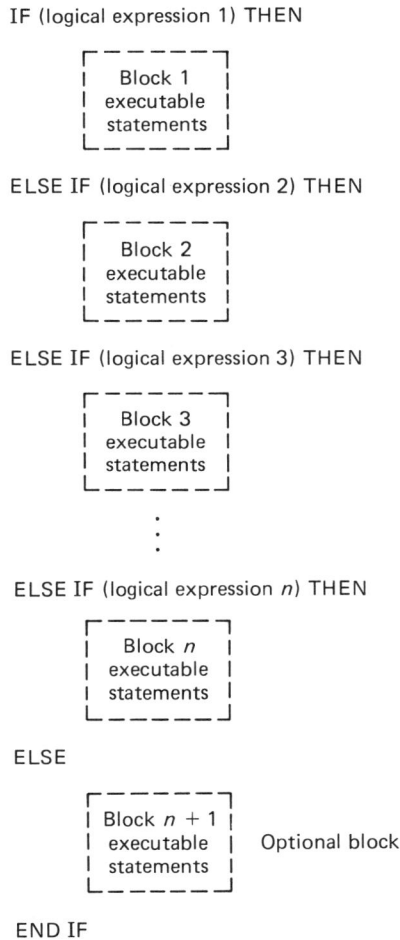

IF (logical expression 1) THEN

```
┌ ─ ─ ─ ─ ─ ┐
│   Block 1   │
│  executable │
│  statements │
└ ─ ─ ─ ─ ─ ┘
```

ELSE IF (logical expression 2) THEN

```
┌ ─ ─ ─ ─ ─ ┐
│   Block 2   │
│  executable │
│  statements │
└ ─ ─ ─ ─ ─ ┘
```

ELSE IF (logical expression 3) THEN

```
┌ ─ ─ ─ ─ ─ ┐
│   Block 3   │
│  executable │
│  statements │
└ ─ ─ ─ ─ ─ ┘
```

$$\vdots$$

ELSE IF (logical expression n) THEN

```
┌ ─ ─ ─ ─ ─ ┐
│   Block n   │
│  executable │
│  statements │
└ ─ ─ ─ ─ ─ ┘
```

ELSE

```
┌ ─ ─ ─ ─ ─ ┐
│ Block n + 1 │    Optional block
│  executable │
│  statements │
└ ─ ─ ─ ─ ─ ┘
```

END IF

Figure 4.17

be printed at the monitor. Provision should be made for real, equal, and complex roots.

Analysis: The quadratic formula is given by

$$X_1 = \frac{-B + \text{Disc}^{1/2}}{2.0A}$$

$$X_2 = \frac{-B - \text{Disc}^{1/2}}{2.0A}$$

$$\text{Disc} = B^2 - 4.0AC$$

When

       Disc > 0 real roots occur

       Disc $= 0$ equal roots occur

       Disc < 0 complex conjugate roots occur

Solution: The algorithm and problem solution are shown in Figures 4.18a and 4.18b. Verification of the program can be accomplished by testing it with known quadratic equations whose roots are real, equal, or complex. What happens if A equals 0?

Algorithm for Sample Problem 4.6. This program solves for the roots of a quadratic equation having the form A*X**2+B*X+C=0 using the quadratic formula. The real coefficients A, B, and C are entered at the terminal.

1. Enter A,B,C.
2. If A=0 then
   a. Calculate
      X=-C/B
   b. Print X
   c. Stop
3. Calculate
   DISC=B*B-4.0*A*C
4. If DISC>0.0 then
   a. Calculate
      SQDISC=SQRT(DISC)
      X1=(-B+SQDISC)/(2.0*A)
      X2=(-B-SQDISC)/(2.0*A)
   b. Print X1,X2
   Else if DISC=0.0 then
   a. Calculate
      X1=-B/(2.0*A)
   b. Print 'Equal roots ',X1
   Else
   a. Calculate
      DISC=-DISC
      SQDISC=SQRT(DISC)
      XR=-B/(2.0*A)
      XI=SQDISC/(2.0*A)
   b. Print 'Complex root: ',XR,XI
   c. Print 'Complex root: 'XR,-XI

Figure 4.18a

```
          PROGRAM ROOTS

     *****************************************************************
     *                                                               *
     *    THIS PROGRAM SOLVES FOR THE ROOTS OF A QUADRATIC EQUATION   *
     *    HAVING THE FORM A*X**2 + B*X + C = 0 USING THE QUADRATIC    *
     *    FORMULA. THE REAL COEFFICIENTS A, B, C ARE ENTERED AT THE   *
     *    TERMINAL.                                                   *
     *                                                               *
     *****************************************************************

          REAL A,B,C,X,DISC,X1,X2,SQDISC,XR,XI

          PRINT*,'ENTER COEFFICIENTS OF THE QUADRATIC EQUATION.'
          PRINT*,'FIRST ENTER A, THEN B, AND THEN C.'
          READ*,A,B,C

          IF(A.EQ.0.)THEN
              X=-C/B
              PRINT 5,X
     5        FORMAT(' THE ROOT IS ',F10.3,'.')
              STOP
          END IF

          DISC=B*B-4.*A*C

          IF(DISC.GT.0.)THEN
              SQDISC=SQRT(DISC)
              X1=(-B+SQDISC)/(2.*A)
              X2=(-B-SQDISC)/(2.*A)
              PRINT 1,X1,X2
     1        FORMAT(' THE ROOTS ARE ',F10.3,' AND ',F10.3,'.')

          ELSE IF(DISC.EQ.0.)THEN
              X1=-B/(2.*A)
              PRINT 2,X1
     2        FORMAT(' BOTH ROOTS ARE ',F10.3,'.')

          ELSE
              DISC=-DISC
              SQDISC=SQRT(DISC)
              XR=-B/(2.*A)
              XI=SQDISC/(2.*A)
              PRINT 3,XR,XI
     3        FORMAT(' THE ROOTS ARE COMPLEX. X1 EQUALS ',F9.3,1X,
        +        F9.3,'I')
              PRINT 4,XR,-XI
     4        FORMAT(' THE ROOTS ARE COMPLEX. X2 EQUALS ',F9.3,1X,
        +        F9.3,'I')

          END IF

          END

     RUN

          ENTER COEFFICIENTS OF THE QUADRATIC EQUATION.
          FIRST ENTER A, THEN B, AND THEN C.
          ? 1,-2,5
          THE ROOTS ARE COMPLEX. X1 EQUALS      1.000      2.000I
          THE ROOTS ARE COMPLEX. X2 EQUALS      1.000     -2.000I
```

Figure 4.18b

4.5 THE WHILE LOOP STRUCTURE

In the beginning of this chapter it was mentioned that another very important means by which the computer can control its own processing is the WHILE loop structure (repetition structure). In this construct a set of actions is repeated indefinitely until a test condition, usually represented by a logical expression, switches value and causes the loop to be exited. The test condition must be initialized (that is, have a value assigned to it before the structure is entered), and it must eventually change inside the loop to allow an exit. The number of ultimate repetitions of the loop is limited by the controlling conditions the programmer establishes.

We met a form of this structure before in Sample Problem 3.7, in which information was to be read in from a file of unknown length. At that time we used

```
10 READ(7,7,END=99) A,S
    .
    .
    .
   GO TO 10
    .
    .
    .
99 STOP
```

As long as data was available, the loop continued and the READ statement was executed repeatedly. When there was no longer any data left, the READ statement caused the loop to be exited immediately and transferred control to statement 99.

Some versions of FORTRAN (for example, WATFIV) include WHILE loops as part of the language, but standard FORTRAN 77 compilers do not provide for its direct use. (The next ANSI [American National Standards Institute] version of FORTRAN will presumably correct this deficiency.) Instead, the WHILE structure must be simulated with an IF-THEN-END IF block that incorporates a GO TO statement.

---

n IF (logical expression) THEN

```
┌ ─ ─ ─ ─ ─ ┐
│  Block of  │
│ executable │
│ statements │
└ ─ ─ ─ ─ ─ ┘
```

GO TO *n*

END IF

Figure 4.19

Its syntax is shown in Figure 4.19. As indicated in the figure, if the logical expression is true, a set of actions is executed, after which control is transferred back to the original logical expression. The logical expression is tested again, and so on. When the logical condition finally turns false, program flow leaves the loop, jumping to the statement that immediately follows the END IF statement. The GO TO statement that effects control transfer back to the beginning of the block references a statement number, which must also appear in columns 1 to 5 of the block IF-THEN statement. Although GO TO statements in general should be avoided, the appearance of one here, transferring control inside a loop, does not affect overall top-down coding.

It should be evident now that we can accomplish the same task of reading the entire data file's contents from Sample Problem 3.7 with an IF-THEN-END IF loop and a **trailer line** in the data. A trailer line is a selected data value that is used as a flag to mark the end of the input data file. When the computer reads this value it knows it must immediately leave the loop. The WHILE structure might then appear as

```
5 IF (X.NE.-999.0) THEN
    READ(8,7)X
    GO TO 5
  END IF
7 FORMAT(F6.0)
```

and the trailer line, the last value of the file in unit number 8, would have the entry -999.0. (The reader should be cautioned to use the .EQ. and .NE. relational operators sparingly when working with real values, since real computations inherently lead to roundoff error and the anticipated results may never appear.)

Sample Problem 4.7

Repeat Sample Problem 2.1, but this time create a data file for the program that incorporates an end-of-data flag. Assume that the number of data values, n, is unknown to begin with. The program must therefore keep track of how many samples have been entered.

Solution: See Figures 4.20a and 4.20b. *Study this algorithm and problem solution carefully.* The solution demonstrates how the computer can be programmed to keep a **running sum** and to count. First SUM, SUMSQ, and the counter variable, N, are initialized (set to 0) before the WHILE loop is entered. Every time a value is read in, it is added to the most recent value of SUM, thus updating it and forming its latest value. This continues each time the loop is traversed, with the value stored in SUM replaced by its old value plus the new value of R that is read in. The same is done for the sum of squared R values; SUMSQ. When the trailer line is read, the loop is immediately exited without SUM being updated. The last values of SUM and SUMSQ attained are the total ones desired. The counter works in the same manner. It is initialized before the loop is entered; then every time a data entry is read, the current value of the counter, N, is increased by 1. This continues until the end-of-data flag is reached. At that time, the loop is exited without the counter changing, and the average, AV, and standard deviation, S, are calculated.

■ **Algorithm for Sample Problem 4.7.** This program reads values from a data file TESTA and then finds the mean value and standard deviation. The data file uses −999 as a trailer line.

1. Read R.
2. Set SUM=SUMSQ=N=0.
3. While R≠−999 then
   a. Calculate
      SUM=SUM+R
      SUMSQ=SUMSQ+R⋆⋆2
      N=N+1
   b. Read R
4. Calculate
   AV=SUM/N
   STO=SUMSQ−N⋆AV⋆⋆2
   S=SQRT(STO/(N−1))
5. Print AV,S,N.

Figure 4.20a

```
          PROGRAM FLAG

     ****************************************************************
     *                                                              *
     *     THIS PROGRAM READS VALUES FROM A DATA FILE 'TESTA' AND THEN *
     *     FINDS THE MEAN VALUE AND STANDARD DEVIATION. THE DATA FILE  *
     *     USES -999 AS A TRAILER LINE.                             *
     *                                                              *
     ****************************************************************

          REAL R,SUM,SUMSQ,STO,S,AV
          INTEGER N

          OPEN(UNIT=8,FILE='TESTA')
          REWIND 8

          READ(8,1)R
        1 FORMAT(F10.0)
          SUM=0.
          SUMSQ=0.
          N=0

     C
     C    WHILE STRUCTURE BEGINS.
     C
        2 IF(R.NE.-999)THEN

              SUM=SUM+R
              SUMSQ=SUMSQ+R**2
              N=N+1
              READ(8,1)R
              GO TO 2
```

Figure 4.20b

```
      END IF

      AV=SUM/N
      STO=SUMSQ-N*AV**2
      S=SQRT(STO/(N-1))

      WRITE(6,4)AV
    4 FORMAT(' ','THE MEAN VALUE WAS ',F10.4)
      WRITE(6,5)S
    5 FORMAT(' ','THE STANDARD DEVIATION WAS ',F10.4)
      WRITE(6,6)N
    6 FORMAT(' ','THE VALUE WAS TAKEN FOR ',I3,' SAMPLES.')

      END


File Testa

2.999
3.001
3.010
3.005
2.995
2.998
3.003
3.003
2.999
3.001
-999.

RUN

THE MEAN VALUE WAS        3.0014
THE STANDARD DEVIATION WAS        .0042
THE VALUE WAS TAKEN FOR  10 SAMPLES.
```

Figure 4.20b   (cont.)

Sample Problem 4.8

Modify the solution program to Sample Problem 4.7 so that a counter is employed to stop the data flow instead of a trailer line. The total number of samples forming the data pool, NDATA, is to be entered from the terminal at execution time. Only every other data value from the pool should be used.

Solution: The algorithm and problem solution are shown in Figures 4.21a and 4.21b. Notice the differences in this program as compared with the previous one. The IF-THEN logical expression for the WHILE loop has been replaced, so that an additional counter, J, now controls the total number of times the loop is cycled and data is read in. The two counters are first initialized (J=1 and N=1), and the final value, test value NDATA, is established before the WHILE loops is entered. The logical expression in the beginning of the block is used to check whether the terminating condition has been met before the loop's execution. If it has not, the loop is performed and the counter J is incremented by 2 and N is incremented by 1 inside the loop. Control is then passed to the test condition again. In effect we are counting up to a limiting number in increments of 2, that is, 1, 3, 5, and so on, to NDATA-1. If the number of samples available to the pool NDATA is 10, five samples will be used (N=5 finally). Note also how the data keeps pace with the counting, by skipping every other data line. (The use of the slash in the FORMAT statement labeled 3 accomplishes this.) NDATA is equal to 20 in this prob-

■   **Algorithm for Sample Problem 4.8.** This program reads every other data value from a data file TESTB that is NDATA long. The mean value for the selected samples and the standard deviation are calculated.

1. Enter NDATA.

2. Read R.

3. Set SUM=R, N=J=1.

4. Calculate
SUMSQ=R∗∗2

5. While J<(NDATA−1)
  a. Read R
  b. Calculate
    SUM=SUM+R
    SUMSQ=SUMSQ+R∗∗2
    N=N+1
    J=J+2

6. Calculate
AV=SUM/N
STO=SUMSQ−N∗AV∗∗2
S=SQRT(STO/(N−1))

7. Print AV,S,N.

Figure 4.21a

```
          PROGRAM COUNT

**************************************************************
*                                                            *
*      THIS PROGRAM READS EVERY OTHER DATA VALUE FROM A DATA FILE  *
*      'TESTB' THAT IS NDATA LONG. THE MEAN VALUE FOR THE SELECTED *
*      SAMPLES AND THE STANDARD DEVIATION ARE CALCULATED.    *
*                                                            *
**************************************************************
          REAL SUM,SUMSQ,R,AV,STO,S
          INTEGER N,J

          OPEN(UNIT=8,FILE='TESTB')
          REWIND 8

          PRINT*,' ENTER THE TOTAL LENGTH OF THE DATA FILE.'
          READ*,NDATA
          READ(8,1)R
        1 FORMAT(F10.0)

          SUM=R
          SUMSQ=R**2
          N=1
          J=1
```

Figure 4.21b

```
2 IF(J.LT.(NDATA-1))THEN

       READ(8,3)R
3      FORMAT(/F10.0)

       SUM=SUM+R
       SUMSQ=SUMSQ+R**2
       N=N+1
       J=J+2
       GO TO 2

   END IF

   AV=SUM/N
   STO=SUMSQ-N*AV**2
   S=SQRT(STO/(N-1))

   WRITE(6,5)AV
5 FORMAT(' ','THE MEAN VALUE WAS ',F10.4)
   WRITE(6,6)S
6 FORMAT(' ','THE STANDARD DEVIATION WAS ',F10.4)
   WRITE(6,7)N
7 FORMAT(' ','THE SAMPLE SIZE WAS ',I3)

   END


File Testb

2.999
55.123
3.001
55.332
3.010
51.555
3.005
52.345
2.995
54.435
2.998
55.100
3.003
56.018
3.003
56.121
2.999
55.000
3.001
54.787

RUN

   ENTER THE TOTAL LENGTH OF THE DATA FILE.
? 20
 THE MEAN VALUE WAS     3.0014
 THE STANDARD DEVIATION WAS      .0042
 THE SAMPLE SIZE WAS  10
```

Figure 4.21b   (cont.)

lem. The student should consider what would happen if the limiting value of J is changed from NDATA-1 to NDATA.

Sample Problem 4.9

One powerful means of finding a root to an equation that is a function of X is by the bisection method. After a single root in a specified region has been isolated, the interval in which it falls can be halved into two smaller intervals. A test can determine whether the root lies in the right or left half-interval. The half-interval that contains the root, in turn, can be halved again, until an acceptably small region containing the root results. At that time, the solution is said to have converged to a root. Write a computer program that employs the bisection method to solve the equation.

$$x^3 + \cos x - 2.9 = 0$$

Analysis: To isolate a root, we will use the incremental-search method. We can start at $x = -3.1415927$ and march along the x-axis, proceeding in steps of $\Delta x = 0.1$. At each x, we test whether a root has been passed by first evaluating

```
F=X**3+COS(X)-2.9
```

Then, after incrementing x by Δx

```
XN=X+DELTAX
```

we also find

```
FN=XN**3+COS(XN)-2.9
```

The product F*FN can tell us if a root has been reached, since when the x-axis is crossed the product will become negative. (It will be zero if we happen to land on a root.) If the axis has not been crossed, we then take

```
F=FN
X=XN
```

find a new XN and FN as before, and test once again. An additional test is included in the loop to prevent x from becoming too large if no root is present. If a root has been passed, we know the region it lies in has

```
XLEFT=X
```

and

```
XRIGHT=XN
```

The bisection method can now be called forth. The method first establishes the midpoint of the region between XLEFT and XRIGHT

```
XMID=(XLEFT+XRIGHT)/2.0
```

To find whether the root lies to the left of the midpoint, we take the product of

```
F=XLEFT**3+COS(XLEFT)-2.9
```

and

    FMID=XMID**3+COS(XMID)-2.9

to see whether it is negative. If it is positive — that is, the root lies to the right — we make

    XLEFT=XMID

and

    F=FMID

A new XMID is now found and the previous steps are repeated. If it lies to the left, we make

    XRIGHT=XMID

and again a new midpoint is found and the test procedure repeated. The process stops when

    ABS(XRIGHT-XLEFT)

is less than some preestablished tolerance value EPSI.

Solution: (Refer to Figures 4.22a and 4.22b.) The solution for Sample Problem 4.9 is a bit complicated. It begins by using an IF-THEN-ELSE IF-END IF structure during the incremental search. Included in this block is a simple IF-THEN-END IF construction to test whether X has become too large. *Every program should have a built in brake such as this one to prevent runaway programs.* This is especially true in the development stage of code when all sorts of unanticipated errors and events are likely to occur.

An IF-THEN-END IF structure with an embedded IF-THEN-ELSE IF-END IF block performs the bisection method. Statement line 200 (statement numbers 100, 200, 300, and 400 are not referenced in the program but were put there as markers for the present discussion) and the one that follows it are there to initialize values before the next loop structure. Interchanging the order in which variables are initialized in FORTRAN can sometimes lead to disastrous results. For instance, note the entirely different result that will occur if statement 100 is interchanged with the one that follows it: X is never altered.

This type of error will not be picked up by the compiler and will prove very difficult to find. When it occurs, one of the basic debugging strategies we have available to us is to *trace each step in the program* to find out how it functions (or fails). We can print out key intermediate values as they are calculated by temporarily including extra PRINT statements during the program development stage. These values can then be compared with what the program should have produced, for example, with a hand-held calculator. Once the errors have been found, the extra PRINT statements are removed.

The interior IF-THEN-ELSE IF-END IF block is used to determine on which side of the region's midpoint the root lies. Statement 300 and the one immediately

Algorithm for Sample Problem 4.9. This program finds a root of a given equation by first using the incremental-search method to isolate its location. The program then switches to the bisection method to find the root's value. A tolerance error of EPSI is accepted in the answer. DELTAX is the step size for X in the incremental search. The starting value of X, DELTAX, and EPSI are given in a data statement. STOPP is the maximum value allowed for X.

1. Initialize X,DELTAX,EPSI,STOPP.
2. Calculate
   ```
   F=X**3+COS(X)-2.9
   XN=X+DELTAX
   FN=XN**3+COS(XN)-2.9
   ```
3. While F*FN>0
   a. Set X=XN
   b. If X>STOPP then
      i. Print 'No answer found.'
      ii. Stop
   c. Calculate
      ```
      XN=X+DELTAX
      ```
   d. Set F=FN.
   e. Calculate
      ```
      FN=XN**3+COS(XN)-2.9
      ```
4. If F*FN=0.0 then
   a. Print XN, 'the root.'
   b. Stop
5. Set XLEFT=X,XRIGHT=XN.
6. While |XRIGHT−XLEFT|>EPSI
   a. Calculate
      ```
      XMID=(XLEFT+XRIGHT)/2.0
      FMID=XMID**3+COS(XMID)-2.9
      ```
   b. If F*FMID>0.0 then
      i. Set XLEFT=XMID,F=FMID
      Else if F*FMID=0.0 then
      i. Set XLEFT=XMID,XRIGHT=XLEFT
      Else
      i. Set XRIGHT=XMID
7. Calculate
   ```
   XMID=(XLEFT+XRIGHT)/2.0
   ```
8. Print XMID, 'the root.'

Figure 4.22a

following it are executed when the root lies to the right of the midpoint. Statement 400 is used when the root lies to the left of the midpoint. The interior block then transfers control back to the WHILE test logical expression, which determines if the

```
            PROGRAM BISECT

      ******************************************************************
      *                                                                *
      *    THIS PROGRAM FINDS A ROOT OF A GIVEN EQUATION BY FIRST       *
      *    USING THE INCREMENTAL-SEARCH METHOD TO ISOLATE ITS           *
      *    LOCATION. THE PROGRAM THEN SWITCHES TO THE BISECTION METHOD  *
      *    TO FIND THE ROOT'S VALUE. A TOLERANCE ERROR OF EPSI IS       *
      *    ACCEPTED IN THE ANSWER. DELTAX IS THE STEP SIZE FOR X IN     *
      *    THE INCREMENTAL SEARCH. THE STARTING VALUE OF X, DELTAX AND  *
      *    EPSI ARE GIVEN IN A DATA STATEMENT. STOPP IS THE MAXIMUM     *
      *    VALUE ALLOWED FOR X.                                         *
      *                                                                *
      ******************************************************************

            REAL X,DELTAX,EPSI,STOPP,F,FN,XN,XLEFT,XMID,XRIGHT,FMID

            DATA X,DELTAX,EPSI,STOPP/-3.1415927,0.1,1.E-5,3.1415927/

            F=X**3+COS(X)-2.9
            XN=X+DELTAX
            FN=XN**3+COS(XN)-2.9
      C
      C     WHILE STRUCTURE BEGINS
      C
          1 IF((F*FN).GT.0.)THEN

        100      X=XN

              IF(X.GT.STOPP)THEN

                  WRITE(6,2)
          2       FORMAT(' ','NO ANSWER FOUND')
                  STOP

              END IF
              XN=X+DELTAX
              F=FN
              FN=XN**3+COS(XN)-2.9
              GO TO 1

          ELSE IF((F*FN).EQ.0.)THEN

                  WRITE(6,3)XN
          3       FORMAT(' ','THE ROOT IS',E14.7)
                  STOP

          END IF
        200 XLEFT=X
            XRIGHT=XN

      C
      C     WHILE STRUCTURE BEGINS
      C
          4 IF(ABS(XRIGHT-XLEFT).GT.EPSI)THEN

                  XMID=(XLEFT+XRIGHT)/2.
                  FMID=XMID**3+COS(XMID)-2.9

              IF((F*FMID).GT.0.0)THEN

        300      XLEFT =XMID
                  F=FMID
```

Figure 4.22b

```
        ELSE IF((F*FMID).EQ.0.0)THEN

            XLEFT=XMID
            XRIGHT=XLEFT

        ELSE

400         XRIGHT=XMID

        END IF
        GO TO 4

    END IF
    XMID=(XLEFT+XRIGHT)/2.

    WRITE(6,5)XMID
  5 FORMAT(' ','THE ROOT IS',E14.7)

    END


RUN

THE ROOT IS  .1397137E+01
```

Figure 4.22b   (cont.)

---

block can be exited, that is, that the value of X has converged close enough to a root so that an answer can be printed out. If X has not converged to the root, the cycle is repeated again for the newly created smaller (halved) region.

Sample Problem 4.10

Repeat Sample Problem 4.9 but this time use the better-known Newton-Raphson method for solution purposes.

Analysis: The Newton-Raphson method converges to the root of the equation $f(x) = 0$ by using the formula

```
XN=X-F/FDERIV
```

where X is the old or first guess for the root, F and FDERIV are the value of the function and its derivative, respectively, evaluated at X, and XN is the new estimate for the root. The new value and the old are compared to see if they have converged. The test for convergence is given by the logical expression

```
ABS(XN-X).LE.EPSI
```

If the estimates are not close enough, X is assigned the value of XN

```
X=XN
```

and the iteration (repetition) continues. The derivative of the function F=X**3+COS(X)-2.9 is given by

```
FDERIV=3.*X**2-SIN(X)
```

Solution: In the solution developed here, the incremental-search method is used to isolate a root, then the method switches over to Newton-Raphson. (See Figures 4.23a and 4.23b.)

■ **Algorithm for Sample Problem 4.10.** This program finds a root of a given equation by first using the incremental-search method to isolate its location. The program then switches to the Newton-Raphson method to find the root's value. A tolerance error of EPSI is accepted in the answer. DELTAX is the step size for X in the incremental search. The starting value of X, DELTAX, and EPSI are given in a data statement. STOPP is the maximum value allowed for X.

1. Initialize X,DELTAX,EPSI,STOPP.
2. Calculate
   ```
   F=X**3+COS(X)-2.9
   XN=X+DELTAX
   FN=XN**3+COS(XN)-2.9
   ```
3. While F*FN>0
   a. Set X=XN
   b. If X>STOPP then
      i. Print 'No answer found.'
      ii. Stop
   c. Calculate
      ```
      XN=X+DELTAX
      ```
   d. Set F=FN.
   e. Calculate
      ```
      FN=XN**3+COS(XN)-2.9
      ```
4. If F*FN=0.0 then
   a. Print XN, 'the root.'
   b. Stop
5. Calculate
   ```
   FDERIV=3.*X**2-SIN(X)
   ```
6. If FDERIV=0.0 then
   a. Print error message
   b. Stop
7. Calculate
   ```
   XN=X-F/FDERIV
   ```
8. While |XN−X|>EPSI
   a. Set X=XN
   b. Calculate
      ```
      FDERIV=3.0*X**2-SIN(X)
      ```
   c. If FDERIV=0.0 then
      i. Print error message
      ii. Stop
   d. Calculate
      ```
      XN=X-F/FDERIV
      ```
9. Print XN, 'the root.'

Figure 4.23a

```
      PROGRAM NEWTON

***************************************************************
*                                                             *
*     THIS PROGRAM FINDS A ROOT OF A GIVEN EQUATION BY FIRST   *
*     USING THE INCREMENTAL-SEARCH METHOD TO ISOLATE ITS       *
*     LOCATION. THE PROGRAM THEN SWITCHES TO THE NEWTON-RAPHSON *
*     METHOD TO FIND THE ROOT'S VALUE. A TOLERANCE ERROR OF EPSI *
*     IS ACCEPTED IN THE ANSWER. DELTAX IS THE STEP SIZE FOR X IN *
*     THE INCREMENTAL SEARCH. THE STARTING VALUE OF X, DELTAX AND *
*     EPSI ARE GIVEN IN A DATA STATEMENT. STOPP IS THE MAXIMUM  *
*     VALUE ALLOWED FOR X.                                      *
*                                                             *
***************************************************************

      REAL X,DELTAX,EPSI,STOPP,F,XN,FN,FDERIV

      DATA X,DELTAX,EPSI,STOPP/-3.1415927,0.1,1.E-5,3.1415927/

      F=X**3+COS(X)-2.9
      XN=X+DELTAX
      FN=XN**3+COS(XN)-2.9

C
C     WHILE STRUCTURE BEGINS
C
    1 IF((F*FN).GT.0.)THEN
         X=XN

         IF(X.GT.STOPP)THEN
            WRITE(6,2)
    2       FORMAT(' ','NO ANSWER FOUND')
            STOP

         END IF

         XN=X+DELTAX
         F=FN
         FN=XN**3+COS(XN)-2.9
         GO TO 1

      ELSE IF((F*FN).EQ.0.)THEN

         WRITE(6,3)XN
    3    FORMAT(' ','THE ROOT IS',E14.7)
         STOP

      END IF

      FDERIV=3.*X**2-SIN(X)

      IF(FDERIV.EQ.0)THEN
         WRITE(6,6)
    6    FORMAT(' ','DERIVATIVE IS ZERO. PROGRAM HAS STOPPED.')
         STOP

      END IF

      XN=X-F/FDERIV

C
C     WHILE STRUCTURE BEGINS
C
```

Figure 4.23b

```
4 IF(ABS(XN-X).GT.EPSI)THEN
     X=XN
     FDERIV=3.*X**2-SIN(X)

     IF(FDERIV.EQ.0)THEN
         WRITE(6,6)
         STOP
     END IF

     F=X**3+COS(X)-2.9
     XN=X-F/FDERIV
     GO TO 4
  END IF

  WRITE(6,5)XN
5 FORMAT(' ','THE ROOT IS',E14.7)

  END


RUN

THE ROOT IS   .1397140E+01
```

Figure 4.23b   (cont.)

4.6 OLD-TIME FAVORITES

The older versions of FORTRAN had no IF-block constructs available. Instead, they relied on other forms of control statements that, although also available in FORTRAN 77, are not recommended today because they lead to very disjointed (nonstructured) programming code. We will list them here for reference purposes, since it is very likely that you will encounter programs written in an earlier version of FORTRAN in your work.

4.6.1 Arithmetic IF

The **arithmetic IF statement** has the syntax

IF(arithmetic expression) n_1, n_2, n_3

If the arithmetic expression is evaluated as positive, control passes to the statement whose number is n_3; if it is zero, control transfers to statement line n_2; and if it is negative, the program branches to statement line number n_1. For example, when executing

    IF(XN-X) 7,10,15

with X = 4.0 and XN = 5.1, control would pass to the statement whose number is 15. If XN = 3.8 instead, the program would then jump to statement number 7.

4.6.2 GO TO

We have already discussed the GO TO statement earlier. Its syntax is

GO TO n

where n is a statement number. The GO TO statement causes an unconditional transfer of control to statement number n to occur.

4.6.3 Computed GO TO

The **computed GO TO statement** has the form

GO TO $(n_1, n_2, \ldots, n_m)$, integer expression

Control is transferred to the statement number $(n_1, n_2, \ldots, n_m)$ pointed to by the result of the integer expression. If the result is 1, control passes to the first statement label. If the result is 5, control passes to the fifth statement label in the list. If the result is negative, or larger than the list, the statement is ignored. For example, with $I = 9$

GO TO (15,27,36), I-7

would have control pass to statement number 27.

4.6.4 Assigned GO TO

The **assigned GO TO statement** works in conjunction with the ASSIGN **statement.** It takes the following forms:

ASSIGN n TO (integer variable)

where n is a statement number, and

GO TO integer variable $(n_1, n_2, \ldots, n_m)$

When executed, control transfers to the statement having the statement number that was assigned to the integer variable. This statement number must be one of the statement numbers included in the GO TO list or an error will result. For example,

ASSIGN 24 TO KART
        .
        .
        .
GO TO KART (15,20,24,30)

would cause control to pass to statement number 24. A simpler form of this command is

GO TO integer variable

One could then substitute the statements

```
KART=24
GO TO KART
```

for the previous two.

REVIEW

Chapter 4 introduces the use of decision structures. Through the use of IF-block construction, it becomes possible to dynamically alter the sequence of program flow online. During program execution, emerging values can be made to control which path the program follows. Several alternative courses of action are usually given. This imparting of decision-making capability to the program requires the introduction of logical elements. The FORTRAN relational and logical operators of Tables 4.1 and 4.2 may be used to form the logical or compound logical expressions needed to effect program decisions. The descending order in which these operators are processed is as follows: relational operators, .NOT., .AND., .OR., .EQV., and .NEQV.

Normal program flow in a program is from top to bottom and is sequential. Decision and repetition (WHILE) structures, introduced via the block IF statement, allow for alteration of sequence. To prevent programs from being disjointed and hard to follow, they are given a standard structural form. This preferred type of program organization has top-down flow, allowing only decision- and repetition-block constructs to modify the succession of events. The blocks themselves may only have one entry and one exit point. They are also sequential, with step movement advancing toward the exit. Programs that adhere to these rules are called structured programs.

The simplest FORTRAN decision structure is the logical IF statement. If the given logical expression contained within the IF statement is true, some simple program action will be performed. The IF-THEN-END IF construction is much more powerful than the IF statement. It frames a block of statements that will be performed if the logical expression (included in the block IF-THEN introductory statement) is true. IF-THEN-END IF structures can also be nested together if desired. To provide for a suitable set of responses to a false condition, IF-THEN-ELSE-END IF structure is needed. Though it is possible to nest IF-THEN-ELSE-END IF structures together, this is not usually done. To accomplish the same multiple tasks an IF-THEN-ELSE IF-END IF construction is preferred because of its simpler format. If several logical expressions are presented in this construct, only the group of statements that belong to the subblock whose logical expression was first evaluated as true is performed.

The repetition of WHILE structure repeats a set of actions indefinitely until a test condition, usually represented by a logical expression, switches values and causes the loop to be exited. WHILE structure can be implemented with a FORTRAN IF-THEN-END IF structure that incorporates a GO TO statement to close the loop. The GO TO statement causes program transfer back to the block IF-THEN conditional (logical) expression. The block is repeatedly executed until the logical expression becomes false. At that point the loop is exited. Therefore, at some point in time the block must be programmed to change the value of the logical expression in order to leave the loop. This particular structural form can be used advantageously for input operation. It would then be used in conjunction with a data file, usually of unknown length, having a trailer line included to mark its end. Sample problems that employ root-finding algorithms (the bisection method and the Newton-Raphson method) are presented in this chapter to demonstrate the WHILE structure.

SUMMARY OF FORTRAN STATEMENTS

IF(logical expression) executable statement

IF (logical expression) THEN
    Block of executable statements
      .
      .

END IF

IF (logical expression) THEN
    Block of executable statements
      .
      .

ELSE
    Block of executable statements
      .
      .

END IF

IF (logical expression 1) THEN
    Block of executable statements
      .
      .

ELSE IF (logical expression 2) THEN
    Block of executable statements
      .
      .

ELSE IF (logical expression i) THEN
    Block of executable statements
      .
      .

ELSE   (ELSE block is optional)
    Block of executable statements
      .
      .

END IF
_____

n IF (logical expression) THEN
    Block of executable statements
      .
      .
      .
    GO TO *n*
  END IF
_____

IF(arithmetic expression) n_1, n_2, n_3
GO TO *n*
GO TO $(n_1, n_2, \ldots, n_m)$, integer expression
_ _ _ _ _ _ _ _ _ _ _ _ _ _

ASSIGN *n* TO integer variable
GO TO integer variable $(n_1, n_2, \ldots, n_m)$
or
GO TO integer variable

KEY TERMS

arithmetic IF statement	**logical expression**
ASSIGN **statement**	**logical IF statement**
assigned GO TO statement	**logical operator**
block IF statement	**nested blocks**
compound logical expression	**relational operator**
computed GO TO statement	**running sum**
END IF **statement**	**selection structure**
GO TO **statement**	**sequential structure**
IF-block structure	**structured programs**
IF structure	**top-down code**
IF-THEN-ELSE-END IF structure	**trailer line**
IF-THEN-ELSE IF-END IF structure	**WHILE loop**
IF-THEN-END IF structure	**WHILE repetition structure**
logical constant	

REVIEW QUESTIONS

1. Write a FORTRAN statement that will set Y equal to 5.0 if X is greater than or equal to 54.1.

2. Write a FORTRAN statement that will set Z equal to the natural logarithm of T if T is less than TIME.

3. Write a FORTRAN statement that will stop the program if K is not equal to a number 4 less than I.

4. Write a compound logical IF statement that will print the value of X at the terminal if T is greater than 5.0 but less than 7.5.

5. Write a compound logical IF statement that will increment RAY by 30.0 if X is equal to 1.E3 or if Y is greater than or equal to the square root of U.

In Problems 6–10, use IF-THEN-END IF structure to program the specified actions.

6. If Q is less than 0.0, print 'THE RADICAL IS COMPLEX' at the terminal and stop the program.

7. If Z is greater than or equal to TIME, increment TIME by DELTA, take the square root of TIME, and then read from the terminal the value of XRAY, in that order.

8. Without resorting to compound logical expressions, write statements that will set BMOM equal to 7500*X and print at the terminal the results if X is greater than or equal to 0 but less than 15. If X is greater than 30, stop the program using a separate IF statement.

9. Repeat Problem 8, using a compound logical expression to achieve the same results.

10. Write program statements that will set STRESS equal to P divided by AREA and then print STRESS if KCASE equals 0 or 4.

11. Use an IF-THEN-ELSE-END IF structure to cause OMEGA to be set to 0 if ZETA is equal to RHO and then print OMEGA at the terminal. Otherwise set OMEGA equal to PHI and increment RHO by EPSI.

12. Replace the given compound logical expression with nested IF-THEN-END IF structures.

```
IF(T.GT.4.0.AND.RHO.LE.ZETA) PRINT *,EPSI
```

13. Replace the compound logical expression given with an appropriate IF-THEN-ELSE IF-END IF structure.

```
IF(ABS(X-X0).LE.0.0001.OR.N.EQ.100) STOP
```

14. Discuss what is wrong with the following sequence of statements:

```
IF(FLOW.NE.Q1) PRESS=100.0
PRESS=ALLOW
```

15. Write FORTRAN statements that will read in values of X, Y, and Z at the terminal and then compare them to find the largest absolute value present. The largest value should be assigned to a fourth variable named RMAX. RMAX should be printed out at the terminal.

16. A force F acting on an object is known to be equal to $3T$ from $T = 0$ to $T = 10$ seconds. For T greater than 10 seconds but less than 15 seconds, the force is constant and equal to 30. For any time greater than or equal to 15 seconds, the force F is 0. Write FORTRAN statements that will assign the proper value to F if T is known.

17. How many times will the variable named VOL be printed out in the following WHILE structure block?

```
      I=0
  7   IF(I.LT.30) THEN
          PRINT *,VOL
          I=2*I+1
          GO TO 7
      END IF
```

18. Complete in terms of I the following WHILE structure block so that the value of X is incremented by DX four times.

```
X=0.0
DX=0.1
I=0
```
7 _ _ _If (I.LT. 15) Then_ _ _ _ _ _ _ _ _
```
      X=X+DX
      I=2*I+1
      GO TO 7
  END  IF
```

Iter

1	0 → 1
2	1 → 3
3	3 → 7
4	7 → 15

COMPUTER PROJECTS

CP1. Modify the solution to Sample Problem 3.4 so that IF-block structure is employed instead.

CP2. Modify the solution to Sample Problem 3.7 so that a WHILE structure is used to enter the data. The data file should now contain a trailer line.

CP3. Write a program that reads in the x- and y-coordinates of a point and then determines which quadrant the point lies in. If the point lies on an axis, the program should identify which one.

CP4. Write a program that will determine whether an integer value entered from the terminal is odd or even.

CP5. Write a FORTRAN program that will add the squares of all odd integers from 0 to 100.

CP6. The "natural number" e is an irrational number whose value is given by

$$e = 2.71828\ldots$$

The quantity e is very important to scientific and engineering calculations because of the many physical processes that can be described by the function e^x. (The FORTRAN intrinsic function EXP(X) can be used to evaluate e^x.) Using the e value given, write a program that will compute e^x for values of x ranging from -4.0 to 4.0 in increments of 0.1. Compare your results to those given by the intrinsic function EXP(X).

CP7. Write a FORTRAN program that will read a series of numbers from the terminal and add them until a flag is reached. The final result should be printed out.

CP8. Repeat CP7, but this time keep track of how many entries were made at the terminal before the flag was encountered. Print the average value of the entries.

CP9. Modify Sample Problem 4.9 to find the root of the equation

$$f(x) = (1 + 70x)e^{-70x} - 0.1 = 0$$

by the bisection method. Start your search at $x = 0$. Let DELTAX = 0.1, EPSI = 1.0E−5, and STOPP = 1.0. (Refer also to CP6.)

CP10. Repeat CP9 but this time use the Newton-Raphson method. Refer to Sample Problem 4.10. The derivative $f'(x)$ of the given function $f(x)$ is

$$f'(x) = -4900xe^{-70x}$$

CP11. Write a FORTRAN program to convert any millimeter data value from 1 to 800, entered at the terminal, to feet and inches. Any remainder should be sorted into half, quarter, eighth, sixteenth, and thirty-second inch parts, in that order. For example, 80 millimeters can be converted to 3 one-inch, 0 half- and quarter-inch, 1 eighth-inch, 0 sixteenth-inch, and 1 thirty-second-inch parts. The output should be presented exactly as shown:

```
    |         |         |         |         |         |
12345678901234567890123456789012345678901234567890123456789012345

         THE  NUMBER  OF  MILLIMETERS:  XXX

         -------------------------------
    NUMBER  OF  FEET:  XX                    NUMBER  OF  INCHES:  XX
    NUMBER  OF  HALF  INCHES:    XX        NUMBER  OF  QUARTER  INCHES:   XX
    NUMBER  OF  EIGHTH  INCHES:    XX    NUMBER  OF  SIXTEENTH  INCHES:   XX
    NUMBER  OF  THIRTY-SECOND  INCHES:   XX
```

CP12. Experimental values for cast steel showing flux density B in kilolines per square inch versus magnetizing force in ampere-turns per inch are given in the following table. Write a program that will take a given value of B, read from the terminal, and linearly interpolate between tabular values to find its corresponding value of H. Output should consist of the values for B and H, and also exactly duplicate the given table. Test your results for the case when $B = 110$ kilolines per square inch.

```
    |         |         |         |         |         |
12345678901234567890123456789012345678901234567890123456789012345

            H                                        B
      AMPERE-TURNS                          KILOLINES  PER
        PER  INCH                            SQUARE  INCH
      ------------                          ------------
            0                                        0
           10                                       44
           20                                       66
           30                                       78
           40                                       86
           70                                       97
          200                                      112

      THE  GIVEN  VALUE  OF  B  IS:  XXX.XX

      THE  CORRESPONDING  VALUE  OF  H  IS:  XXX.XX
```

CP13. A 230-volt shunt motor is designed to operate with constant armature current of 100 amp. Since the armature resistance is known to be 0.1 ohm, the armature terminal voltage E_a is 220 volts. Using the motor's characteristics given in the accompanying table, write a FORTRAN program to find the necessary shunt field current I_f that will cause the motor to operate at any desired speed (to be entered at the terminal) by linearly interpolating between the given values. Besides printing your answer, the program should also exactly duplicate the given table. Test your results at 1500 rpm.

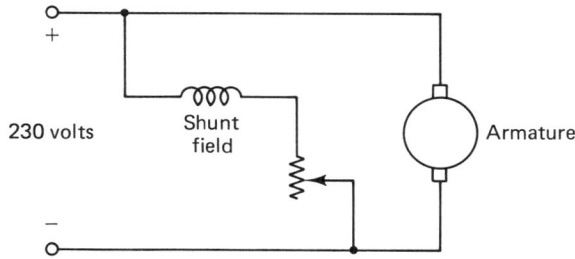

CP4.13

```
    |          |          |          |          |          |
1 2 3 4 5 6 7 8 9 0 1 2 3 4 5 6 7 8 9 0 1 2 3 4 5 6 7 8 9 0 1 2 3 4 5 6 7 8 9 0 1 2 3 4 5 6 7 8 9 0 1 2 3 4 5
```

EA/N VOLTS/RPM	IF SHUNT–FIELD CURRENT, AMP
. 0 0 0	. 0 0
. 0 8 5	. 5 0
. 1 5 5	1 . 0 0
. 2 0 5	1 . 5 0
. 2 2 0	1 . 7 5
. 2 3 0	2 . 0 0

```
   THE  NUMBER  OF  RPM  IS:  XXXX.X        THE  FIELD  CURRENT  IS:  X.XX
```

CP14. Maintenance costs for a machine are estimated to be $400 each year, payable at the end of the year. Write a computer program to find the present worth of its total maintenance expense over a ten year period if the interest rate is 8 percent. Present worth can be calculated using the formula

$$P = F(1 + I)^{-n}$$

where F is the future value, I is the interest rate, and n is the number of years. (*Hint:* There are ten individual worths that must be found and added together. Use WHILE loop structure in your program. The values of F, I, and n are to be entered at the terminal. Your output should have the following form:

```
    |          |          |          |          |          |
1 2 3 4 5 6 7 8 9 0 1 2 3 4 5 6 7 8 9 0 1 2 3 4 5 6 7 8 9 0 1 2 3 4 5 6 7 8 9 0 1 2 3 4 5 6 7 8 9 0 1 2 3 4 5

THE  PRESENT  WORTH  OF  XX  YEARS  OF  MAINTENANCE  COSTING  $XXX.XX

ANNUALLY,  AND  WITH  A  CONSTANT  INTEREST  RATE  OF  X.XX %  IS
$XXXX.XX.
```

CP15. The flow of heat through a pipe's insulation covering, per linear foot, is given by the relationship

$$q = \frac{2\pi(T_1 - T_n)}{\displaystyle\sum_{m=1}^{n-1} \frac{\ln(r_{m+1}/r_m)}{K_m}} \quad \text{BTU/hr}$$

where T_1 is the pipe's surface temperature, T_n is the temperature at the outside surface with $n - 1$ layers of insulation acting, r_m and r_{m+1} are inner and outer radii, respectively, of layer m, and K_m is the value of its thermal conductivity. For the values given in the figure, write a computer program that will calculate the outside temperature given a heat flow, q, of 323 BTU/hr-ft. All given values can be read from a data file or the terminal. The program should be written using WHILE loop structure in a general manner so that an arbitrary number of layers can be employed. Output should have the form shown.

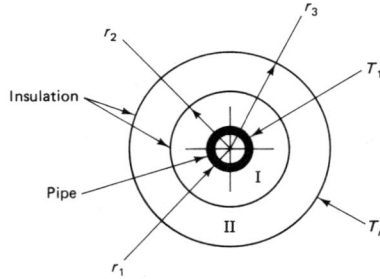

CP4.15

$T_1 = 700°F$

$r_3 = 17.75$ in

$r_2 = 13.75$ in

$r_1 = 10.75$ in

$K_I = 0.050$ BTU/hr-ft-F $K_{II} = 0.089$ BTU/hr-ft-F

```
      |         |         |         |         |         |
12345678901234567890123456789012345678901234567890123456789012345

           INNER RADIUS    OUTSIDE RADIUS     CONDUCTIVITY
   LAYER       IN.              IN.          BTU/HR-FT-F
   ------------------------------------------------------------
    XX       XXX.XXX          XXX.XXX           X.XXX

     "          "                "                 "

     "          "                "                 "

     "          "                "                 "


  THE OUTSIDE SURFACE TEMPERATURE AT COVERING INSULATION LAYER
  XX IS XXX.X.
```

CP16. The height, expressed as a function of time, of a sounding rocket launched vertically in a constant gravitational field with drag forces neglected is given by the simple relationship

$$Y = \left(-9.81\frac{T}{2.0} + V\right)T \quad \text{meters}$$

where T is time, V is the launch velocity, and Y is the height. Using WHILE loop structure, write a FORTRAN program that will calculate and print out values of Y starting at $T = 0$, and continuing output for every one second afterward until a maximum time of 20 seconds is reached. Let the computer also calculate the highest height attained and the time at which it occurs. (*Hint:* Initialize a variable name that will be used to store the maximum value of height YMAX. At each pass through the WHILE loop structure, the maximum value of Y, YMAX, should be tested to see if it is less than the current value of Y. If it is greater than the most recent value of Y, no change is required and the program should continue. If it is less than the current value of Y, the two values should be interchanged — after printing Y, so that YMAX will still contain the largest value. At the same time, interchange the values of T and TMAX so that the proper TMAX is also available for printout. No comparisons of time magnitudes are needed, however. The variable name that contains TMAX will not store the largest value of T, but the value of T that is associated with YMAX.) Take as your initial value $V = 100$ m/s. Your output should follow the form shown.

```
           |           |           |           |           |           |
 1234567890123456789012345678901234567890123456789012345678901234567890123456789012345

                   T SEC                        Y METERS
                 -------------------------------------------------
                     .0                         .000EXXX
                   XX.X                         .XXXEXXX
                    "                             "
                    "                             "
                    "                             "
                   20.0                         .XXXEXXX


     THE MAXIMUM HEIGHT ATTAINED FOR THE ROCKET WAS  .XXXEXXX
     METERS.   THIS OCCURRED AT XX.X SECONDS.
```

CHAPTER 5

The DO Loop Structure

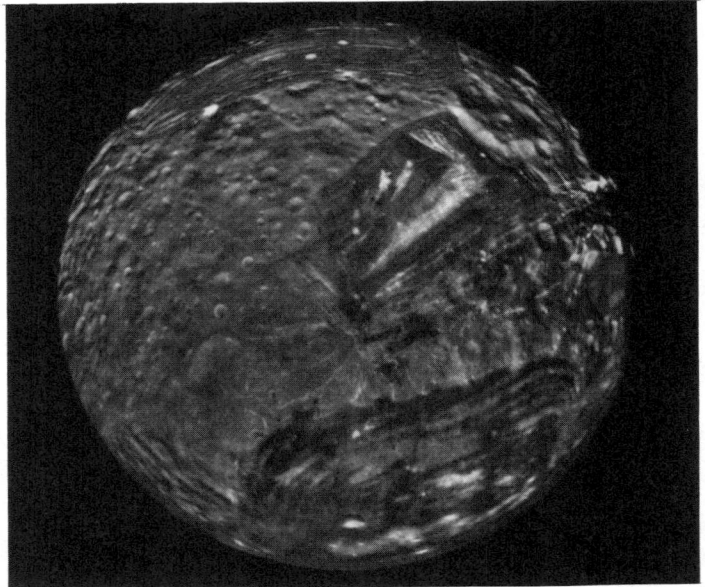

Uranus' moon Miranda is shown in this computer-assembled and enhanced mosaic of images obtained by the *Voyager 2* spacecraft. *(Photo courtesy of JPL/NASA.)*

5.1 INTRODUCTION

Another widely used form of the WHILE repetition structure provided for in the FORTRAN language is the DO loop. The DO syntax allows the programmer a simple means of forming counter-controlled repetition loops. This type of structure is so important to proper programming style that this present chapter will be devoted exclusively to its discussion.

5.2 THE DO LOOP STRUCTURE

The DO Loop structure has the general syntax

DO *n* index variable=initial value, limit, increment
    Block of executable statements
n CONTINUE

The DO **statement** instructs the computer to perform all subsequent statements up to and including the statement having label n for a specified number of counted times (see Figure 5.1). The named **index variable** is the variable to be counted. Its initial value and its limit must be assigned. If no increment is specified for the counter, a default value of 1 is assumed. The final statement of the loop, having statement label n, can be any executable statement except a GO TO, assigned GO TO, IF, RETURN, STOP, END, or another DO statement. Statements belonging to the IF-block syntax — IF–THEN, ELSE IF, ELSE, and END IF — are not allowed either.

Usually, for clarity's sake, a CONTINUE **statement** is used to close each DO loop. The author strongly recommends this usage. For example, the statements

```
    A=0.0
    DO 5  I=1,4
        READ *,R
        A=A+R
  5 CONTINUE
```

would sum the next four values of R entered at the terminal. Though it is not mandatory, for presentation purposes, the body of the loop should be indented. The statements

```
    A=0.0
    DO 75  J=1,4,2
        A=A+6.0
 75 CONTINUE
```

would produce a numeric value of 12.0 for the variable name A when the loop was finally exited. The index variable J takes on the values 1, 3, and 5, being incremented by 2. When J becomes 5, which is larger than the limiting value, the loop is immediately exited and control passes to the statement immediately following the statement numbered 75. In other words, testing *precedes* execution of the block's

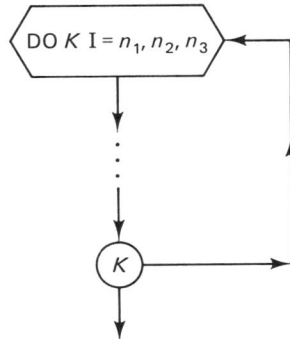

Figure 5.1

statements. Upon exiting, J retains the value of 5. The statements

```
A=0.0
DO 101 S=1.1,4.0,2.1
    A=A+6.0
101 CONTINUE
```

would produce the same final result for A. Upon exiting the loop, the variable index S would have a value of 5.3. One can also write

```
A=0.0
DO 55 I=3,-1,-3
    A=A+6.0
55 CONTINUE
```

which would produce A = 12.0 and I = −3 upon termination. As yet another example, the statements

```
N=2
DO 4 KAT=3,-1,-2
    N=N*N
4 CONTINUE
```

would produce N = 256 and KAT = −3 when the loop was exited.

One can summarize this discussion by saying that *the number of times that a loop will be executed, K, is equal to the value (integer) of the following expression:*

$$K = \frac{\text{limit} - \text{initial value} + \text{increment}}{\text{increment}} \text{, integer part}$$

If K is less than zero, the loop is not performed. That means that for a positive increment, the specified limit value must be equal to or greater than the initial value. The opposite holds true for a negative increment. A zero increment is not allowed.

Variable names and even arithmetic expressions are permissible for the DO statement parameters. For instance, one can write

```
READ*,I,N,INC
A=0.0
DO 55 J=2*I,N,INC
    .
    .
    .
55 CONTINUE
```

If I=1, N=5 and INC=2 are entered at the keyboard, the DO loop block statements will be executed twice.

Once the structure has been activated and its parameters set, the control variables may be changed inside the loop without having any effect on the number of times the loop is actually executed. For instance, the statements

```
        READ *,N
        SUM=N
        DO 7  I1=N,2,-1
            N=I1-1
            SUM=SUM+N
    7  CONTINUE
        PRINT *,SUM
```

would produce and print for N = 8 the sum (SUM) of 36

```
    8+7+6+5+4+3+2+1
```

as a result.

The *index variable*, however, may not have its value changed inside the loop. This type of fatal error is commonly committed by beginning programmers. It leads to an error message during compilation time that fortunately can readily be interpreted and corrected, namely,

```
    "REDEFINES A DO CONTROL INDEX"
```

For example, one cannot write

```
        READ *,N
        SUM=N
        DO 7  I1=N,2,-1
            I1=I1-1
            SUM=SUM+I1
    7  CONTINUE
```

since this redefines the index variable I1 inside the loop, that is, I1 appears to the left of an equal sign within the loop.

Sample Problem 5.1

Rework Sample Problem 4.8 using the DO-loop form of the WHILE structure.

Solution: The algorithm and problem solution are shown in Figures 5.2a and 5.2b. Almost identical to the solution of Sample Problem 4.8, the program shows clearly how a DO loop clarifies and shortens program development, leaving less chance for the introduction of programming errors. A single DO statement replaces several program steps. The DO statement causes a counter variable J to be initialized at the start of the loop, and then increments this variable by 2, comparing its current value with a predefined limit, NDATA, each time the loop is traversed. When the limiting value of the counter is reached, the loop is exited. Note how the DO loop block statements are indented in the source listing for clarity. Some programmers prefer even to separate the entire DO loop from the rest of the program by using blank comment lines before and after the loop.

■ **Algorithm for Sample Problem 5.1.** This program reads every other data value from a data file TESTB that is NDATA long. The mean value for the selected samples and the standard deviation are calculated.

1. Enter NDATA.
2. Set SUM=SUMSQ=N=0.
3. Do, with index J ranging from 1 to NDATA in increments of 2, the following:
   a. Read R
   b. Calculate
      SUM=SUM+R
   c. SUMSQ=SUMSQ+R**2
   d. N=N+1
4. Calculate
   AV=SUM/N
   STO=SUMSQ-N*AV**2
   S=SQRT(STO/(N-1))
5. Print AV,S,N.

Figure 5.2a

```
            PROGRAM COUNT

      ****************************************************************
      *                                                              *
      *     THIS PROGRAM READS EVERY OTHER DATA VALUE FROM DATA FILE  *
      *     'TESTB' THAT IS NDATA LONG. THE MEAN VALUE FOR THE SELECTED *
      *     SAMPLES AND THE STANDARD DEVIATION ARE CALCULATED.        *
      *                                                              *
      ****************************************************************

            REAL SUM,SUMSQ,R,AV,STO,S
            INTEGER N,NDATA

            OPEN(UNIT=8,FILE='TESTB')
            REWIND 8

            PRINT*,' ENTER THE TOTAL LENGTH OF THE DATA FILE.'
            READ*,NDATA

            SUM=0.
            SUMSQ=0.
            N=0.

            DO 2 J=1,NDATA,2

                READ(8,1)R

                SUM=SUM+R
                SUMSQ=SUMSQ+R**2
                N=N+1

          2 CONTINUE
```

Figure 5.2b

```
    AV=SUM/N
    STO=SUMSQ-N*AV**2
    S=SQRT(STO/(N-1))

    WRITE(6,5)AV
  1 FORMAT(F10.0/)
  5 FORMAT(' ','THE MEAN VALUE WAS ',F10.4)
    WRITE(6,6)S
  6 FORMAT(' ','THE STANDARD DEVIATION WAS ',F10.4)
    WRITE(6,7)N
  7 FORMAT(' ','THE SAMPLE SIZE WAS ',I3)

    END


File Testb

2.999
55.123
3.001
55.332
3.010
51.555
3.005
52.345
2.995
54.435
2.998
55.100
3.003
56.018
3.003
56.121
2.999
55.000
3.001
54.787


RUN

  ENTER THE TOTAL LENGTH OF THE DATA FILE.
? 20
  THE MEAN VALUE WAS      3.0014
  THE STANDARD DEVIATION WAS      .0042
  THE SAMPLE SIZE WAS  10
```

Figure 5.2b   (cont.)

Sample Problem 5.2

Write a FORTRAN program that will read in a positive integer value n from the terminal and then calculate n factorial. Output should be displayed on the monitor.

Analysis: n factorial is defined by

$$n! = n \times (n - 1) \times (n - 2) \times \cdots \times 1$$

with

$$0! = 1$$

Solution: The program's flowchart is illustrated in Figure 5.3, and the algorithm and problem solution are shown in Figures 5.4a and 5.4b. In Sample Problems 4.7 and 4.8

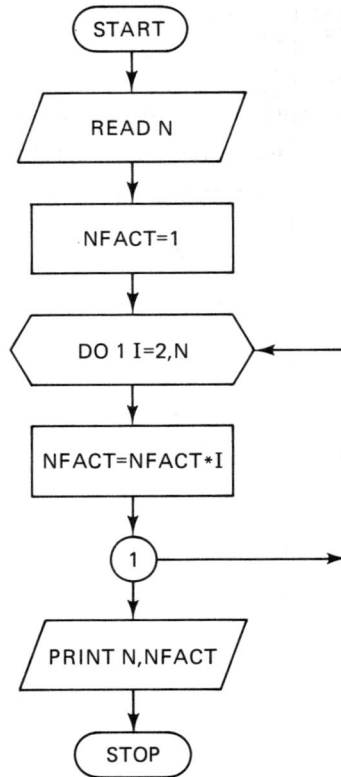

Figure 5.3

■ **Algorithm for Sample Problem 5.2.** This program reads in a positive integer value N from the terminal and then calculates N factorial. Results are displayed at the terminal.

1. Enter N.
2. Set NFACT=1.
3. Do, for index I ranging from 2 to N, the following:
   a. Calculate
      NFACT=NFACT * I
4. Print NFACT.

Figure 5.4a

```
          PROGRAM FACTOR

*******************************************************************
*                                                                 *
*     THIS PROGRAM READS IN A POSITVE INTEGER VALUE OF N FROM THE *
*     TERMINAL AND THEN CALCULATES N FACTORIAL. RESULTS ARE       *
*     DISPLAYED AT THE TERMINAL.                                  *
*                                                                 *
*******************************************************************

          INTEGER N,NFACT

          PRINT*,'ENTER THE VALUE OF N.'
          READ*,N
          NFACT=1

          DO 1 I=2,N

              NFACT=NFACT*I

        1 CONTINUE

          PRINT*,'THE VALUE FOR ',N,' FACTORIAL IS ',NFACT

          END


   RUN

    ENTER THE VALUE OF N.
    ? 8
     THE VALUE FOR 8 FACTORIAL IS 40320
```

Figure 5.4b

we introduced a running sum. In this problem we develop a **running product.** Note how this product must be initialized with a nonzero value. What key values should be included in any test of the program?

Sample Problem 5.3

Write a FORTRAN program that will develop a table of sin x for the values of x ranging from 0 to 1 in increments of 0.1 radians. Use a Taylor's series expansion to approximate sin x, and keep only the first four terms of the expansion. Comparc your results with the value of sin x returned by the computer's intrinsic function $SIN(X)$.

Analysis: From Calculus, we learn that the Taylor's series expansion for sin x, keeping only the first four terms, is given by

$$\sin x = \frac{x}{1!} - \frac{x^3}{3!} + \frac{x^5}{5!} - \frac{x^7}{7!}$$

where, for example,

$$7! = 7 \times 6 \times 5 \times 4 \times 3 \times 2 \times 1$$

Solution: The algorithm and problem solution are shown in Figures 5.5a and 5.5b. Note how the FORMAT statements are optionally placed at the end of the program this time.

■ **Algorithm for Sample Problem 5.3.** This program develops the values of SIN(X) using a Taylor's series expansion with only the first four terms. A table is drawn up for the values of SIN(X) with X ranging from 0 to 1 in increments of 0.1. Results are compared with the SIN(X) value found using the intrinsic FORTRAN SIN function.

1. Do, with index X ranging from 0 to 1 in increments of 0.1, the following:
    a. Calculate
       ```
       SIN1=X
       SIN2=-X*X*SIN1/(3.0*2.0)
       SIN3=-X*X*SIN2/(5.0*4.0)
       SIN4=-X*X*SIN3/(7.0*6.0)
       SINE=SIN1+SIN2+SIN3+SIN4
       Y=SIN(X)
       ```
    b. Print X,SINE,Y.

Figure 5.5a

```
        PROGRAM SINX

        ****************************************************************
        *                                                              *
        *    THIS PROGRAM DEVELOPS THE VALUES OF SIN(X) USING A TAYLOR'S *
        *    SERIES EXPANSION WITH ONLY THE FIRST FOUR TERMS. A TABLE   *
        *    IS DRAWN UP FOR THE VALUES OF SIN(X) WITH X RANGING FROM 0 *
        *    TO 1 IN INCREMENTS OF 0.1. RESULTS ARE COMPARED TO THE     *
        *    SIN(X) VALUE FOUND USING THE INTRINSIC FORTRAN SIN         *
        *    FUNCTION.                                                  *
        *                                                              *
        ****************************************************************

        REAL SIN1,SIN2,SIN3,SIN4,SINE,Y,X

        PRINT 1
        PRINT 2
        PRINT 3
        PRINT 2

        DO 5 X=0.,1.0,0.1

            SIN1=X
            SIN2=-X*X*SIN1/(3.*2.)
            SIN3=-X*X*SIN2/(5.*4.)
            SIN4=-X*X*SIN3/(7.*6.)
            SINE=SIN1+SIN2+SIN3+SIN4
            Y=SIN(X)

            PRINT 4,X,SINE,Y

      5 CONTINUE

      1 FORMAT(T27,'TABLE FOR SIN(X)',//)
      2 FORMAT(65('-'))
```

Figure 5.5b

```
3 FORMAT(T14,'X',TR18,'SINX',TR15,'SIN(X)')
4 FORMAT(T6,E14.7,2(TR6,E14.7))

  END


RUN
```

                              TABLE FOR SIN(X)


```
----------------------------------------------------------------
        X                    SINX                    SIN(X)
----------------------------------------------------------------
    .0000000E+00          .0000000E+00            .0000000E+00
    .1000000E+00          .9983342E-01            .9983342E-01
    .2000000E+00          .1986693E+00            .1986693E+00
    .3000000E+00          .2955202E+00            .2955202E+00
    .4000000E+00          .3894183E+00            .3894183E+00
    .5000000E+00          .4794255E+00            .4794255E+00
    .6000000E+00          .5646424E+00            .5646425E+00
    .7000000E+00          .6442176E+00            .6442177E+00
    .8000000E+00          .7173557E+00            .7173561E+00
    .9000000E+00          .7833258E+00            .7833269E+00
    .1000000E+01          .8414683E+00            .8414710E+00
```

Figure 5.5b   (cont.)

Sample Problem 5.4

The roller cam follower shown in Figure 5.6 attains the measured vertical motion described presently in the first quarter of the cam's rotation starting from rest. It is desired to use this given information, read from a file, to estimate the maximum value of the follower's speed and the time that that occurs. Write a FORTRAN program that analyzes the cam follower's motion to find these values.

T (seconds)	S (centimeters)
0.0	0.0000
0.05	0.0373
0.10	0.1387
0.15	0.288
0.20	0.4693
0.25	0.6667
0.30	0.864
0.35	1.0453
0.40	1.1947
0.45	1.2960
0.50	1.3333

Analysis: The speed of the cam follower, v, can be numerically estimated by

$$v \simeq \frac{\Delta S}{\Delta t}$$

which is the time average of the change in position. As Δt becomes smaller, the accuracy of this expression increases, but only up to a limit, after which roundoff error be-

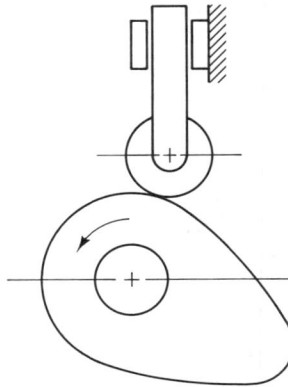

Figure 5.6

comes predominant. Note that the instantaneous average, where Δt becomes so small that it is practically zero, is the mathematical time derivative of S with respect to t. From physics we know this derivative as the instantaneous speed.

$$v = \lim_{\Delta t \to 0} \frac{\Delta S}{\Delta t} = \frac{dS}{dt}$$

In this problem, not only must the derivative, v, be *approximated* at each step (remember the computer cannot divide by numbers approaching zero), but its current value must also be compared with the largest value previously calculated, v_1. The larger of these two quantities is then kept as the new reference for the next comparison, and so on. At the end of the program, the value finally retained will be the largest speed, and its accompanying time, t_1, will be the answer for the moment when it occurs.

Solution: See Figures 5.7a and 5.7b for the algorithm and problem solution. Note that not only are the initial displacement and time assigned before entering the DO loop, but the variables that are used to store the maximum speed value, V1, and its associated time, T1, are also initialized. In addition, note that in the logical expression absolute values are used to accommodate the presence of any negative speeds. This IF-block is nested (embedded) in the DO loop and is shown indented relative to the DO-loop block for legibility. This example also demonstrates the problems associated with using a real index variable. The DO-loop limit was purposely set to 0.50001 to accommodate the roundoff errors introduced by the computer when performing real-variable operations.

■ **Algorithm for Sample Problem 5.4.** This program estimates the speed that a roller cam follower attains, starting from rest, given its displacement values every DT seconds. The calculated speeds are printed out at the terminal, as well as the values of the maximum speed and the time when it occurred.

1. Read DT,S0.

2. Set T0=V0=T1=V1=0.

3. Print T0,V0.

4. Do, with index T ranging from 0.05 to 0.50001 in increments of DT, the following:
   a. Read S
   b. Calculate
      V=(S-S0)/DT
   c. Print T,V
   d. Set S0=S
   e. If |V|>|V1| then
      i. Set V1=V,T1=T

5. Print V1,T1.

Figure 5.7a

```
        PROGRAM CAM

****************************************************************
*                                                              *
*    THIS PROGRAM ESTIMATES THE SPEED THAT A ROLLER CAM FOLLOWER *
*    ATTAINS, STARTING FROM REST, GIVEN ITS DISPLACEMENT VALUES *
*    EVERY DT SECONDS. THE CALCULATED SPEEDS ARE PRINTED OUT AT  *
*    THE TERMINAL AS WELL AS THE VALUES OF THE MAXIMUM SPEED AND *
*    THE TIME WHEN IT OCCURRED.                                 *
*                                                              *
****************************************************************

        REAL DT,SO,TO,VO,T1,V1,V,T,S

        OPEN(UNIT=8,FILE='DISP')
        REWIND 8

        READ(8,*)DT
        READ(8,*)SO

        TO=0.
        VO=0.
        T1=TO
        V1=VO

        PRINT 1,TO,VO

        DO 2 T=0.05,0.50001,DT
```

Figure 5.7b

```
        READ(8,*)S
        V=(S-S0)/DT
        PRINT 1,T,V
        S0=S

        IF(ABS(V).GT.ABS(V1))THEN
            V1=V
            T1=T
        END IF

    2 CONTINUE

        PRINT 3,V1,T1

    1 FORMAT(5X,'THE SPEED AT T = ',F4.2,' S IS = ',F9.4,' CM/S')
    3 FORMAT(//5X,'THE MAXIMUM SPEED IS = ',F9.4,' CM/S AND',
      + /,5X,'OCCURS AT T = ',F4.2,' S')

        END


    File Disp

    0.
    0.0373
    0.1387
    0.288
    0.4693
    0.6667
    0.864
    1.0453
    1.1947
    1.296
    1.333


    RUN

        THE SPEED AT T =   .00 S IS =       .0000 CM/S
        THE SPEED AT T =   .05 S IS =       .7460 CM/S
        THE SPEED AT T =   .10 S IS =      2.0280 CM/S
        THE SPEED AT T =   .15 S IS =      2.9860 CM/S
        THE SPEED AT T =   .20 S IS =      3.6260 CM/S
        THE SPEED AT T =   .25 S IS =      3.9480 CM/S
        THE SPEED AT T =   .30 S IS =      3.9460 CM/S
        THE SPEED AT T =   .35 S IS =      3.6260 CM/S
        THE SPEED AT T =   .40 S IS =      2.9880 CM/S
        THE SPEED AT T =   .45 S IS =      2.0260 CM/S
        THE SPEED AT T =   .50 S IS =       .7400 CM/S


        THE MAXIMUM SPEED IS =     3.9480 CM/S AND
        OCCURS AT T =   .25 S
```

Figure 5.7b   (cont.)

5.3 TRANSFER TO AND FROM A DO LOOP

A DO loop is only activated if and when its DO statement is processed; otherwise, it lies dormant. Transferring control into the range of an inactive DO loop is not legal and would produce a compilation time error that might be signaled on the monitor via the message "ILLEGAL TRANSFER TO INSIDE A CLOSED DO LOOP OR

IF BLOCK." When transferring from the loop prematurely, that is, before the index counter is full, the index variable retains its current value.

A DO loop is made inactive either through a normal exit, a transfer of control, or when a RETURN, a STOP, or an END statement appears in the program while the loop is being executed. If one DO loop is nested within the range of another DO loop (see Sample Problem 5.9), it would also become inactive when the outer external DO loop becomes inactive or terminated. Note that if an IF-block structure is introduced inside a DO loop, it must be entirely contained within that DO loop's range (see Sample Problem 5.4). Similar rules apply to a DO loop structure introduced within an IF-block element; that is, it must have its terminating statement appear before the structure's END IF statement (see Figure 5.8).

Sample Problem 5.5

For experimentally obtained measurements, it is sometimes desirable to draw a representational curve for the data. A drawn curve cannot ordinarily pass smoothly through all the measured data points; therefore, some criteria of "best fitness" must be accepted. Usually, the method of least squares, or least squared error, it used to establish the curve. To start with, one must have some general knowledge of the expected shape of the curve. In this problem assume that the data can be represented by a *straight line* (see Figure 5.9). Write a FORTRAN program that will read the seven following data points from a file and then generate the straight-line parameters that will "best fit" the data.

X	Y
0.20	54.0
0.25	48.8
0.30	45.3
0.35	40.1
0.40	35.2
0.45	32.3
0.50	27.5

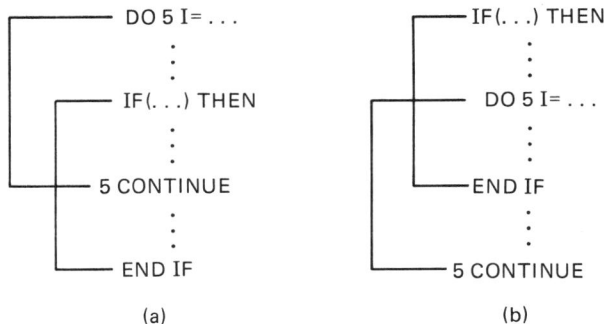

Figure 5.8 Operations not allowed

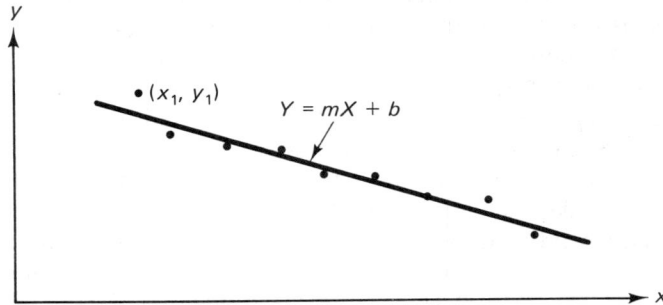

Figure 5.9

Analysis: The method of least squares best fits a curve to the data points having the general shape that the user predefines. When the curve is a straight line that has the equation

$$y = mx + b$$

the method of least squares will yield m and b as

$$m = \frac{n \sum\limits_{i=1}^{n} y_i x_i - \sum\limits_{i=1}^{n} x_i \sum\limits_{i=1}^{n} y_i}{n \sum\limits_{i=1}^{n} x_i^2 - \left(\sum\limits_{i=1}^{n} x_i \right)^2}$$

$$b = \frac{\sum\limits_{i=1}^{n} y_i - m \sum\limits_{i=1}^{n} x_i}{n}$$

Here n is the number of data points, and y_i and x_i are the respective paired data values.

Solution: See the algorithm and problem solution given in Figures 5.10a and 5.10b. The program should be run with other values for verification purposes. Points that are known to fall on a specific straight line may be used as fictitious data to check whether the program can correctly find the line's known slope and y-intercept.

Sample Problem 5.6

One way to numerically find the area that lies beneath a given curve contained between two fixed endpoints is to substitute a series of rectangular shapes for the given area. The sum of the areas of the rectangles approximates the area sought. The bases for each of these area elements are usually chosen to have equal length, Δx. Their heights have the y-ordinate values that are associated with the curve for each x. (See Figure 5.11.) Accuracy in this method increases as Δx gets smaller, but only up to a limit, after which roundoff error predominates. From calculus, the mathematical definition of a definite integral has Δx become vanishingly small while the number of boxes, n, simul-

■ **Algorithm for Sample Problem 5.5.** This program fits a straight line to a set of N data points using the method of least squares. Data points are entered from file FIT. The answers appear at the terminal.

1. Read N.
2. Set SUM1=SUM2=SUM3=SUM4=0.
3. Do, with index I ranging from 1 to N, the following:
   a. Read X,Y
   b. Calculate
      SUM1=SUM1+X
      SUM2=SUM2+Y
      SUM3=SUM3+X∗Y
      SUM4=SUM4+X∗X
4. Calculate
   SLOPE=(N∗SUM3−SUM1∗SUM2)/(N∗SUM4−SUM1∗∗2)
   B=(SUM2−SLOPE∗SUM1)/N
5. Print SLOPE,B.

Figure 5.10a

```
          PROGRAM FIT

      ****************************************************************
      *                                                              *
      *   THIS PROGRAM FITS A STRAIGHT LINE TO A SET OF N DATA POINTS *
      *   USING THE METHOD OF LEAST SQUARES. DATA POINTS ARE ENTERED  *
      *   FROM FILE 'FIT'. THE ANSWERS APPEAR AT THE TERMINAL.        *
      *                                                              *
      ****************************************************************

          REAL SUM1,SUM2,SUM3,SUM4,X,Y,SLOPE,B
          INTEGER N

          OPEN(UNIT=8,FILE='FIT')
          REWIND 8

          READ(8,*)N

          SUM1=0.
          SUM2=0.
          SUM3=0.
          SUM4=0.

          DO 1 I=1,N

             READ(8,*)X,Y

             SUM1=SUM1+X
             SUM2=SUM2+Y
             SUM3=SUM3+X*Y
             SUM4=SUM4+X*X
```

Figure 5.10b

```
1 CONTINUE

  SLOPE=(N*SUM3-SUM1*SUM2)/(N*SUM4-SUM1*SUM1)
  B=(SUM2-SLOPE*SUM1)/N

  PRINT 2,SLOPE,B

2 FORMAT('THE SLOPE OF THE STRAIGHT LINE THAT BEST FITS THE',
 + ' DATA IS = ',E10.3,/,'ITS INTERCEPT IS = ',E10.3)

  END


File Fit

7
0.2,54.0
0.25,48.8
0.3,45.3
0.35,40.1
0.4,35.2
0.45,32.3
0.5,27.5

RUN

THE SLOPE OF THE STRAIGHT LINE THAT BEST FITS THE DATA IS =   -.876E+02
```

Figure 5.10b   (cont.)

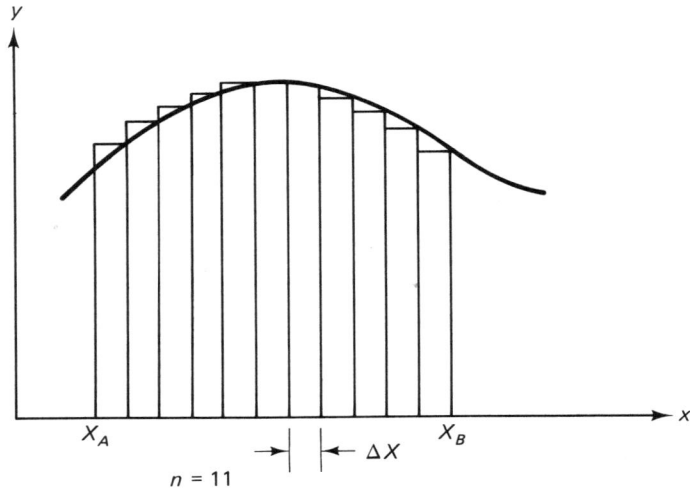

Figure 5.11

taneously goes to an infinite value during the summing process:

$$\int_{x_A}^{x_B} f(x)\, dx = \lim_{\substack{\Delta x \to 0 \\ n \to \infty}} \sum_{i=1}^{n} f(x_i)\, \Delta x_i$$

$$\Delta x_i = (x_B - x_A)/n$$

With this background information, write a FORTRAN program that will read the data from the following table and find the area under the curve.

X	Y
.0	.0
.05	−.425
.10	−.700
.15	−.825
.20	−.800
.25	−.625
.30	−.300
.35	.175
.40	.800

Analysis: If values for the curve are given in tabular form, with equally spaced x intervals, n is given by

$$n = \text{number of data points} - 1$$

and the area by

$$\text{Area} = \sum_{i=1}^{n} (y_{i+1})\, \Delta x = \Delta x \left(\sum_{i=1}^{n} y_{i+1} \right)$$

Solution: See Figures 5.12a and 5.12b for the algorithm and problem solution.

Sample Problem 5.7

The accuracy of the integration method introduced in Sample Problem 5.6 can be improved greatly if trapezoidal elements instead of rectangular ones are used to approximate the area. The approximated curve will then no longer appear step-wise irregular but will have straight line segments joining the given, equally spaced data points (see Figure 5.13). Write a FORTRAN program that uses the trapezoidal rule to solve for the area under the curve whose data points were given in Sample Problem 5.6.

Analysis: The area of a trapezoidal element is given by

$$(\text{Area})_i = \Delta x \frac{(y_{i+1} + y_i)}{2}$$

and the total area is

$$\text{Area} = \sum_{i=1}^{n} (\text{Area})_i = \left(y_i + y_{n+1} + 2 \sum_{i=2}^{n} y_i \right) \frac{\Delta x}{2}$$

Algorithm for Sample Problem 5.6. This program calculates the area under a given curve by approximating the given region with a series of rectangles with equal widths. The heights of the rectangles are the individual ordinates associated with each X value of the curve. The areas of the rectangles are summed.

1. Read NDATA,DX.
2. Set SUM=0.
3. Do, with index I ranging from 1 to NDATA, the following:
   a. Read Y
   b. If I>1 then
      i. Calculate
         SUM=SUM+Y
4. Calculate
   AREA=SUM*DX
5. Print AREA.

Figure 5.12a

```
        PROGRAM AREAS

        ****************************************************************
        *                                                              *
        *    THIS PROGRAM CALCULATES THE AREA UNDER A GIVEN CURVE BY    *
        *    APPROXIMATING THE GIVEN REGION WITH A SERIES OF RECTANGLES *
        *    WITH EQUAL WIDTHS. THE HEIGHTS OF THE RECTANGLES ARE THE   *
        *    INDIVIDUAL ORDINATES ASSOCIATED WITH EACH X VALUE OF THE   *
        *    CURVE. THE AREAS OF THE RECTANGLES ARE SUMMED.             *
        *                                                              *
        ****************************************************************

        REAL SUM,DX,Y,AREA
        INTEGER NDATA

        OPEN(UNIT=8,FILE='YVAL')
        REWIND 8

        READ(8,*)NDATA,DX

        SUM=0.

        DO 2 I=1,NDATA

            READ(8,*)Y

            IF(I.GT.1)THEN
                SUM=SUM+Y
            END IF

      2 CONTINUE

        AREA=SUM*DX
```

Figure 5.12b

```
      PRINT 1,AREA
    1 FORMAT(//'THE AREA UNDER THE GIVEN CURVE IS = ',E10.3)

      END


File Yval

11,.1
0.0000
0.1105
0.2443
0.4050
0.5967
0.8244
1.0933
1.4096
1.7804
2.2136
2.7183

RUN


THE AREA UNDER THE GIVEN CURVE IS =    .114E+01
```

Figure 5.12b   (cont.)

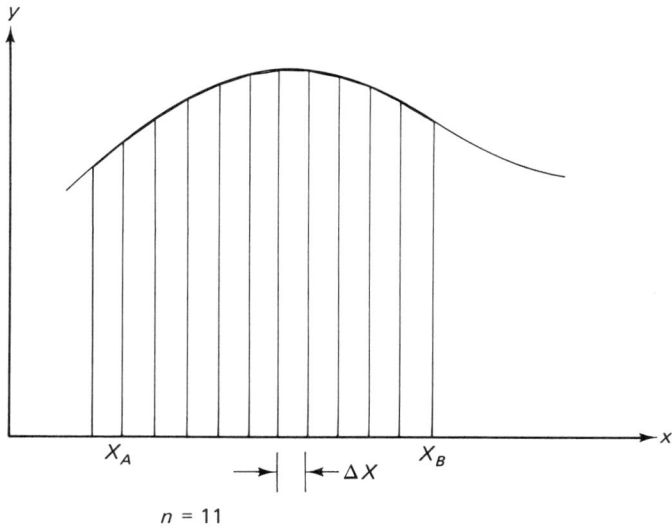

$n = 11$

Figure 5.13

Once more

$$n = \text{number of data points} - 1$$

and

$$\Delta x = \frac{x_B - x_A}{n}$$

Solution: See Figures 5.14a and 5.14b for the algorithm and problem solution. Will the program work for an even number of data points as well? What happens when n equals 2?

Algorithm for Sample Problem 5.7. This program uses the trapezoidal rule to calculate the area under a given curve that lies between points XA and XB. Data for the curve at equally spaced intervals is read from a file. The file also provides information concerning NDATA and DX.

1. Read NDATA, DX.
2. Read Y.
3. Calculate
   N=NDATA−1
4. Set SUM=Y.
5. Do, with index I ranging from 2 to N, the following:
   a. Read Y
   b. Calculate
      SUM=SUM+2.0*Y
6. Read Y.
7. Calculate
   SUM=SUM+Y
   AREA=DX*SUM/2.0
8. Print AREA.

Figure 5.14a

```
        PROGRAM TRAP

*****************************************************************
*                                                               *
*     THIS PROGRAM USES THE TRAPEZOIDAL RULE TO CALCULATE THE   *
*     AREA UNDER A GIVEN CURVE THAT LIES BETWEEN POINTS XA AND  *
*     XB. DATA FOR THE CURVE AT EQUALLY SPACED INTERVALS IS READ *
*     FROM A FILE. THE FILE ALSO PROVIDES INFORMATION CONCERNING *
*     NDATA AND DX.                                             *
*                                                               *
*****************************************************************
```

Figure 5.14b

```
      REAL DX,SUM,Y,AREA
      INTEGER N,NDATA

      OPEN(UNIT=8,FILE='YVAL')
      REWIND 8

      READ(8,*)NDATA,DX
      READ(8,*)Y

      N=NDATA-1
      SUM=Y

      DO 3 I=2,N
          READ(8,*)Y
          SUM=SUM+2.*Y
    3 CONTINUE

      READ(8,*)Y

      SUM=SUM+Y
      AREA=DX*SUM/2.

      PRINT 2,AREA
    2 FORMAT(//'THE AREA UNDER THE GIVEN CURVE IS = ',E10.3)

      END


File Yval

11,.1
0.0000
0.1105
0.2443
0.4050
0.5967
0.8244
1.0933
1.4096
1.7804
2.2136
2.7183

RUN


THE AREA UNDER THE GIVEN CURVE IS =   .100E+01
```

Figure 5.14b   (cont.)

Sample Problem 5.8

For the simply supported beam (see Sample Problem 4.2) and data presented in Figure 5.15, write a FORTRAN program that will search for the position, x, and value, y_{max}, of the largest deflection, along the beam. Since the slope of the beam must be zero at the position of maximum deflection, locate this position of zero slope first. Use the incremental search method to find the region where the slope changes sign (a turning point). Then use the bisection method discussed in Sample Problem 4.9 ten times to close in on the x value where the slope is zero. Note that for an initial search increment

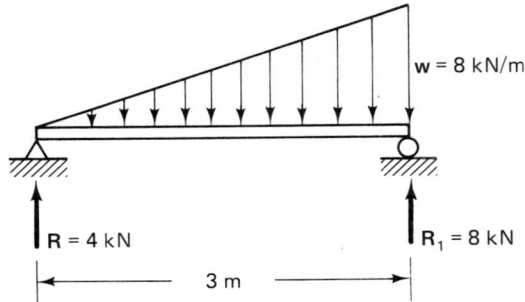

Figure 5.15

of $\Delta x = 0.1$, halving the increment ten times will result in a new Δx,

$$\Delta x = 0.1(0.5)^{10} = 0.0001$$

which is small enough to reveal the answer.

Analysis: It is known that the deflection of a beam with this specific loading can be written as

```
Y=(R*X*(X*X-BL*BL)/6.0-W*X*(X**4-BL**4)/(120.0*BL))/EI
```

where

BL	is the beam's length in meters
R	is the reaction force at the left support in knewtons
X	is the distance along the beam measured from the left in meters
W	is the maximum value of the uniformly varying load in knewtons/meter
EI	is a constant of the beam in knewtons/meter² or kPa

Therefore, the slope or derivative of this function is given by

```
SLOPE=(R*(3.0*X*X-BL*BL)/6.0-W*(5.0*X**4-BL**4)/
       (120.0*BL))/EI
```

Solution: See Figures 5.16a and 5.16b for the algorithm and problem solution. The algorithm starts at X=0.0 with DX=0.1. Note how the IF-THEN-ELSE IF-END IF structure nested inside the DO loop is totally contained inside the DO loop's range. Similarly, the DO loop itself is nested, and totally enclosed, in an outer IF-THEN-ELSE IF-END IF block. Instead of exiting from this outer block whenever a condition of zero slope is found and thus introducing unnecessary branches, the answer is printed while inside the boundary of the block, and the program is then halted. The outer loop is a WHILE structure and provides for the eventual braking of the program in the event of a programming logic error. The maximum value that X can take on is BL. Rather than specify the numerical value of BL in the block IF statement, it is represented by its variable name. This is done so that one can easily change its value through the DATA statement later on if it becomes necessary.

■ **Algorithm for Sample Problem 5.8.** This program solves for the maximum deflection of a simply supported beam. The equation for the beam's slope is employed to find the point of maximum displacement where the slope equals 0. The incremental-search method followed by the bisection method repeated ten times is used to find the X value where the slope is zero.

1. Initialize DX, R, BL, W, EI.

2. Set X=0.

3. Calculate
```
FX=(R*(3.0*X*X-BL*BL)/6.0-W*(5.0*X**4-BL**4)/
    (120.0*BL))/EI
X1=X+DX
```

4. While X1≤BL then

a. Calculate
```
FX1=(R*(3.0*X1*X1-BL*BL)/6.0-W*(5.0*X1**4-BL**4)/
    (120.0*BL))/EI
```

b. If FX*FX1<0.0 then

i. Do, with index I ranging from 1 to 10, the following:

(a) Calculate
```
XM=(X+X1)/2.0
FXM=(R*(3.0*XM*XM-BL*BL)/6.0-W*(5.0*XM**4-BL**4)/
    (120.0*BL))/EI
```

(b) If FX*FXM=0.0 then

(1) Calculate
```
YMAX=(R*(XM**3-XM*BL*BL)/6.0-W*(XM**5-XM*BL**4)/
    (120.0*BL))/EI
```

(2) Print YMAX, XM

(3) Stop

Else if FX*FXM<0.0 then

(1) Set X1=XM

Else

(1) Set X=XM,FX=FXM

ii. Calculate
```
XM=(X+X1)/2.0
YMAX=(R*(XM**3-XM*BL*BL)/6.0-W*(XM**5-XM*BL**4)/
    (120.0*BL))/EI
```

iii. Print YMAX, XM

iv. Stop

Else if FX*FX1=0.0 then

i. Set XM=X1

ii. Calculate
```
YMAX=(R*(XM**3-XM*BL*BL)/6.0-W*(XM**5-XM*BL**4)/
    (120.0*BL))/EI
```

Figure 5.16a

      iii. Print YMAX, XM

      iv. Stop

   Else

       i. Set X=X1, FX=FX1

      ii. Calculate

         X1=X1+DX

5. Print 'No answer found.'

Figure 5.16a   (cont.)

```
         PROGRAM DEFL

   *****************************************************************
   *                                                               *
   *     THIS PROGRAM SOLVES FOR THE MAXIMUM DEFLECTION OF THE      *
   *     SIMPLY SUPPORTED BEAM OF SAMPLE PROBLEM 5.8. THE EQUATION  *
   *     FOR THE BEAM'S SLOPE IS EMPLOYED TO FIND THE POINT OF      *
   *     MAXIMUM DISPLACEMENT WHERE THE SLOPE EQUALS ZERO.          *
   *     THE INCREMENTAL SEARCH METHOD FOLLOWED BY THE BISECTION    *
   *     METHOD REPEATED 10 TIMES IS USED TO FIND THE X VALUE WHERE *
   *     THE SLOPE IS ZERO.                                         *
   *                                                               *
   *****************************************************************

         REAL DX,R,BL,W,EIX,FX,X1,FX1,XM,FXM,YMAX

         DATA DX,R,BL,W,EI/.1,4.E3,3.,8.E3,1.84E6/

         X=0.
         FX=(R*(3.*X*X-BL*BL)/6.-W*(5.*X**4-BL**4)/(120.*BL))/EI
         X1=X+DX

   C
   C     WHILE STRUCTURE BEGINS
   C
       2 IF(X1.LE.BL)THEN
             FX1=R*(3.*X1*X1-BL*BL)/(6.*EI)
             FX1=FX1-W*(5.*X1**4-BL**4)/(120.*BL*EI)

             IF(FX*FX1.LT.0.)THEN

                 DO 1 I=1,10
                     XM=(X+X1)/2.
                     FXM=R*(3.*XM*XM-BL*BL)/(6.*EI)
                     FXM=FXM-W*(5.*XM**4-BL**4)/(120.*BL*EI)

                     IF(FX*FXM.EQ.0.)THEN
                         YMAX=R*XM*(XM*XM-BL*BL)/(6.*EI)
                         YMAX=YMAX-W*XM*(XM**4-BL**4)/(120.*BL*EI)
                         PRINT*,'  THE MAXIMUM VALUE OF Y IS = ',YMAX
                         PRINT*,'  IT OCCURS AT X = ',XM
                         PRINT*,'  ANSWERS ARE IN METERS.'
                         STOP

                     ELSE IF(FX*FXM.LT.0.)THEN
                         X1=XM
```

Figure 5.16b

```
                        ELSE
                            X=XM
                            FX=FXM

                        END IF

1               CONTINUE

                XM=(X+X1)/2.
                YMAX=R*XM*(XM*XM-BL*BL)/(6.*EI)
                YMAX=YMAX-W*XM*(XM**4-BL**4)/(120.*BL*EI)

                PRINT*,'   THE MAXIMUM VALUE OF Y IS = ',YMAX
                PRINT*,'   IT OCCURS AT X = ',XM
                PRINT*,'   ANSWERS ARE IN METERS.'
                STOP

            ELSE IF(FX*FX1.EQ.0.)THEN
                XM=X1
                YMAX=R*XM*(XM*XM-BL*BL)/(6.*EI)
                YMAX=YMAX-W*XM*(XM**4-BL**4)/(120.*BL*EI)
                PRINT*,'   THE MAXIMUM VALUE OF Y IS = ',YMAX
                PRINT*,'   IT OCCURS AT X = ',XM
                PRINT*,'   ANSWERS ARE IN METERS.'
                STOP

            ELSE
                X=X1
                FX=FX1
                X1=X+DX
C
C     END OF WHILE BLOCK
C
                GO TO 2

            END IF

        END IF

        PRINT*,'   NO MAXIMUM FOUND.'

        END


RUN

    THE MAXIMUM VALUE OF Y IS = -.002296943141441
    IT OCCURS AT X = 1.557958984375
    ANSWERS ARE IN METERS.
```

Figure 5.16b   (cont.)

5.4 NESTED DO LOOPS

Nested DO loops are DO loops that are completely contained within the range of an outer DO loop. They may share a common terminal statement, such as CONTINUE, but this is not advisable. Instead, it is recommended that every DO loop have its own terminal CONTINUE statement. This becomes mandatory when there is more than one inner loop sharing the same level of nesting. Since the range of the inner-most DO loop must be completely contained within the next outer loop, its CONTINUE statement must appear before the CONT I NUE statement from the outer loop: the loops closing from the inside out. As discussed in Section 5.3 a DO loop can only be activated by executing its DO statement. Transfer of control from an outer DO loop to the midst of an inner DO loop is not allowed. Once a loop has been activated, control can transfer out of the loop to activate another loop that, like itself, is completely contained within the same outer DO range; or control can transfer entirely out of the nested structure. Some of the possible transfers, both valid and invalid, are shown in Figure 5.17. In Figure 5.18 it may be seen that overlapping of loops is not permitted. Also, note that if two inner loops sharing the same level of nesting are

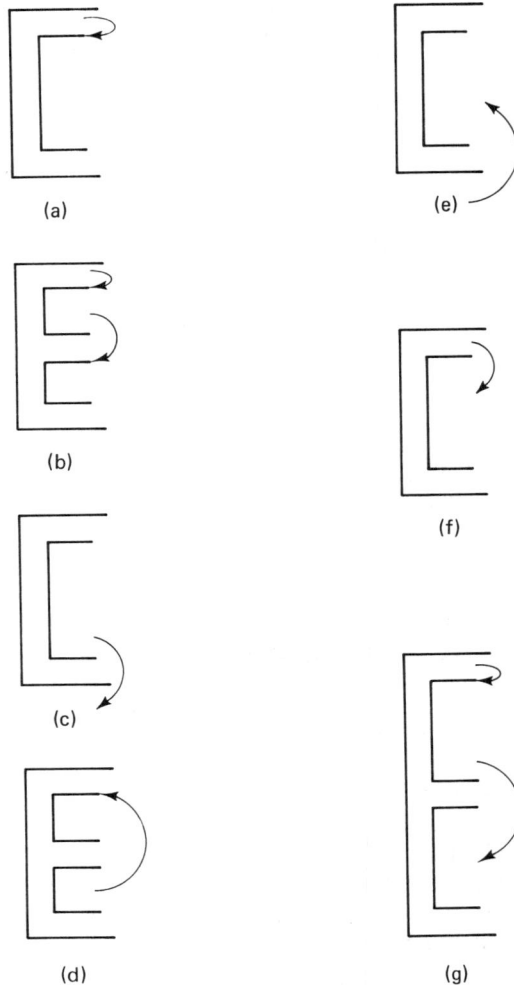

Figure 5.17   (a) through (d) are legal transfers; (d) through (g) are illegal transfers.

(a)

```
┌─ DO 1 I=...    · · ·
│   ┌─ DO 2 J=...    · · ·
│   │   ┌─ DO 3 K=...    · · ·
│   │   └──────── CONTINUE
│   │          3
│   └──────────────────── CONTINUE    · · ·
│                      2
└────────────────────────── 1 CONTINUE    · · ·
```

(b)

```
┌─ DO 1 I=...    · · ·
│   ┌─ DO 2 J=...    · · ·
│   └──────── CONTINUE    · · ·
│          2
│   ┌─ DO 3 K=...    · · ·
│   └──────── CONTINUE
│          3
└──────────── 1 CONTINUE    · · ·
```

(c)

```
┌─ DO 1 I=...
│   ┌─ DO 2 J=...    · · ·
│   └──────── CONTINUE
│          2
│   ┌─ DO 3 J=...    · · ·
│   └──────── CONTINUE
│          3
└──────────── 1 CONTINUE    · · ·
```

(d)

```
┌─ DO 1 I=...    · · ·
│   ┌─ DO 1 J=...    · · ·
│   │   ┌─ DO 1 K=...    · · ·
│   │   │
└───┴───┴──── 1 CONTINUE
```

(e)

```
┌─ DO 1 J=...
│   ┌─ DO 2 J=...    · · ·
│   └──────── CONTINUE    · · ·
│          2
│   ┌─ DO 3 I=...    · · ·
│   └──────── CONTINUE    · · ·
│          3
└──────────── 1 CONTINUE
```

(f)

```
┌─ DO 1 I=...
│   ┌─ DO 2 J=...    · · ·
│   └──────── 1 CONTINUE    · · ·
└────────────── 2 CONTINUE
```

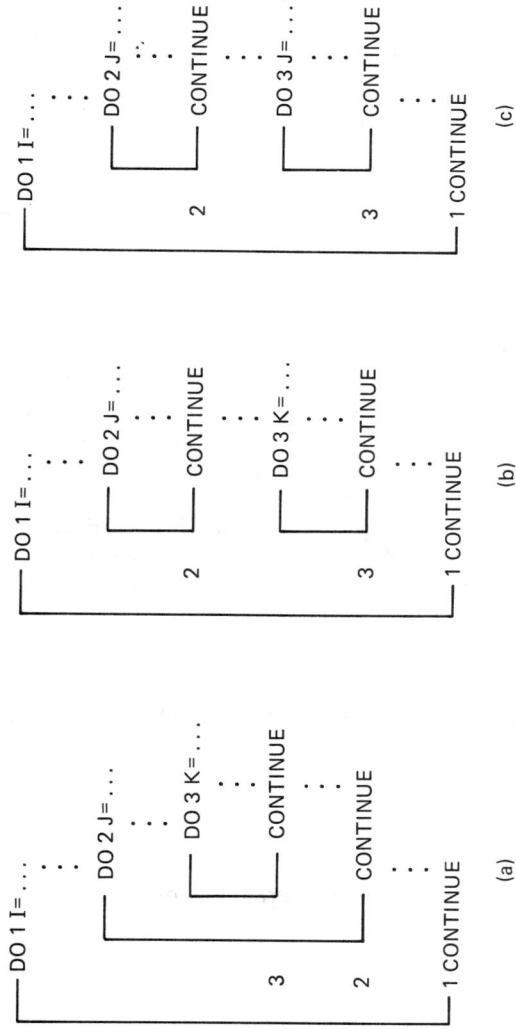

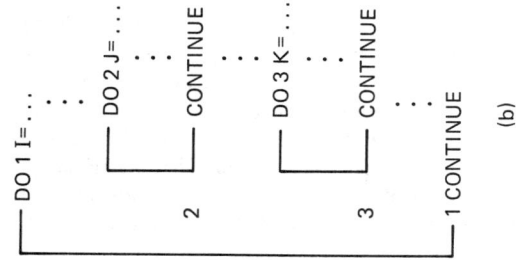

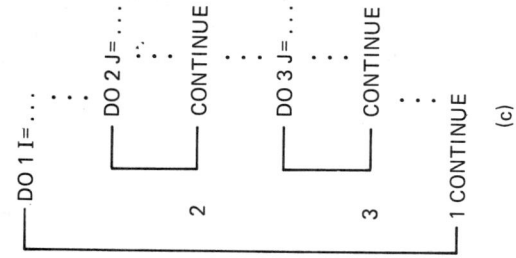

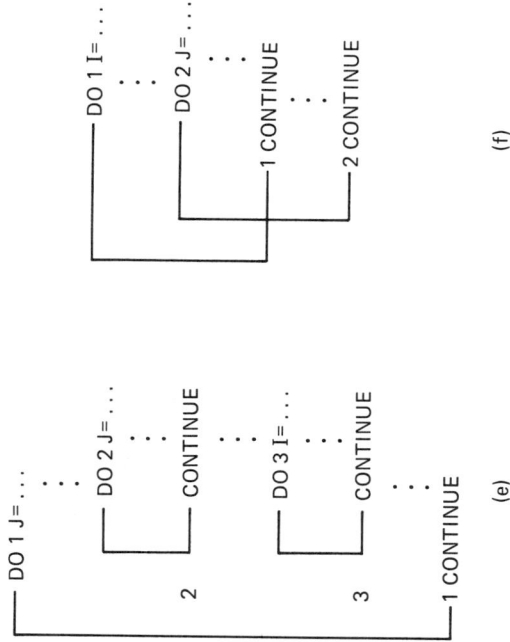

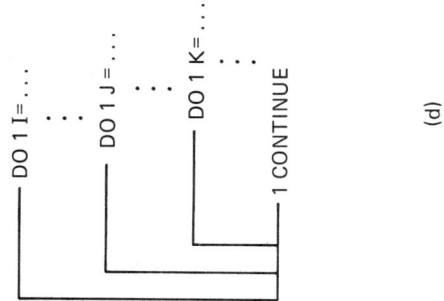

Figure 5.18 (a) through (c) are valid; (d) is valid but not recommended; (e) and (f) are invalid.

163

independent, they can share the same index variable name. In general, an inner loop must be given an entirely different index variable from the other loop indices.

Sample Problem 5.9

Write a FORTRAN program that will estimate the value of the natural number *e*. (See Chapter 4, CP6.) Use the sum of the first ten terms of its infinite series representation. The number *e* is the base of the system of natural logarithms.

Analysis: The number *e* is given by the relationship

$$e = \sum_{i=0}^{\infty} \frac{1}{i!}$$

where *i*! means *i* factorial

$$i \times (i - 1) \times (i - 2) \times \cdots \times 1$$

and

$$0! = 1.$$

Solution: See Figures 5.19a and 5.19b for the algorithm and problem solution. Note how the inner DO loop uses the index variable I from the outer loop as an initial parameter. Do not interchange I with the current loop index variable J. For example, do not write NFACT=NFACT*I, which would be a difficult error to find. The END IF statement placed in juxtaposition to the CONTINUE statement eliminates the need for a GO TO statement to transfer control from the first block of the IF-THEN-ELSE-END IF structure to the CONTINUE statement of the outer loop. It may not be possible at all times to completely avoid using a GO TO statement when nesting is present; it should, however, be a standard to aim for.

Algorithm for Sample Problem 5.9. This program estimates the value of the number E by summing the first ten terms of its infinite-series representation.

1. Set E=0
2. Do, with index I ranging from 0 to 9, the following:
   a. If I≥2 then
      i. Set NFACT=1
      ii. Do, with index J ranging from I to 1 in decrements of 1, the following:
         a. Calculate
            NFACT=NFACT*J
      iii. Calculate
         E=E+1.0/NFACT
      Else
      i. Calculate
         E=E+1
3. Print E.

Figure 5.19a

```
       PROGRAM EBASE

*******************************************************************
*                                                                 *
*  THIS PROGRAM ESTIMATES THE VALUE OF THE NUMBER E BY SUMMING    *
*  THE FIRST 10 TERMS OF ITS INFINITE SERIES REPRESENTATION.      *
*                                                                 *
*******************************************************************
       REAL E
       INTEGER NFACT,J

       E=0.0

       DO 1 I=0,9

           IF(I.GE.2)THEN
               NFACT=1

               DO 2 J=I,1,-1
                   NFACT=NFACT*J
2              CONTINUE

               E=E+1.0/NFACT

           ELSE
               E=E+1.0

           END IF
1      CONTINUE

       PRINT*,' THE NUMBER E EQUALS ',E

       END


   RUN

   THE NUMBER E EQUALS 2.718281525573
```

Figure 5.19b

Sample Problem 5.10

Another well-known means of finding the area that lies beneath a curve between two fixed endpoints that is more accurate than the trapezoidal method employs Simpson's rule. Write a FORTRAN program that uses Simpson's rule to evaluate the area under the given curve of Sample Problem 5.6.

Analysis: The expression for the integral, or area, using Simpson's rule can be written as

$$\text{Area} = \left(y_1 + 4 \sum_{i=2}^{n} y_{\text{even}} + 2 \sum_{i=3}^{n-1} y_{\text{odd}} + y_{n+1} \right) \frac{\Delta x}{3}$$

In this expression, n is equal to the total number of area elements and must be even. The value of n is equal to the total number of data points, *NDATA*, minus 1. The given data points must also be specified at equally spaced distances, Δx, along the x-axis.

$$\Delta x = \frac{x_B - x_A}{NDATA - 1}$$

Solution: The algorithm and problem solution are shown in Figures 5.20a and 5.20b. What values of NDATA would you use to verify the program? What is the minimum value that NDATA can be?

Algorithm for Sample Problem 5.10. This program uses Simpson's rule to evaluate the area lying under a given curve between the end data points XA and XB. The data is taken from NDATA points on the curve equally spaced from each other. NDATA must be odd. Information is entered from the file YVAL1.

1. Read NDATA,XA,XB,YA.
2. Calculate
   ```
   N=NDATA-1
   DX=(XB-XA)/N
   ```
3. Set SUM=YA.
4. Do, with index I ranging from 2 to N in increments of 2, the following:
   a. Read YEVEN
   b. Calculate
      ```
      SUM=SUM+4.0*YEVEN
      ```
   c. If I≤N−2 then
      i. Read YODD
      ii. Calculate
         ```
         SUM=SUM+2.0*YODD
         ```
5. Read YB.
6. Calculate
   ```
   SUM=SUM+YB
   AREA=SUM*DX/3.0
   ```
7. Print AREA.

Figure 5.20a

```
      PROGRAM SIMPS

**********************************************************************
*                                                                    *
*   THIS PROGRAM USES SIMPSON'S RULE TO EVALUATE THE AREA LYING      *
*   UNDER A GIVEN CURVE BETWEEN THE END DATA POINTS XA AND XB.       *
*   THE DATA IS TAKEN FROM NDATA POINTS ON THE CURVE EQUALLY         *
*   SPACED FROM EACH OTHER. NDATA MUST BE ODD. INFORMATION IS        *
*   ENTERED FROM THE FILE 'YVAL1'.                                   *
*                                                                    *
**********************************************************************

      REAL XA,XB,YA,DX,SUM,YEVEN,YODD,YB,AREA
      INTEGER NDATA,N

      OPEN(UNIT=8,FILE='YVAL1')
      REWIND 8
```

Figure 5.20b

```
      READ(8,*)NDATA,XA,XB
      READ(8,*)YA

      N=NDATA-1
      DX=(XB-XA)/N
      SUM=YA

      DO 1 I=2,N,2
          READ(8,*)YEVEN
          SUM=SUM+4.*YEVEN

          IF(I.LE.N-2)THEN
              READ(8,*)YODD
              SUM=SUM+2.*YODD
          END IF

    1 CONTINUE

      READ(8,*)YB

      SUM=SUM+YB
      AREA=SUM*DX/3.

      PRINT 2,AREA
    2 FORMAT(//'THE AREA UNDER THE GIVEN CURVE IS = ',E10.3)

      END


  File Yval1

  11,0.,1.
  0.0000
  0.1105
  0.2443
  0.4050
  0.5967
  0.8244
  1.0933
  1.4096
  1.7804
  2.2136
  2.7183


  RUN


  THE AREA UNDER THE GIVEN CURVE IS =    .100E+01
```

Figure 5.20b   (cont.)

Sample Problem 5.11

Shown in Figures 5.21 is a stepped shaft having two different diameters that is driven by a motor delivering constant output power. For the three paired cases of loading given in the table, find the shaft's output speed as well as the relative angular displacements of twenty equally spaced points along the shaft relative to end *A* located near the motor's armature. Neglect vibrational effects. The motor's output power is equal to 100 kw.

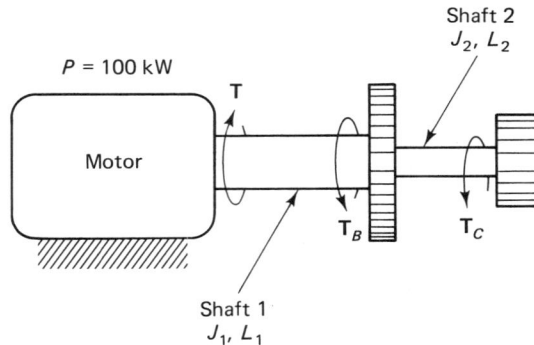

Figure 5.21

Case	T_B	T_C	R_1	R_2	L_1	L_2
1	1500	1000	.05	.04	.20	.30
2	1000	1500	.04	.05	.30	.20
3	1200	1200	.04	.04	.25	.25

Analysis: The shaft's output power, P, is given by the mechanical relationship

$$P = \frac{2\pi n T}{60.0}$$

where

$$\pi = 3.141593$$

$$T = \text{torque delivered} = T_B + T_C \quad \text{newton-meters}$$

$$n = \text{shaft revolutions per minute (RPM)}$$

Shaft angular deflection, ϕ, relative to end A is found using the two formulas

$$\phi_1 = \frac{Tx}{J_1 G} \quad \text{radians,} \quad \text{for } 0 \le x \le L_1$$

and

$$\phi_2 = \frac{TL_1}{J_1 G} + \frac{T_C(x - L_1)}{J_2 G} \quad \text{radians,} \quad \text{for } L_1 \le x \le L_1 + L_2$$

The shaft is made of steel and the constant G, the material's modulus of rigidity, equals 80.0×10^9 Pa. The quantities J_1 and J_2, the polar moments of inertia for area, are calculated using the general expression

$$J = \frac{\pi R^4}{2.0} \quad \text{in m}^4$$

where R, the shaft radius, is either R_1 or R_2.

Solution: The algorithm and problem solution are shown in Figures 5.22a and 5.22b. The reader's attention is drawn to the use of the REAL type statements for J1, J2, L1, L2, and N. Note also that gear *B* is assumed to lie on the portion of the shaft having polar moment of inertia J1.

REVIEW

In this important chapter, the FORTRAN DO loop structure is introduced. The DO loop is controlled by a counter that has an index variable name, an initial value, a final value, and an increment or decrement. The DO statement that activates the loop contains all the pertinent information concerning the counter, and a statement num-

Algorithm for Sample Problem 5.11. This program calculates the angular speed and deflection of a stepped shaft, given the shaft's length, radii, and loading. Displacements are found at twenty equally spaced points along its length. Motor power is 100 kw and G equals 80.0E9 Pa. The output torque is TB at length L1, and is TC at the very end. The number of cases studied is M.

1. Initialize PI,G,P.
2. Read M.
3. Do, with index I ranging from 1 to M, the following:
    a. Read TB,TC,R1,R2,L1,L2
    b. Calculate
        ```
        T=TB+TC
        J1=PI*R1**4/2.0
        J2=PI*R2**4/2.0
        N=60.0*P/(2.0*PI*T)
        DX=(LI+L2)/20.0
        ```
    c. Print TB,TC,R1,R2,L1,L2,J1,J2,N,P.
    d. Set X=PHI=0.
    e. Print X,PHI.
    f. Do, with index J ranging from 1 to 20, the following:
        i. Calculate
            ```
            X=X+DX
            ```
        ii. If X≤L1 then
            (a) Calculate
                ```
                PHI=T*X/(J1*G)
                ```
            Else
            (a) Calculate
                ```
                PHI=T*L1/(J1*G)+TC*(X-L1)/(J2*G)
                ```
        iii. Print X,PHI.

Figure 5.22a

```
        PROGRAM SHAFT

****************************************************************
*                                                              *
*   THIS PROGRAM CALCULATES THE ANGULAR SPEED AND DEFLECTION OF *
*   A STEPPED SHAFT GIVEN THE SHAFT'S LENGTH, RADII, AND        *
*   LOADING. DISPLACEMENTS ARE FOUND AT 20 EQUALLY SPACED POINTS *
*   ALONG ITS LENGTH. MOTOR POWER IS 100 KW AND G EQUALS 80.E9  *
*   PA. THE OUTPUT TORQUE IS TB AT LENGTH L1, AND IS TC AT THE  *
*   VERY END. THE NUMBER OF CASES STUDIED IS M. SEE THE FIGURE  *
*   GIVEN.                                                      *
*                                                              *
****************************************************************
        REAL J1,J2,L1,L2,N,PI,G,P,TB,TC,R1,R2,T,DX,X,PHI
        INTEGER M,CASE,LENGTH

        OPEN(UNIT=8,FILE='SHAFT')
        REWIND 8

        DATA PI,G,P/3.141593,80.E9,1.E5/
        READ(8,*)M

        DO 1 CASE=1,M

            READ(8,*)TB,TC,R1,R2,L1,L2

            T=TB+TC
            J1=PI*R1**4/2.
            J2=PI*R2**4/2.
            N=60.*P/(2.*PI*T)
            DX=(L1+L2)/20.

            PRINT 2,CASE
            PRINT 3
            PRINT 4,TB,TC
            PRINT 5,R1,R2,L1,L2
            PRINT 6,J1,J2
            PRINT 7,N,P
            PRINT 3

            X=0.
            PHI=0.

            PRINT 9,X,PHI

            DO 8 LENGTH=1,20
               X=X+DX

               IF(X.LE.L1)THEN
                   PHI=T*X/(J1*G)

               ELSE
                   PHI=T*L1/(J1*G)+TC*(X-L1)/(J2*G)
               END IF

               PRINT 9,X,PHI

8       CONTINUE
1 CONTINUE

2 FORMAT(//T30,'CASE ',I1)
3 FORMAT(/65('-'))
4 FORMAT(/T5,'TB = ',E10.3,TR13,'TC = ',E10.3,' NEWTN-METRS')
5 FORMAT(T5,'R1 = ',F5.2,TR3,'R2 = ',F5.2,TR3,'L1 = ',F5.2,
 + TR3,'L2 = ',F5.2,'  METERS')
```

Figure 5.22b

```
6 FORMAT(T5,'J1 = ',E10.3,TR13,'J2 = ',E10.3,'   METERS**4')
7 FORMAT(T5,'N = ',E10.3,' RPM',TR10,'P = ',E10.3,' KWATTS')
9 FORMAT(T13,'X = ',E10.3,TR8,'PHI = ',E10.3,' RAD')

    END


File Shaft

3
1500,1000,.05,.04,.2,.3
1000,1500,.04,.05,.3,.2
1200,1200,.04,.04,.25,.25

RUN
```

CASE 1

```
---------------------------------------------------------------------

    TB =    .150E+04              TC =    .100E+04 NEWTN-METRS
    R1 =    .05   R2 =    .04  L1 =    .20   L2 =    .30   METERS
    J1 =    .982E-05             J2 =    .402E-05   METERS**4
    N =     .382E+03 RPM         P =     .100E+06 KWATTS

---------------------------------------------------------------------

            X =     .000E+00        PHI =    .000E+00 RAD
            X =     .250E-01        PHI =    .796E-04 RAD
            X =     .500E-01        PHI =    .159E-03 RAD
            X =     .750E-01        PHI =    .239E-03 RAD
            X =     .100E+00        PHI =    .318E-03 RAD
            X =     .125E+00        PHI =    .398E-03 RAD
            X =     .150E+00        PHI =    .477E-03 RAD
            X =     .175E+00        PHI =    .557E-03 RAD
            X =     .200E+00        PHI =    .637E-03 RAD
            X =     .225E+00        PHI =    .714E-03 RAD
            X =     .250E+00        PHI =    .792E-03 RAD
            X =     .275E+00        PHI =    .870E-03 RAD
            X =     .300E+00        PHI =    .947E-03 RAD
            X =     .325E+00        PHI =    .103E-02 RAD
            X =     .350E+00        PHI =    .110E-02 RAD
            X =     .375E+00        PHI =    .118E-02 RAD
            X =     .400E+00        PHI =    .126E-02 RAD
            X =     .425E+00        PHI =    .134E-02 RAD
            X =     .450E+00        PHI =    .141E-02 RAD
            X =     .475E+00        PHI =    .149E-02 RAD
            X =     .500E+00        PHI =    .157E-02 RAD
```

CASE 2

```
---------------------------------------------------------------------

    TB =    .100E+04              TC =    .150E+04 NEWTN-METRS
    R1 =    .04   R2 =    .05  L1 =    .30   L2 =    .20   METERS
    J1 =    .402E-05             J2 =    .982E-05   METERS**4
    N =     .382E+03 RPM         P =     .100E+06 KWATTS

---------------------------------------------------------------------

            X =     .000E+00        PHI =    .000E+00 RAD
            X =     .250E-01        PHI =    .194E-03 RAD
            X =     .500E-01        PHI =    .389E-03 RAD
            X =     .750E-01        PHI =    .583E-03 RAD
            X =     .100E+00        PHI =    .777E-03 RAD
```

Figure 5.22b   (cont.)

```
X =    .125E+00        PHI =   .971E-03 RAD
X =    .150E+00        PHI =   .117E-02 RAD
X =    .175E+00        PHI =   .136E-02 RAD
X =    .200E+00        PHI =   .155E-02 RAD
X =    .225E+00        PHI =   .175E-02 RAD
X =    .250E+00        PHI =   .194E-02 RAD
X =    .275E+00        PHI =   .214E-02 RAD
X =    .300E+00        PHI =   .233E-02 RAD
X =    .325E+00        PHI =   .238E-02 RAD
X =    .350E+00        PHI =   .243E-02 RAD
X =    .375E+00        PHI =   .247E-02 RAD
X =    .400E+00        PHI =   .252E-02 RAD
X =    .425E+00        PHI =   .257E-02 RAD
X =    .450E+00        PHI =   .262E-02 RAD
X =    .475E+00        PHI =   .267E-02 RAD
X =    .500E+00        PHI =   .271E-02 RAD
```

CASE 3

```
-------------------------------------------------------------------

TB =   .120E+04             TC =   .120E+04 NEWTN-METRS
R1 =   .04    R2 =    .04   L1 =   .25   L2 =    .25   METERS
J1 =   .402E-05             J2 =   .402E-05    METERS**4
N =    .398E+03 RPM         P =    .100E+06 KWATTS

-------------------------------------------------------------------

X =    .000E+00        PHI =   .000E+00 RAD
X =    .250E-01        PHI =   .187E-03 RAD
X =    .500E-01        PHI =   .373E-03 RAD
X =    .750E-01        PHI =   .560E-03 RAD
X =    .100E+00        PHI =   .746E-03 RAD
X =    .125E+00        PHI =   .933E-03 RAD
X =    .150E+00        PHI =   .112E-02 RAD
X =    .175E+00        PHI =   .131E-02 RAD
X =    .200E+00        PHI =   .149E-02 RAD
X =    .225E+00        PHI =   .168E-02 RAD
X =    .250E+00        PHI =   .187E-02 RAD
X =    .275E+00        PHI =   .196E-02 RAD
X =    .300E+00        PHI =   .205E-02 RAD
X =    .325E+00        PHI =   .214E-02 RAD
X =    .350E+00        PHI =   .224E-02 RAD
X =    .375E+00        PHI =   .233E-02 RAD
X =    .400E+00        PHI =   .242E-02 RAD
X =    .425E+00        PHI =   .252E-02 RAD
X =    .450E+00        PHI =   .261E-02 RAD
X =    .475E+00        PHI =   .270E-02 RAD
X =    .500E+00        PHI =   .280E-02 RAD
```

Figure 5.22b   (cont.)

ber, n, to reference the loop's end, that is, the last statement. The last statement of the loop should be a CONTINUE statement.

        DO n index variable=initial value, limit, increment
          Block of executable statements
   n CONTINUE

     The index variable name can be of the real or the integer type. If no increment is given in the DO statement, a value of 1 is assumed. Immediately following the DO statement, and preceding the CONTINUE statement, are the block statements to be performed. Before these statements are executed (the loop is traversed), the current value of the index variable is compared with its limit value. If it is greater than (or less than when decrements are employed) the final value, the loop is exited and control is transferred to the statement immediately following the CONTINUE statement. Each time the loop is passed through, the index variable increases by the increment. The opposite occurs when decrements are used. Variable names and arithmetic expressions are permissible for the DO statement parameters. Once the loop is activated, the control variables may change value; however, the index variable may not have its value changed in the loop by an assignment statement. The total number of passes through a DO loop is given by

$$K = \frac{\text{limit} - \text{initial value} + \text{increment}}{\text{increment}}, \text{ integer part}$$

A DO loop remains dormant until its DO statement is processed. Transferring control into the middle of a DO loop is not permitted. Once a DO loop has been initiated, however, it becomes possible to leave the loop at any time and then reenter it anywhere, providing that the index variable has not changed its value outside the loop. A loop may even contain a STOP statement to end the program. If the DO loop is nested within an outer DO loop, control transfer back to the inner loop is not possible if the outer loop has been deactivated, for example, by a normal exit.

IF-block structure, if present, must be completely contained within the DO loop. A DO loop nested inside an outer IF structure must also be completely enclosed and not extend beyond the END IF statement. When DO loops are nested, it is recommended that each loop have its own individual CONTINUE statement. Nested loops close from the inside out; therefore, the CONTINUE statement associated with the innermost loop should appear first in the program. Separate CONTINUE statements must have different statement numbers. Similarly, for nested loops, each loop must have an individual index-variable name assigned to it unless they are at the same level, that is, one is not contained within the other. In any event, it is usually a good rule to assign separate index-variable names within nested DO loops.

KEY TERMS

DO **statement**   **index variable name**
CONTINUE **statement**   **running product**

REVIEW QUESTIONS

1. In the following problems, determine how many record lines must be present in an accompanying data file (UNIT = 9) to supply the information requested by the program.

   (a)   DO 7 I=1,20,2   (b)   DO 7 I=15,3,-2
                   READ(9,1) R   READ(9,1) R
           7 CONTINUE   7 CONTINUE
           1 FORMAT(F4.2)   1 FORMAT(F4.2)

(c) DO 7 X=0.1,3.2,0.3
         READ(9,1) R
     7 CONTINUE
     1 FORMAT(F4.2)

(d) DO 7 I=4,16,2
         DO 9 J=1,5
             READ(9,1) R
     9   CONTINUE
         R=R+10.0
     7 CONTINUE

2. What are the values of the index I after the DO loops are exited in Problems 1a, 1b, and 1d?

3. What is the value of the index X after the DO loop is exited in Problem 1c?

4. What is the value of RESULT after the following operations have been performed?

(a) RESULT=0.0
         DO 7 I=9,1
             RESULT=RESULT+I
     7 CONTINUE
         PRINT *,RESULT

(b) RESULT=0.0
         DO 7 I=9,1,-3
             RESULT=RESULT+I
     7 CONTINUE
         PRINT *,RESULT

(c) RESULT=0.0
         DO 7 I=5,5
             RESULT=RESULT+I
     7 CONTINUE
         PRINT *,RESULT

(d) RESULT=1.0
         DO 7 X=3.1,0.0,-0.2
             IF(X.GT.1.) RESULT=RESULT*X
     7 CONTINUE
         PRINT *,RESULT

5. Describe any errors that may be present in the following program segments. Assume that all data files have already been opened.

(a) SUM=0.0
         DO 5 I=1,10
             READ(9,2) R
     2 FORMAT(F4.2)
             SUM=SUM+R
             IF(I.EQ.10) PRINT *,SUM
     5 STOP

(b) SUM=0.0
         DO 11 X=100.0,2.0,-1.0
             SUM=X*(X-1.0)+SUM
             IF (X.EQ.2.) THEN
                 PRINT *,SUM
             END IF
    11 CONTINUE

(c) NFACT=1
         DO 7 K1=1,9
             K1=K1+1
             NFACT=NFACT*K1
     7 CONTINUE
         DO 9 K1=1,5
             READ(7,2) R
     2   FORMAT(F4.2)
             PRINT *,R
     9 CONTINUE
         PRINT *,NFACT

(d) SUM=0.0
         X=0.0
         DO 11 I=1,10,3
             READ *,R
             WRITE(7,2) R
     2   FORMAT(F4.2)
             DO 8 K=1,5
                 SUM=SUM+R
     8   CONTINUE
             DO 9 I=1,5
                 X=X-R
     9   CONTINUE
             WRITE(7,2) SUM,X
    11 CONTINUE

[handwritten annotations] Can't use I because it's your iterator

have 2 values But format only has one

6. Convert the following WHILE loops to DO loop structures.

(a)
```
    SUM=0.0
    I=1
  7 IF (I.LT.12) THEN
        READ *,X
        SUM=SUM+X
        I=I+1
        GO TO 7
    END IF
```

(b)
```
    PROD=1.0
    I=12
 15 IF (I.GT.1) THEN
        PROD=PROD*I
        I=I-1
        GO TO 15
    END IF
```

[handwritten:]
```
Do 15  I = 2,12
   If(I.GT.1) Then
      Prod = Prod*I
15 Continue
```

7. Fill in and complete the FORTRAN block shown, which is designed to read the contents of a file named DATA and determine the average value of the contained records. It is known that the total number of records in the file is not greater than 100.

```
    OPEN(UNIT=9,FILE='DATA')
    K=0
    SUM=0.0

    DO _ _ I=_ _ _ _ _ _ _ _ _ _

        READ(_ _, _ _,END=_ _) R

        SUM=_ _ _ _ _ _ _ _ _ _

        K=_ _ _ _ _ _ _ _ _
 14 CONTINUE

 11 AV=_ _ _ _ _ _ _ _ _ _
    PRINT *,'THE AVERAGE VALUE IS',AV
  5 FORMAT(F4.2)
```

8. Fill in and complete the FORTRAN block shown, which is designed to read the contents of a file named DATA to determine the largest algebraic value, RMAX, present and then display it at the terminal. It is known that the total number of records in the file is not greater than 100.

```
    OPEN(UNIT=9,FILE='DATA')
    RMAX=0.0

    DO 14 K=_ _ _1,100_ _

        READ(9, 2,END=5) R

        IF (R.GT.RMAX) THEN
            RMAX=R
        END IF
 14 CONTINUE
  2 FORMAT(F4.2)
  5 PRINT *,RMAX
```

9. Fill in and complete the FORTRAN block shown, which is designed to sum the squares of the first twenty positive odd integers until the difference between a squared integer and the previous squared term is greater than 40. The sum should then be printed at the terminal.

```
    ISQ=0
    NSUM=0

    DO 5  I=_1_,40_,2__

      _I1_=I**2

      IF((I1-ISQ).GT.40) GO TO _11___

      ISQ=__I1_____

      NSUM=_NSUM_+_ISQ_
  5 CONTINUE
 11 PRINT *,NSUM
```

10. Fill in and complete the FORTRAN block shown, which is written to evaluate a given function of both X and Y. Values of X range from 0.1 to 5.7 in increments of 0.2, while values of Y start at 0.3 and go to 3.9 in steps of 0.3.

```
    DO _____
      DO _____
        F=X*X+3.0*Y
        PRINT *,F
  7   CONTINUE
  5 CONTINUE
```

11. Write a FORTRAN program segment using DO loop structure that will calculate the following values:
(a) 9!
(b) The sum of the first eight terms of the sequence

$$1/2 - 1/4 + 1/8 - 1/16 + \ldots$$

(c) The sum of the first five terms of the sequence

$$1/(3 \times 1) + 1/(5 \times 3) + 1/(7 \times 5) + 1/(9 \times 7) + \ldots$$

COMPUTER PROJECTS

CP1. Modify Sample Problem 3.7 so that a DO loop replaces the WHILE structure.
CP2. Rework Problem CP16 of Chapter 4 using a DO loop to generate the time increments. In the following problems, use DO loop structure only.

CP3. The value of π, the ratio of a circle's circumference to its diameter, can be estimated by the series

$$\frac{\pi}{4} = 1 - \frac{1}{3} + \frac{1}{5} - \frac{1}{7} + \frac{1}{9} - \cdots$$

Write a FORTRAN program that will sum this series. Stop the program and print the result at the terminal when an individual term becomes less than 1.0×10^{-3}. Approximately how many terms are included in the result?

CP4. Write a program to evaluate the function $z = x \ln(A)$ for values of x from 0 to 10 in increments of 0.1. Take A as 2.4. Print your results in tabular form. Use the intrinsic function, ALOG(A).

CP5. Write a program to evaluate $y = 2.1 \sin(3.1416t) + 1.05 \sin(6.2832t)$ for values of t from 0 to 2 in increments of 0.1. Print your results in tabular form.

CP6. Write a program to evaluate $z = e^x \cos(y) \sin(x)$ for values of x ranging from 0 to 1 in increments of 0.1, and values of y ranging from 0 to 0.4 in increments of 0.2. Print your results in tabular form.

CP7. Use the program written for Sample Problem 5.5 to fit a straight line to the following data points:

x	y
0.1	0.64
0.2	0.79
0.3	0.99
0.4	1.25
0.5	1.47
0.6	1.71
0.7	1.85
0.8	2.10
0.9	2.25
1.0	2.45

CP8. Use the program written for Sample Problem 5.6 to find the area beneath the curve that is described by the following data points:

x	y
0.0	0.0000
0.1	0.1105
0.2	0.2443
0.3	0.4050
0.4	0.5967
0.5	0.8244
0.6	1.0933
0.7	1.4096
0.8	1.7804
0.9	2.2136
1.0	2.7183

CP9. Use the trapezoidal rule to find the area beneath the curve described in CP8 (see Sample Problem 5.7).

CP10. Use Simpson's rule to find the area beneath the curve described in CP8 (see Sample Problem 5.10).

CP11. In later advanced work, you will learn that the majority of periodic functions can be represented by a series of sine and cosine terms. Fourier (1768–1830) was the great French mathematician who introduced this concept, and the resulting series bears his name: Fourier series. For example, the periodic triangular voltage wave shown in the figure, can be represented by the following Fourier series:

$$f(t) = \frac{8A}{\pi^2} \sum_{k=1}^{\infty} (-1)^{k-1} \frac{1}{(2k-1)^2} \sin(2k-1)\omega t$$

Write a FORTRAN program that will sum the first five terms of the series and print out the answers for $t = 0$ to $t = 1.2$ seconds in time increments of 0.2. Take $A = 10.0$ volts, and $\omega = 20.0$ rad/sec. Present your output as shown.

```
    |        |        |        |        |        |
1234567890123456789012345678901234567890123456789012345

--------------------------------------------------------------
         T(SEC)                        V(VOLTS)
--------------------------------------------------------------
         F3.1                          E10.3
           .                              .
           .                              .
           .                              .
--------------------------------------------------------------
```

CP12. The fuel economy (miles per gallon) of a certain automobile may be approximated by the formula

$$\text{F.E.} = \frac{1200v}{v^2 + 1600}$$

where v is given in miles per hour (mph). Write a FORTRAN program to find the

CP5.11

model's fuel economy for values of v ranging from 0 to 80 mph in increments of 5 mph.

CP13. The *binomial coefficient* $\binom{n}{k}$ is defined as

$$\binom{n}{k} = \frac{n!}{k!\,(n-k)!}$$

where

$$n! = n \times (n-1) \times (n-2) \times (n-3) \times \cdots \times 1$$

and

$$0! = 1$$

Write a FORTRAN program that will evaluate and print out specific binomial coefficients given the n and k values as input. Test your program for values of $n = 30$ and $k = 10$.

CP14. The *binomial series* is defined by

$$(1+x)^n = \sum_{k=0}^{\infty} \binom{n}{k} x^k \quad \text{for } -1 < x < 1$$

where $\binom{n}{k}$ is a binomial coefficient (see CP13). Write a FORTRAN program to evaluate the series for the case when $x = 0.5$ and $n = 5$. Note that when $k > n$, the coefficients $\binom{n}{k}$ vanish.

CP15. The infinite series for arcsin x valid for $|x| < 1$ is given by

$$\arcsin x = x + \sum_{n=1}^{\infty} \frac{(2n-1)(2n-3)\ldots(1)}{2^n n!}\,\frac{x^{2n+1}}{2n+1}$$

Write a FORTRAN program that will evaluate and sum the first five terms of the series. Test your results for the case where $x = 0.5$.

CP16. The infinite series for arcsinh x valid for $|x| < 1$ is given by

$$\operatorname{arcsinh} x = x - \sum_{n=1}^{\infty} \frac{(-1)^{n+1}(2n-1)(2n-3)\ldots(1)}{2^n n!}\,\frac{x^{2n+1}}{2n+1}$$

Write a FORTRAN program that will evaluate and sum the first five terms of the series. Test your results for the case where $x = 0.4$.

CP17. The volume of a rectangular package is to be maximized while its surface area is to be kept constant at 0.54 square meters. Plan and program a systematic search to find the right dimensions of the package's sides by varying the lengths of two sides from 0.01 to 0.5 meters in 0.01 increments. Derive, using calculus, the exact answer.

CP18. The acceleration of the piston in an automobile engine operating at constant angular speed is approximated by the following design formula:

$$a_p = -R\omega^2\left(\cos\theta + \frac{R}{L}\cos 2\theta\right)$$

CP5.18

In this expression ω is the engine speed in rad/sec, R is the crank length, R/L is the ratio of crank length to connecting rod length, and θ is the crank angle in radians. For the two cases, where $R/L = 0.25$ and where $R/L = 0.2$, find the acceleration of the piston for one complete revolution of the crank at 15-degree increments starting at 0 degrees. Take the engine speed to be 1500 rev/min and R to be 7 cm. Express your answer in cm/sec.[2] Form two tables for your results having the form indicated.

```
     |         |         |         |         |         |
1234567890123456789012345678901234567890123456789012345
```

```
                         R/L =0.25
-----------------------------------------------------------------
          THETA                      ACCELERATION
        (DEGREES)                     (CM/SEC**2)
-----------------------------------------------------------------
           I3                          E10.3
            .                            .
            .                            .
            .                            .
-----------------------------------------------------------------

                         R/L =0.20
-----------------------------------------------------------------
          THETA                      ACCELERATION
        (DEGREES)                     (CM/SEC**2)
-----------------------------------------------------------------
           I3                          E10.3
            .                            .
            .                            .
            .                            .
-----------------------------------------------------------------
```

CP19. The magnitude of the current i in a series RLC circuit with an alternating current of $V \sin(\omega t)$ is given by the following relationship:

$$i = \frac{V \sin(\omega t)}{[R^2 + (\omega L - 1/\omega C)^2]^{1/2}}$$

For the two cases $R = 50$ ohms and $R = 100$ ohms, determine the largest magnitude of the circuit current for values of ω equal to 0.2, 0.4, 0.6, 0.8, 1.0, 1.2, and 1.4 times the value of $1.0/\sqrt{L*C}$. Take $L = 10.0$ millihenrys, $C = 2.0$ microfarads, and $V = 110.0$ volts. Express your answer in milliamperes. Form two tables for your results having the form indicated.

CP5.19

```
    I         I              I           I            I             Î
123456789012345678901234567890123456789012345678901234567890123456789012345
```

                              R = 50

            OMEGA   CURRENT
          (RAD/SEC)   (MILLIAMPS)

           E10.3   E10.3
             .   .
             .   .
             .   .

                              R = 100

            OMEGA   CURRENT
          (RAD/SEC)   (MILLIAMPS)

           E10.3   E10.3
             .   .
             .   .
             .   .

CP20. Electric intensity at a point is given by the vector summation

$$\mathbf{E} = \frac{1}{4\pi k} \sum_{l=1}^{n} \frac{Q_l}{r_l^2} \mathbf{i}_l$$

CP5.20

In newtons/coulomb. Here Q_l stands for the magnitude of point charge l of the field, r_l is the distance from the point charge to the field position where the intensity is to be calculated, and $k = 9.0 \times 10^9$ newton-meters squared per coulomb squared, a constant. The unit vector $\mathbf{i}_l$ is a vector of magnitude 1 that points away from the charge l if it is positive and toward the charge if it is negative. The total intensity of the field at point A in the x direction is then given by

$$E_x = \frac{\frac{1}{4\pi k}\sum_{l=1}^{n} Q_l \cos \beta_l}{[(x_A - x_l)^2 + (y_A - y_l)^2]^{1/2}}$$

and in the y direction by

$$E_y = \frac{\frac{1}{4\pi k}\sum_{l=1}^{n} Q_l \sin \beta_l}{[(x_A - x_l)^2 + (y_A - y_l)^2]^{1/2}}$$

where x_l and y_l are the coordinates of the point charge l, x_A and y_A are the coordinates of the point in the field to be studied, and β_l is given by

$$\beta_l = \tan^{-1}\frac{(y_A - y_l)}{(x_A - x_l)}$$

Write a FORTRAN program that will calculate the field intensity at a point for the two cases described in the following table. Data should be read in from a data file with information about each individual charge ($x, y,$ and Q) placed in one record line of the data file. The data file will then be four lines long. The innermost DO loop of the nested DO loop structure should execute a total of four times and contain the following read statement:

```
READ(7,35) X,Y,Q
```

Output from the program should be the completed table.

```
   I         I         I         I         I         I
12345678901234567890123456789012345678901234567890123456789012345
                    ELECTRIC  INTENSITY
         ------------------------------------------
             CASE  1 :   XA  =  .06    YA  =  .08
         ------------------------------------------
         X1  =   -.06    Y1  =   .0    Q1  =     .12E-07
         X2  =    .04    Y2  =   .0    Q2  =   -.12E-07
         ------------------------------------------
         EX  =  XX.XXXEXXX        EY   =  XX.XXXEXXX
         ------------------------------------------
             CASE  2 :   XA  =  .08    YA  =  .06
         ------------------------------------------
         X1  =   -.05    Y1  =   .01   Q1  =     .12E-07
         X2  =    .05    Y2  =  -.01   Q2  =   -.12E-07
         ------------------------------------------
         EX  =  XX.XXXEXXX        EY   =  XX.XXXEXXX
         ------------------------------------------
```

CP21. The first law of thermodynamics states that the heat, Q, supplied to a system equals the work, W, done by the system plus the increase in its internal energy, ΔE. If an engine is known to perform the four-process cycle described in the accompanying table, use the first law of thermodynamics to complete the table's entries. Also calculate the total net rate of work output in kilowatts. Two cases should be studied. The first has Q_4 equal to -229.0 kilowatts; the second case has Q_4 equal to -115.0 kilowatts. Duplicate the following tables filling in the missing entries using the E10.3 format.

```
       |          |          |          |          |          |          |
123456789012345678901234567890123456789012345678901234567890123456789012345
```

```
                                 CASE  1
---------------------------------------------------------------------------
                    Q                DELTA E              W
PROCESS          (KWATTS)           (KWATTS)           (KWATTS)
---------------------------------------------------------------------------
   1             .000E+00           .362E+02          XX.XXXEXXX
   2             .352E+03           .352E+03          XX.XXXEXXX
   3            -.352E+02          -.615E+03          XX.XXXEXXX
   4            -.229E+03           .211E+03          XX.XXXEXXX
---------------------------------------------------------------------------
                                    TOTAL  W          XX.XXXEXXX

                                 CASE  2
---------------------------------------------------------------------------
                    Q                DELTA E              W
PROCESS          (KWATTS)           (KWATTS)           (KWATTS)
---------------------------------------------------------------------------
   1             .000E+00           .362E+02          XX.XXXEXXX
   2             .352E+03           .352E+03          XX.XXXEXXX
   3            -.352E+02          -.615E+03          XX.XXXEXXX
   4            -.115E+03           .211E+03          XX.XXXEXXX
---------------------------------------------------------------------------
                                    TOTAL  W          XX.XXXEXXX
```

CP22. The vertical hydrostatic force acting on a submerged surface, in the upward or downward direction, is equal to the weight of fluid that *could* be found directly above the surface divided by the liquid's free area. In the present exercise, it is desired to find the

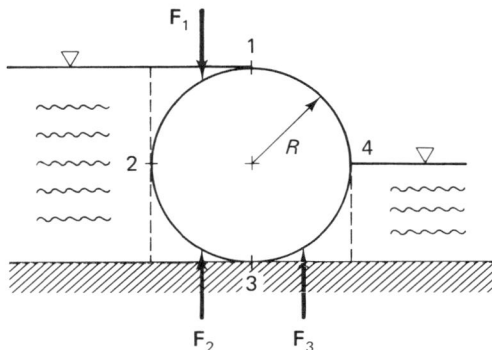

CP5.22

total force per unit length acting on a submerged solid cylinder having various radii. Two situations are studied: the first has the solid barely covered; the second case has the fluid level on the right-hand side lowered to half height. The fluid is water which has a specific weight of 9807 newtons per meter cubed. The weight of water per unit length above surface 1–2 is given by

$$F_1 = 9807.0\left(1.0 - \frac{\pi}{4.0}\right)R^2 \quad \text{N/m downward}$$

The force acting upward on surface 2–3 is equal to

$$F_2 = 9807.0\left(1.0 + \frac{\pi}{4.0}\right)R^2 \quad \text{N/m upward}$$

Then for the first case, the total force acting equals $2.0(F_2 - F_1)$. For the second case, the total force acting equals $F_2 + F_3 - F_1$, where the force on surface 3–4, F_3, is

$$F_3 = \frac{9807.0\pi R^2}{4.0} \quad \text{N/m upward}$$

Write a complete FORTRAN program that will fill in the following table and then calculate the total force. The program should be automatically performed twice (use outer DO loop control) to cover both cases. Output should appear as shown using the E10.3 format.

```
          |         |         |         |         |         |
12345678901234567890123456789012345678901234567890123456789012345


-----------------------------------------------------------------
CASE  1    R=1.0       R=1.5       R=2.0       R=2.5       R=3.0
-----------------------------------------------------------------
    F1     XX.XXXEXXX  XX.XXXEXXX  XX.XXXEXXX  XX.XXXEXXX  XX.XXXEXXX
    F2     XX.XXXEXXX  XX.XXXEXXX  XX.XXXEXXX  XX.XXXEXXX  XX.XXXEXXX
    F3     XX.XXXEXXX  XX.XXXEXXX  XX.XXXEXXX  XX.XXXEXXX  XX.XXXEXXX
-----------------------------------------------------------------
FTOTAL XX.XXXEXXX  XX.XXXEXXX  XX.XXXEXXX  XX.XXXEXXX  XX.XXXEXXX


-----------------------------------------------------------------
CASE  2    R=1.0       R=1.5       R=2.0       R=2.5       R=3.0
-----------------------------------------------------------------
    F1     XX.XXXEXXX  XX.XXXEXXX  XX.XXXEXXX  XX.XXXEXXX  XX.XXXEXXX
    F2     XX.XXXEXXX  XX.XXXEXXX  XX.XXXEXXX  XX.XXXEXXX  XX.XXXEXXX
    F3     XX.XXXEXXX  XX.XXXEXXX  XX.XXXEXXX  XX.XXXEXXX  XX.XXXEXXX
-----------------------------------------------------------------
FTOTAL XX.XXXEXXX  XX.XXXEXXX  XX.XXXEXXX  XX.XXXEXXX  XX.XXXEXXX
```

CHAPTER 6

Arrays

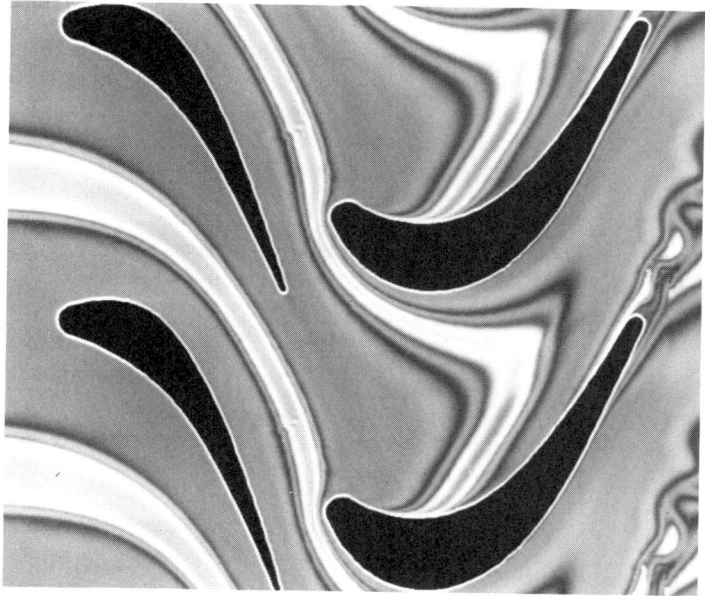

Simulated flow past blades of
a turbojet engine. The
computer analysis was done
on a CRAY-2 supercomputer.
*(Photo courtesy of NASA
Ames Research Center.)*

6.1 INTRODUCTION

A FORTRAN **array** is a group of allotted sequential storage locations in memory
that share a common name. The name assigned to the group is referred to as the
array name. Each **element** of the array, or individual storage location, is given an
integer value **subscript** to differentiate it from the other elements of the group. Array
elements are therefore also known as **subscripted variables.** The subscripts, enclosed within parentheses, are assigned in consecutive order. Because the subscript
can be represented by an integer variable or integer variable expression, arrays are
ideally suited for the DO loop operations introduced in Chapter 5. More important,
they also make it possible to process large quantities of related data using one collective name rather than tediously assigning a different variable name for each
data entry.

6.2 ONE-DIMENSIONAL ARRAYS

An array with a single subscript is called a **one-dimensional array.** It can be thought of as storing a row or column of information. To inform the compiler of an array's size so that it can allot the proper amount of storage space, a DIMENSION **statement** or a type specification statement is given. A DIMENSION statement is a nonexecutable specification statement containing information about the array's name and size. It must be given before the program's first executable statement. The process of reserving storage space for the program's arrays is called **dimensioning.** A DIMENSION statement for one-dimensional arrays appears as follows:

DIMENSION $array_1(ll:ul)$, $array2(ll:ul)$, . . .

where ll and ul are the lower and upper limits, respectively, of the subscript values. These values are constants. If the lower limit is not specified, it is assumed to be 1. The DIMENSION statement would then appear as

DIMENSION $array_1(ul)$, $array_2(ul)$, . . .

Once an array name has been specified in a DIMENSION statement, it cannot appear later without a subscript in an assignment statement, nor can a simple variable be given the name of a dimensioned array. More than one DIMENSION statement may appear in a program.

All values in an array must be of the same type. If not otherwise specified, arrays are typed implicitly using the first letter of the array name. Typing is done for the entire array. When explicit typing is used, the size of the array may be specified at the same time. In that instance, the array should not be listed in a DIMENSION statement as well. For example, the two statements

```
DIMENSION A(15)
INTEGER A
```

can be replaced by the statement

```
INTEGER A(15)
```

However, the statement pair

```
DIMENSION A(15)
INTEGER A(15)
```

is not allowed.

Like DIMENSION statements, type statements are nonexecutable specification statements (see Chapter 2) and must appear before the first executable statement of the program. It is possible to have a lower and an upper limit given on the subscript range in a type statement as well. For example, the two statements

```
DIMENSION A(4:9)
REAL A
```

can be replaced by the single statement

```
REAL A(4:9)
```

Because it is good programming practice to explicitly type every array that appears in a program unit, many programmers prefer to simply use the type specification statement for dimensioning purposes.

Array subscript values are assigned consecutively. Even negative subscript values are acceptable as long as the lower limit on the range of subscript values is less than or equal to the upper range limit. For example, the statement

```
INTEGER A(-12:-1),B(-5:6)
```

specifies two integer arrays, A and B, both having twelve elements.

Array elements can be assigned values individually,

```
I(5)=2
A(1)=2.0
A(2)=4.0*SIN(T)
```

or in groups using DO loop structure. Subscripted variables are ideally suited for DO loop operations. For example, study the following short examples.

```
      PROGRAM TEST
      REAL X(10)
      DO 7 I=1,4
          X(I)=0.0
7 CONTINUE
```

In the foregoing program statements, the value 0 is assigned to the first four elements of array X. The other six elements are left undefined. (Dimensioning an array does not initialize it.) In the next example, the segment

```
      PROGRAM TEXT 1
      REAL X(6)
      DO 7 I=1,3
          X(2*I)=3.0*I
          X(2*I-1)=X(2*I)-1.0
7 CONTINUE
```

assigns

```
      X(1)=2.0
      X(2)=3.0
      X(3)=5.0
      X(4)=6.0
      X(5)=8.0
      X(6)=9.0
```

6.2.1 Input/Output Operations of One-dimensional Arrays

READ and DATA statements may be used to input array data. Strictly speaking, DATA statements do not input information; they assign numerical values to variable names during compilation (see Section 2.7). DATA statements must follow type declaration or DIMENSION specifications so that the variable names included in the DATA statement can be processed properly by the computer during compilation time. For example, the program statements

```
DIMENSION X(4),NINT(3)
DATA X,NINT/3.0,2.0,2*1.0,3*2/
```

would assign

```
X(1)=3.0
X(2)=2.0
X(3)=1.0
X(4)=1.0
NINT(1)=2
NINT(2)=2
NINT(3)=2
```

A DATA statement having an array name given without subscripts will have values assigned to *all* elements of the array. There must be a correct number of data values provided in the statement to do that. For example, the statements

```
REAL X(40)
DATA X/30*1.0/
```

would produce a compiler error message, since ten numerical values for the array are missing. It is possible, however, to write

```
REAL X(4)
DATA X(1),X(2)/1.0,2.0/
```

where only specific elements of the array are initialized.

Sample Problem 6.1

Referring to the solution for Sample Problem 2.1 given in Figure 2.3b, modify the source code so that a ten element one-dimensional array, X, is used for the data. Use a DATA statement.

Solution: The algorithm and problem solution are shown in Figures 6.1a and 6.1b.

To read in data for an array, one can list the specific array elements that are to receive values. For example,

```
READ *,A(1),A(2),A(3)
```

or

```
READ 7,A(1),A(2),A(3)
```

Algorithm for Sample Problem 6.1. This program modifies the solution to Sample Problem 2.1, so that an array, X, may be used for the calculations.

1. Initialize N,X(1),X(2), . . . , X(10).
2. Set SUM=SUMSQ=0.
3. Do, with index I ranging from 1 to N, the following:
   a. Calculate
      SUM=SUM+X (I)
      SUMSQ=SUMSQ+X (I) ∗∗2
4. Calculate
   XAV=SUM/N
   STO=SUMSQ−N∗XAV∗∗2
   S=SQRT (STO/ (N−1))
5. Print XAV,S.

Figure 6.1a

```
        PROGRAM STAT

****************************************************************
*                                                              *
*      THIS PROGRAM MODIFIES THE SOLUTION TO SAMPLE PROBLEM 2.1 *
*      SO THAT AN ARRAY, X, MAY BE USED FOR THE CALCULATIONS.   *
*                                                              *
****************************************************************
        REAL X(10),XAV,STO,SUM,SUMSQ,S
        INTEGER N

        DATA N,X/10,2.999,3.001,3.010,3.005,2.995,2.998,2*3.003,
       + 2.999,3.001/

        SUM=0.
        SUMSQ=0.

        DO 4 I=1,N
            SUM=SUM+X(I)
            SUMSQ=SUMSQ+X(I)**2
      4 CONTINUE

        XAV=SUM/N
        STO=SUMSQ-N*XAV**2
        S=SQRT(STO/(N-1))

        PRINT*
        PRINT*,' THE AVERAGE IS ',XAV
        PRINT*,' THE STANDARD DEVIATION IS ',S

        END

 RUN

    THE AVERAGE IS 3.0014
    THE STANDARD DEVIATION IS .004168666204058
```

Figure 6.1b

or

```
READ(9,7)A(1),A(2),A(3)
```

The last statement reads from a file. One can alternatively specify just the array's name in the READ statement—the **short-list technique.** For example,

```
READ *,A
```

The data is entered into the array in order of increasing subscript value. When read from a file, the data for every element must be available. For example, when using a short list, the statements

```
REAL A(3)
READ *,A
```

are equivalent to writing

```
REAL A(3)
READ *,A(1),A(2),A(3)
```

When using data files, the READ statement's list receives data values from one record line of the file (see Section 3.8). If the list is greater than the number of fields specified for the record line, the next record is begun. For instance, with the data file ZETA,

```
1.0,2.0
3.0,4.0
5.0,6.0
```

the entries

```
REAL A(3)
READ(9,*)A
```

or

```
REAL A(3)
READ(9,*)A(1),A(2),A(3)
```

would assign

```
A(1)=1.0
A(2)=2.0
A(3)=3.0
```

Note that with the same data file, the statements

```
REAL A(3),B(3)
READ(9,*)A,B
```

would assign

```
A(1)=1.0
A(2)=2.0
```

```
A(3)=3.0
B(1)=4.0
B(2)=5.0
B(3)=6.0
```

Data for an array may also be entered using a DO loop. For example,

```
REAL A(3)
DO 5 I=1,3
    READ(9,*),A(I)
5 CONTINUE
```

With this method the data file is accessed three separate times. The single READ statement effectively becomes three READ statements. This means that for the previously given data file, ZETA, the values A receives in the DO loop will be

```
A(1)=1.0
A(2)=3.0
A(3)=5.0
```

The READ statement only refers to one element at a time; therefore, only one field of each record (one record per READ statement) is used. The second field of the record line is ignored. This also means, on the other hand, that a minimum of three separate data lines must be included in the file. If only one record line appears in the file, for example

```
1.0,2.0,3.0
```

an error message "END-OF-FILE encountered" will appear during program execution.

The final way to enter data for an array is by using a READ statement with an **implied-DO list.** This is by far the most common method for effecting input/output operations. Any amount of data (that does not exceed the array's size) can be read in for the array in any order designated by the programmer. Some examples of implied-DO-list READ statements follow.

```
READ *,(A(I),I=1,3)

READ 7,(A(I),I=1,3),(B(I),I=1,3)

READ (9,7),(A(I),A(I+1),I=1,9,2)
```

Note that the entire expression for the list of variables is in parentheses and that a comma follows the right parenthesis for each index variable.

The implied-DO-list READ statement receives its values from one record line of a data file. If the list is greater than the number of fields specified for the record line, the next record is begun. Again, for the data file ZETA previously cited,

```
1.0,2.0
3.0,4.0
5.0,6.0
```

the entries

```
REAL  A(3)
READ(9,*)(A(I),I=1,3)
```

would result in

```
A(1)=1.0
A(2)=2.0
A(3)=3.0
```

DATA statements may also use implied-DO lists. For example,

```
REAL  X(40)
DATA  (X(I),I=1,30)/30*1.0/
```

is permissible. Note that in the foregoing example, ten values were not assigned.

Sample Problem 6.2

Referring to the solution of Sample Problem 6.1 given in Figure 6.1b, rewrite the source code so that X is entered from a file using an implied-DO list.

Solution: The algorithm and problem solution are shown in Figures 6.2a and 6.2b.

Output operations with arrays are accomplished in a similar manner. One can list separately, for example, the specific elements that are to be printed out:

```
PRINT  *,A(1),A(2),A(3)
```

or

```
PRINT  7,A(1),A(2),A(3)
```

Algorithm for Sample Problem 6.2. This program modifies the solution to Sample Problem 6.1, so that the array, X, is entered from a file.

1. Read N,X(1),X(2),...,X(10).
2. Set SUM=SUMSQ=0.
3. Do, with index I ranging from 1 to N, the following:
   a. Calculate
      ```
      SUM=SUM+X(I)
      SUMSQ=SUMSQ+X(I)**2
      ```
4. Calculate
   ```
   XAV=SUM/N
   STO=SUMSQ-N*XAV**2
   S=SQRT(STO/(N-1))
   ```
5. Print XAV,S.

Figure 6.2a

```
         PROGRAM STAT

*******************************************************************
*                                                                 *
*      THIS PROGRAM MODIFIES THE SOLUTION TO SAMPLE PROBLEM 6.1    *
*      SO THAT THE ARRAY, X, IS ENTERED FROM A FILE.              *
*                                                                 *
*******************************************************************
         REAL X(10),SUM,SUMSQ,XAV,STO,S
         INTEGER N

         OPEN(UNIT=8,FILE='TEST')
         REWIND 8

         READ(8,*) N,(X(I),I=1,N)

         SUM=0.
         SUMSQ=0.

         DO 4 I=1,N
             SUM=SUM+X(I)
             SUMSQ=SUMSQ+X(I)**2
       4 CONTINUE

         XAV=SUM/N
         STO=SUMSQ-N*XAV**2
         S=SQRT(STO/(N-1))

         PRINT*
         PRINT*,' THE AVERAGE IS ',XAV
         PRINT*,' THE STANDARD DEVIATION IS ',S

         END

File Test

10
2.999
3.001
3.010
3.005
2.995
2.998
3.003
3.003
2.999
3.001


RUN

  THE AVERAGE IS 3.0014
  THE STANDARD DEVIATION IS .004168666204058
```

Figure 6.2b

or

```
     WRITE(9,4),A(1),A(2),A(3)
```

One can alternatively use the short-list technique,

```
     PRINT *,A
```

which will effect ouput of all elements in the array in order of increasing subscript.

The implied-DO list is also available for data output,

```
    PRINT *,(A(I),I=1,3)
    PRINT 5,(I,X(I),Y(I),Z(I+3),I=8,2,-1)
  5 FORMAT(5X,I2,2(5X,F5.2),E(14.7))
```

as well as the DO loop structure:

```
    DO 5 I=1,3
        PRINT *,A(I)
  5 CONTINUE
```

The DO loop structure, however, is restricted to separate lines of output. Each line of output matches the individual PRINT operations specified within the loop during one pass. Notice that the statements

```
    REAL A(3)
    DO 7 I=1,3
        PRINT 5,A(I)
  7 CONTINUE
  5 FORMAT (3F6.1)
```

will produce output on three separate lines instead of one.

Since the control of output offered by implied-DO lists is far superior to other means of listing output variables, the use of implied-DO lists is preferred by most programmers.

Sample Problem 6.3

In a proposed machine's design, the four-bar linkage shown in Figure 6.3 is used to convert rotary motion of the driver link *AB* to oscillatory motion of the output link *CD*. (If you are wondering, the ground, link *AD*, is considered the fourth link.) The ouput angle, ϕ, was computed as a function of the input angle, θ, which rotated from 0 to 360 degrees. The angle ϕ, corresponding to each 30-degree rotation of the input, was stored in a file called PHILE. Write a program that will read in the data from this file as an array and find the largest output angle present. Both the element's value and its index should be printed out.

```
    24.4
    38.8
    61.6
    86.3
   110.0
   129.0
   131.7
   108.3
    78.9
    55.3
    37.8
    26.1
```

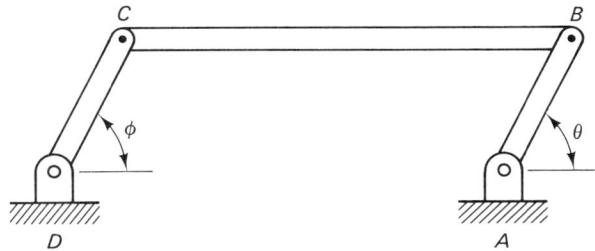

Figure 6.3

Solution: The algorithm and problem solution are shown in Figures 6.4a and 6.4b. How would you modify the program to calculate the corresponding driver position for the largest output angle?

Sample Problem 6.4

Referring to Sample Problem 6.3, write a program that will serially search for a specific angular value in the PHI array. If the value is found, the element's position (index) is to be printed out. If it is not found, a suitable message should be printed at the terminal.

Solution: The algorithm and problem solution are shown in Figures 6.5a and 6.5b. Note that because PHI and VALUE are real, the equal comparison may not be exact and some tolerance (1.E-8) must be provided for. Test your program with the values 86.3 and 131.701.

Sample Problem 6.5

One very important type of engineering structure that is commonly used in the construction of buildings and bridges is the truss. A truss consists of straight structural members connected by joints located at the member's extremities; that is, no truss member is continuous at a joint. Shown in Figure 6.6 is a plane bridge truss supporting

Algorithm for Sample Problem 6.3.   This program finds the largest element of a given one-dimensional array. The array's elements contain the output angle of a four-bar linkage. The element's value and index are printed at the terminal.

1. Read PHI(1),PHI(2), . . . , PHI(12).
2. Set PLARGE=PHI(1),ILARGE=1.
3. Do, with index I ranging from 2 to 12, the following:
    a. If PHI(I)>PLARGE then
        i. Set PLARGE=PHI(I),ILARGE=I
4. Print PLARGE,ILARGE.

Figure 6.4a

```
            PROGRAM ANGLE

******************************************************************
*                                                                *
*       THIS PROGRAM FINDS THE LARGEST ELEMENT OF A GIVEN ONE-    *
*       DIMENSIONAL ARRAY. THE ARRAY'S ELEMENTS CONTAIN THE       *
*       OUTOUT ANGLE OF A FOUR-BAR LINKAGE. THE ELEMENT'S VALUE   *
*       AND INDEX ARE PRINTED AT THE TERMINAL.                    *
*                                                                *
******************************************************************
            REAL PHI(12),PLARGE
            INTEGER ILARGE,I

            OPEN(UNIT=7,FILE='PHILE')
            REWIND 7

            READ(7,*) (PHI(I),I=1,12)

            PLARGE=PHI(1)
            ILARGE=1

            DO 1 I=2,12

               IF(PHI(I).GT.PLARGE)THEN
                  PLARGE=PHI(I)
                  ILARGE=I
               END IF

      1 CONTINUE

            PRINT*
            PRINT*,' THE LARGEST OUTPUT ANGLE IN THE ARRAY HAS THE',
          + ' VALUE ',   PLARGE
            PRINT*,' ITS INDEX VALUE IS ',ILARGE

            END

File Phile

24.4
38.8
61.6
86.3
110.0
129.0
131.7
108.3
78.9
55.3
37.8
26.1

RUN

   THE LARGEST OUTPUT ANGLE IN THE ARRAY HAS THE VALUE 131.7
   ITS INDEX VALUE IS 7
```

Figure 6.4b

three vertical loads of 5000 pounds. Using the method of joints (learned in statics), it is possible to solve for the load in each member of the truss. The resulting values are shown in the accompanying table for half the structure and are assumed to reside in a file, TRUSS. Because the loading and structure are symmetric, there is no need to form a solution for the right-hand side. Write a program that will read in the values from

■ **Algorithm for Sample Problem 6.4.** This program performs a serial search of an array to find a specified value. The given array contains the output angles from Sample Problem 6.3. The value sought is entered at the terminal, and if the value is found in the array, the position or index of the element is printed at the terminal.

1. Read PHI(1),PHI(2), . . . , PHI(12).

2. Enter VALUE.

3. Set K=0.

4. Do, with index I ranging from 1 to 12, the following:
   a. If |PHI(I)−VALUE|<1.0E−8 then
      i. Set K=I
      ii. Print K

5. If K=0 then
   a. Print 'Value was not found.'

Figure 6.5a

```
        PROGRAM SEARCH

*******************************************************************
*                                                                 *
*   THIS PROGRAM PERFORMS A SERIAL SEARCH OF AN ARRAY TO FIND A   *
*   SPECIFIED VALUE. THE GIVEN ARRAY CONTAINS THE OUTPUT ANGLES   *
*   FROM SAMPLE PROBLEM 6.3. THE VALUE SOUGHT IS ENTERED AT THE   *
*   TERMINAL, AND IF THE VALUE IS FOUND IN THE ARRAY, THE         *
*   POSITION OR INDEX OF THE ELEMENT IS PRINTED AT THE TERMINAL.  *
*                                                                 *
*******************************************************************
        REAL PHI(12),VALUE
        INTEGER K,I

        OPEN(UNIT=7,FILE='PHILE')
        REWIND 7

        READ(7,*) (PHI(I),I=1,12)
        PRINT*,' ENTER THE ANGULAR VALUE SOUGHT IN THE ARRAY.'
        READ*,VALUE

        K=0

        DO 1 I=1,12

            IF(ABS(PHI(I)-VALUE).LT.1.E-8)THEN
                K=I
                PRINT*,' THE VALUE SOUGHT IS AT INDEX POSITION ',K
            END IF

  1     CONTINUE

        IF(K.EQ.0)THEN
            PRINT*,' THE VALUE SOUGHT WAS NOT FOUND.'
        END IF
```

Figure 6.5b

```
        END

File Phile

 24.4
 38.8
 61.6
 86.3
110.0
129.0
131.7
108.3
 78.9
 55.3
 37.8
 26.1


RUN

   ENTER THE ANGULAR VALUE SOUGHT IN THE ARRAY.
? 131.7
   THE VALUE SOUGHT IS AT INDEX POSITION 7
```

Figure 6.5b   (cont.)

TRUSS as an array and find the largest stress acting on a truss member. (Structural steel will permanently set, for example, if the internal stress level becomes greater than 36×10^3 pounds per inch square — psi.) The index value of the element corresponding to the truss member's number from the figure should also be printed out. The cross-sectional area of each member is 30 square inches.

Member	Internal load
1	−7,500
2	0
3	12,500
4	−10,000
5	−2,500
6	−13,333
7	4,167
8	10,000
9	0

Analysis: The stress in any member is given by

$$\text{stress} = \frac{\text{load}}{\text{area}}, \text{psi}$$

Positive stresses are called tensile stresses; negative stresses are called compressive stresses.

Solution: The algorithm and problem solution are shown in Figures 6.7a and 6.7b. The maximum number of array elements that the program can process is 100. Notice how the READ statement handles a file smaller than that. After entering the data, the program searches for, and identifies, the largest number value. It then looks for all

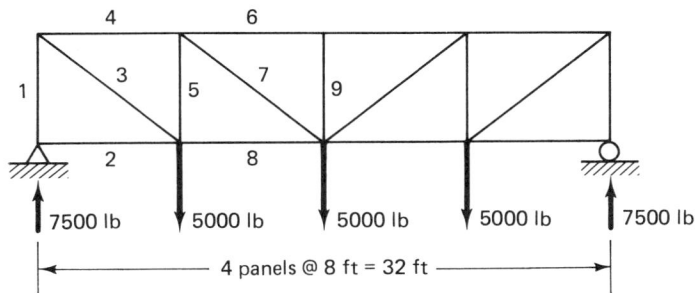

Figure 6.6

the members that carry the same load in either compression or tension. Stresses are printed out.

Sample Problem 6.6

Assuming now that the values from Sample Problem 6.5 are stored in ascending order — that is, the first element contains the smallest numerical value, and so on — write a program that employs a binary search to find a specific load value in the array. If the value

Algorithm for Sample Problem 6.5. This program reads in bridge-member load values from the file TRUSS. The values are stored in real array LOADS. Array LOADS is limited in size to 100 elements. The largest element of the array is then found, PMAX, and its load value is divided by the standard bridge-member cross-sectional area of 30.0 s.i. to yield the stress. The index numbers of all bridge members that have the same absolute stress level are printed out. Negative stresses are identified as compressive; positive stresses are called tensile stresses.

1. Initialize A.
2. Set PMAX=0,N=0.
3. Do, with index I ranging from 1 to 100, the following:
   a. Read LOADS(I) until the file is empty. When that occurs, leave the loop
   b. Set N=N+1
   c. If $|LOADS(I)|>PMAX$ then
      i. Set $PMAX=|LOADS(I)|$
4. Do, with index I ranging from 1 to N, the following:
   a. If $||LOADS(I)|-PMAX|<1.0E-8$ then
      i. S=LOADS(I)/A
      ii. If S<0.0, print I,S. Label as compressive stress
      iii. If S>0.0, print I,S. Label as tensile stress

Figure 6.7a

```
        PROGRAM BRIDGE

****************************************************************
*                                                              *
*   THIS PROGRAM READS IN BRIDGE MEMBER LOAD VALUES FROM FILE  *
*   'TRUSS'. THE VALUES ARE STORED IN REAL ARRAY LOADS. ARRAY  *
*   LOADS IS LIMITED IN SIZE TO 100 ELEMENTS. THE LARGEST      *
*   ELEMENT OF THE ARRAY IS THEN FOUND, PMAX, AND ITS LOAD VALUE *
*   IS DIVIDED BY THE STANDARD BRIDGE MEMBER CROSS-SECTIONAL   *
*   AREA OF 30. S.I. TO YIELD THE STRESS. THE INDEX NUMBERS OF *
*   ALL BRIDGE MEMBERS THAT HAVE THE SAME ABSOLUTE STRESS LEVEL *
*   ARE PRINTED OUT. NEGATIVE STRESSES ARE IDENTIFIED AS       *
*   COMPRESSIVE; POSITIVE STRESSES ARE CALLED TENSILE STRESSES. *
*                                                              *
****************************************************************

        REAL LOADS(100),A,PMAX,S
        INTEGER N,I

        OPEN(UNIT=9,FILE='TRUSS')
        REWIND 9

        DATA A/30./

        PMAX=0.
        N=0

        DO 9 I=1,100

            READ(9,*,END=30)LOADS(I)
            N=N+1

            IF(ABS(LOADS(I)).GT.PMAX) THEN
                PMAX=ABS(LOADS(I))
            END IF

      9 CONTINUE

C
C     THE SECOND PART OF THE PROGRAM FINDS ALL MEMBERS HAVING
C     IDENTICAL STRESS LEVELS.
C

     30 DO 7 I=1,N

            IF(ABS(ABS(LOADS(I))-PMAX).LT.1.E-8) THEN
                S=LOADS(I)/A
                IF(S.LT.0.) PRINT 4,I,ABS(S)
                IF(S.GT.0.) PRINT 5,I,S
            END IF

      7 CONTINUE

      4 FORMAT(' THE MAXIMUM COMPRESSIVE STRESS OCCURS IN MEMBER',
     + I3,' AND EQUALS',E10.3,' PSI')
```

File Truss

```
-13333
-13333
-7500
-2500
0
0
```

Figure 6.7b

```
4167
12500
12500


RUN

     THE MAXIMUM COMPRESSIVE STRESS OCCURS IN MEMBER   1 AND EQUALS   .444E+03 PSI
     THE MAXIMUM COMPRESSIVE STRESS OCCURS IN MEMBER   2 AND EQUALS   .444E+03 PSI
```

Figure 6.7b   (cont.)

---

is found, the members (element indices) that bear that load are to be printed out. If it is not found, a suitable error message should be printed at the terminal.

Analysis: In a binary search, the middle element of an array is inspected first. If it is larger than the value sought, the value, if present, must lie in the first half of the array, since it is given in ascending order. If it is smaller than the value sought, the value could be present in the second half. The half region in which the value may be found is then halved itself. The middle value is once again checked and the process repeats until the value is found, or until the region cannot be reduced further (the value sought is not present). Provision should also be made in program logic to allow for the possibility that a middle element might exactly equal the value sought, or that several values may repeat. Because the binary search is carried out in regions that are consecutively halved, it is a much faster search method than the serial search method described previously.

Solution: The algorithm and problem solution appear in Figures 6.8a and 6.8b. Notice how the range between ILOW and IHIGH is halved each time: either ILOW becomes MID+1, or IHIGH becomes MID−1, depending on which half the value lies in. The program stops if VALUE has not been found. This occurs when ILOW becomes greater than IHIGH. If the value is found, the array LOADS is checked to see whether identical values exist in the neighboring elements. Two FORTRAN WHILE loops are employed for that purpose. The first loop has the index value I increase from MID+1 (remember the first appearance of the value was at index MID) to IHIGH. After the first succeeding element that does not equal VALUE appears in the array, the array is searched once more in the second WHILE loop, with the index J descending from MID−1 toward ILOW. If a preceding element does not equal VALUE, the program is stopped. (*Caution:* when testing this program, do not forget that the member indices differ in this problem from those in the last one, since they have been placed in ascending order in the file TRUSS.)

Sample Problem 6.7

In Sample Problem 6.6, it was assumed that the given array was found in ascending order. If this is not the case, the array will have to be sorted first, for the binary search method to work. Write a utility program that takes a general one-dimensional array and sorts it into ascending order.

Analysis: One straightforward means of sorting an array is the bubble sort method. The bubble sort first passes through the array, comparing each adjacent pair of values. The values of each pair are interchanged, if necessary, so that the pair is reestablished in ascending order. After all consecutive pairs are exhausted, the largest value of the array should be at the bottom. Another pass is then taken through the array, which will

Algorithm for Sample Problem 6.6. This program performs a binary search of an array to find a specified load VALUE. The array is assumed to be already sorted in ascending order. The value sought is entered at the keyboard, and if the value is found, the position or index of the element (member) is printed at the terminal.

1. Read LOADS(1),LOADS(2), . . . , LOADS(9).
2. Enter VALUE.
3. Set ILOW=1, IHIGH=9.
4. While ILOW≤IHIGH then
    a. Calculate
       i. MID=(ILOW+IHIGH)/2
      ii. If LOADS(MID)=VALUE then
         (a) Print MID
         (b) Set I=MID+1
         (c) While $|$LOADS(I)−VALUE$|<1.0E-8$ and I≤HIGH then
           (1) Print I
         (d) Set J=MID−1
         (e) While $|$LOADS(J)−VALUE$|<1.0E-8$ and J≥ILOW then
           (1) Print J
         (f) Stop
        Else if LOADS(MID)<VALUE then
        (a) Calculate
          I LOW=M I D+1
        Else
        (a) Calculate
          I H I GH=I H I GH−1
5. Print 'Value not found.'

Figure 6.8a

```
                PROGRAM BINARY

    ****************************************************************
    *                                                              *
    *   THIS PROGRAM PERFORMS A BINARY SEARCH OF AN ARRAY TO FIND A *
    *   SPECIFIED LOAD VALUE. THE ARRAY IS ASSUMED TO BE ALREADY    *
    *   SORTED IN ASCENDING ORDER. THE VALUE SOUGHT IS ENTERED AT   *
    *   THE KEYBOARD, AND IF THE VALUE IS FOUND, THE POSITION OR    *
    *   INDEX OF THE ELEMENT (MEMBER) IS PRINTED AT THE TERMINAL.   *
    *                                                              *
    ****************************************************************

                REAL LOADS(9),VALUE
                INTEGER ILOW,IHIGH,MID,I,J

                OPEN(UNIT=7,FILE='TRUSS')
                REWIND 7
```

Figure 6.8b

```
      READ(7,*) (LOADS(I), I=1,9)
      PRINT*, ' ENTER THE LOAD VALUE SOUGHT IN THE STRUCTURE.'
      READ*,VALUE

      ILOW=1
      IHIGH=9
C
C  IHIGH HAS THE VALUE OF THE LARGEST INDEX PRESENT
C
    3 IF(ILOW.LE.IHIGH)THEN
          MID=(ILOW+IHIGH)/2
          IF(ABS(LOADS(MID)-VALUE).LT.1.E-8)THEN
              PRINT*,' THE VALUE SOUGHT IS AT INDEX POSITION ',MID
C
C  THE VALUE SOUGHT IS AT INDEX MID. THE PROGRAM WILL NOW
C  SEARCH FROM THAT POSITION IN THE ARRAY TOWARDS IHIGH, AND THEN
C  TOWARDS ILOW, TO FIND ANY OTHER EQUAL-VALUED ELEMENTS.
C
              I=MID+1

   15         IF(ABS(LOADS(I)-VALUE).LT.1.E-8.AND.I.LE.IHIGH)THEN
                  PRINT*,' THE VALUE SOUGHT IS AT INDEX',
     +                ' POSITION ',I
                  I=I+1
                  GO TO 15
              END IF

              J=MID-1

   17         IF(ABS(LOADS(J)-VALUE).LT.1.E-8.AND.J.GE.ILOW)THEN
                  PRINT*,' THE VALUE SOUGHT IS AT INDEX',
     +                ' POSITION ',J
                  J=J-1
                  GO TO 17
              END IF

              STOP

          ELSE IF(LOADS(MID).LT.VALUE)THEN
              ILOW=MID+1

          ELSE
              IHIGH=MID-1
          END IF

          GO TO 3

      END IF

      PRINT*,' THE VALUE SOUGHT WAS NOT FOUND.'

      END


File Truss

-13333
-13333
-7500
-2500
0
0
4167
12500
12500
```

Figure 6.8b   (cont.)

RUN

```
ENTER THE LOAD VALUE SOUGHT IN THE STRUCTURE.
? 4167
THE VALUE SOUGHT IS AT INDEX POSITION 7
```

Figure 6.8   (cont.)

eventually place the next largest element in the next-to-last position. For an *n* element array, *n* − 1 such passes will be needed to sort all the values. After each pass, the larger values "sink" to the bottom, while the smaller values seem to "bubble up" (thus the name bubble sort). Note that for each successive pass through the array, one less pair of values needs to be inspected. This is because the previous passes through the array positioned the largest element at the bottom, followed by the next largest element, and so forth. The bubble sort method is illustrated numerically for a five-element array in Figure 6.9.

Solution: The algorithm and a general bubble-sort program are shown in Figures 6.10a and 6.10b. Note especially the need for a temporary storage position, ATEMP, to effect a correct interchange; otherwise, the original value of A(I) will be lost.

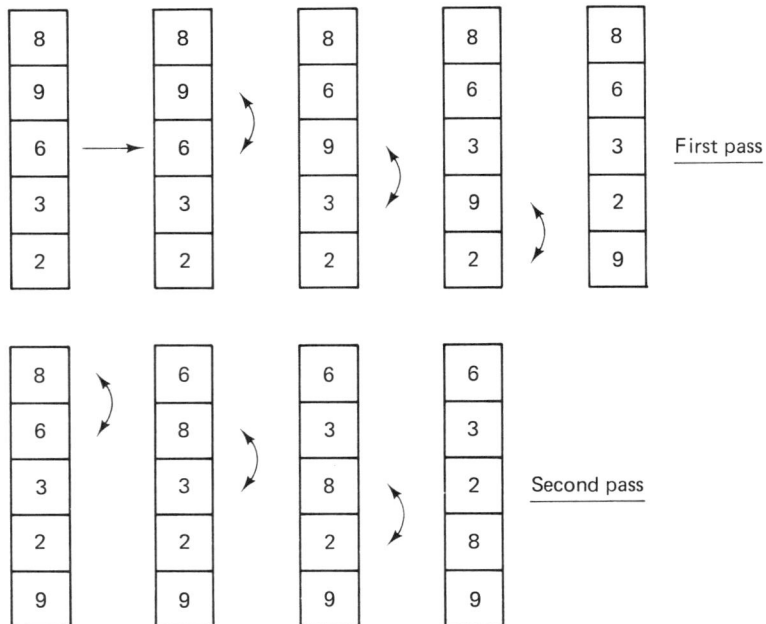

Figure 6.9

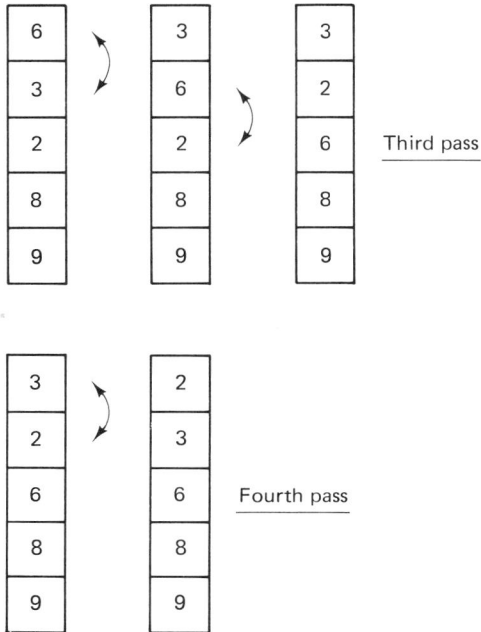

Figure 6.9   (cont.)

Algorithm for Sample Problem 6.7. This program uses a bubble sort to arrange elements of a general one-dimensional array in ascending order.

1. Read NMAX,A(1),A(2), . . . , A(10).
2. Do, with index NFINAL ranging from NMAX−1 to 1 in decrements of 1, the following:
   a. Do, with index I ranging from 1 to NFINAL, the following:
      i. If A(I+1)<A(I) then
         (a) Set ATEMP=A(I)
         (b) Set A(I)=A(I+1)
         (c) Set A(I+1)=ATEMP
3. Print A(1),A(2), . . . , A(10).

Figure 6.10a

```
      PROGRAM BUBBLE

****************************************************************
*                                                              *
*     THIS PROGRAM USES A BUBBLE SORT TO ARRANGE ELEMENTS OF A *
*     GENERAL ONE-DIMENSIONAL ARRAY IN ASCENDING ORDER.        *
*                                                              *
****************************************************************
      REAL A(100),ATEMP
      INTEGER NMAX,NFINAL,I

      OPEN(UNIT=7,FILE='ARRAY')
      REWIND 7

      READ(7,*) NMAX,(A(I),I=1,NMAX)
C
C  NMAX IS THE LARGEST INDEX VALUE PRESENT.
C
      DO 2 NFINAL=NMAX-1,1,-1

         DO 1 I=1,NFINAL

            IF(A(I+1).LT.A(I))THEN
               ATEMP=A(I)
               A(I)=A(I+1)
               A(I+1)=ATEMP

            END IF

1           CONTINUE
2 CONTINUE

      PRINT 3,(A(I),I=1,NMAX)
3 FORMAT(5X,F9.2)

      END

File Array

10
10.9
33.3
29.2
44.1
14.3
12.3
15.8
27.4
22.3
18.

RUN

10.90
12.30
14.30
15.80
18.00
22.30
27.40
29.20
33.30
44.10
```

Figure 6.10b

6.3 TWO-DIMENSIONAL ARRAYS

An array with two subscripts is called a **two-dimensional array.** Instead of containing a single column of information, a two-dimensional array can be thought of as storing a group of columns. The position of an element in a two-dimensional array requires two subscripts for identification. The first integer subscript, or index, denotes the element's row number. The second index, separated from the first by a comma, specifies the element's column position. These indices are enclosed in parentheses. A two-dimensional array with m rows and n columns is called an $m \times n$ array.

A DIMENSION statement for a two-dimensional array has the general form

DIMENSION array name($ll_r : ul_r, ll_c : ul_c$), . . .

where ll_r and ul_r refer to the lower and upper limits for the array's row numbers, and ll_c and ul_c refer to the corresponding limits for the array's column numbers. If no lower limit is specified, a default value of 1 is assigned. Some dimensioning examples follow. Note that a single DIMENSION or type statement can refer both to one- and two-dimensional arrays.

```
DIMENSION X(0:5,-1:1),Z(3,0:4)
DIMENSION Y(10),X(6,3)
```

A two-dimensional array may not be specified in both a DIMENSION and a type statement. For instance, the statements

```
DIMENSION X(6,3)
REAL X(6,3)
```

paired together are not allowed.

Two-dimensional array elements can be assigned values individually

```
I(3,4)=2
A(2,7)=54.0
A(2*J,K+2)=17.0
```

or in groups using the DO loop structure. When using DO loops, one may choose between processing the array in a **row-wise** or in a **column-wise** fashion. To access each element of an array, the DO loops must be nested. For example, to assign zero values to all elements of a 4×5 array A, the following program segment may be used,

```
    PROGRAM EXTRA
    REAL A(4,5)
    DO 3 I=1,4
        DO 4 J=1,5
            A (I,J)=0.0
4       CONTINUE
3 CONTINUE
```

which processes the array by rows. To process the array by columns one may write

```
PROGRAM EXTRA
REAL A(4,5)
DO 3 J=1,5
    DO 4 I=1,4
        A (I,J)=0.0
4       CONTINUE
3 CONTINUE
```

6.3.1 Input/Output Operations of Two-dimensional Arrays

For the input of numerical values to a two-dimensional array, DATA and READ statements may be used to enter or assign values to specific terms of the array, to the entire array, or to a select group of terms of the array. If the entire array is to be entered, the short-list technique may be employed. One may also use a nested DO loop structure or an implied-DO list for greater control of input operations. The short-list technique may not be used when only a partial group of values are to be read from a file.

When using the short-list technique, it is important to keep in mind that values are entered into the array in a *column-wise manner,* that is, they are assigned to the array's columns in successive order beginning with column 1. For a 3×3 array A, for example, the DATA statement

```
DATA A/3*1.0,3*2.0,3*3.0/
```

will assign

```
A(1,1)=1.0
A(2,1)=1.0
A(3,1)=1.0
A(1,2)=2.0
A(2,2)=2.0
A(3,2)=2.0
A(1,3)=3.0
A(2,3)=3.0
A(3,3)=3.0
```

The statements

```
PROGRAM INPUT
REAL A(3,3)
OPEN (UNIT=9,FILE='TESTS')
REWIND 9
READ(9,*) A
```

will also cause data to be entered in a column-wise order. If the data file TESTS appears as follows,

```
1.0,1.0,1.0,2.0,2.0,2.0,3.0,3.0,3.0
```

the values assigned to the array's elements will be the same as those assigned in the preceding DATA statement example.

   To use a DO loop structure to read the data from the file TESTS, one can write

```
      DO 2 J=1,3
         DO 1 I=1,3
            READ(9,*) A(I,J)
1        CONTINUE
2 CONTINUE
```

However, the data file must be changed to appear as shown:

```
      1.0
      1.0
      1.0
      2.0
      2.0
      2.0
      3.0
      3.0
      3.0
```

Nine lines of data are needed because of the nine separate times the READ statement within the DO loop is executed.

   Implied-DO lists could have been used instead in the program INPUT to control data entry. For example, the statements

```
      DO 2 J=1,3
         READ(9,*)(A(I,J),I=1,3)
2 CONTINUE
```

or the full implied-DO list

```
      READ(9,*)((A(I,J),I=1,3),J=1,3)
```

could have been employed (with the original given file TESTS) to effect the same results.

   Note that when using the implied-DO list with more than one index, the inner index — that which appears immediately after the innermost right parenthesis — is filled first (varies the fastest). It acts like an inner nested DO loop. The following READ statement uses a different order,

```
      READ(9,*)((A(I,J),J=1,3)I=1,3)
```

and would read and assign array values by *rows*. When used with the original given file TESTS, the values assigned would be

```
      A(1,1)=1.0
      A(1,2)=1.0
      A(1,3)=1.0
      A(2,1)=2.0
```

```
A(2,2)=2.0
A(2,3)=2.0
A(3,1)=3.0
A(3,2)=3.0
A(3,3)=3.0
```

Using implied-DO lists is generally preferred for input operations because it allows for full or partial entry of the array in any order, and it diminishes the chances of placement errors that might present themselves when using short-list representation. Because implied-DO lists also allow several data entries to be read per record line from a data file, they offer the user a more compact file arrangement. A DATA statement having an implied-DO list representing the same information as before might appear as follows:

```
DATA((A(I,J),J=1,3),I=1,3)/3*1.0,3*2.0,3*3.0/
```

Output operations have the same syntax. When writing the full array one can use a short-list, a nested-DO-loop, or an implied-DO-list structure. When only partial groups are printed out, a short-list procedure will not work. With the use of implied-DO lists, either partial or full arrays may be written in any desired order and several values per output record line are possible. For these reasons, implied-DO lists are also preferred for array output operations. The following are some typical output statements with implied-DO lists:

```
PRINT *,((A(I,J),I=1,3),J=1,5)

PRINT 5,((X(L,M),M=2,8),L=1,5)

WRITE(6,5)((I,J,X(I,J)),I=1,4),J=1,7,2)

WRITE(9,5)(Y(I),I=1,4),((X(I,J),I=1,3),J=7,1,-1)
```

If the short-list form is used, array elements will be printed out in a column-wise order beginning with the first column. Using short-lists for input/output operations can lead to confusion. To see what problems can arise, study the following program segment, data file, and output:

```
    PROGRAM SHORT
    REAL A(3,3)
    OPEN(UNIT=9,FILE='TESTER')
    REWIND 9
    READ(9,*) A
    PRINT 1,A
1   FORMAT(3(5X,F4.1))
    PRINT 2,A(3,1),A(1,3)
2   FORMAT(//2(5X,F4.1))
```

The given data file TESTER is

```
1.0,2.0,3.0
4.0,5.0,6.0
7.0,8.0,9.0
```

Program output then appears as

```
1.0        2.0        3.0
4.0        5.0        6.0
7.0        8.0        9.0

3.0        7.0
```

Although it appears from the initial output that array A was read in a row-wise manner and then printed out in the same way, this was not the case. Element A(3,1) — the third row, first column element — does not equal 7.0 but equals 3.0! By using the short-list data input form, the data from the input file was assigned to the array elements in a column-wise order; thus, the value assigned to A(3,1) was 3.0. When the elements of the array were specified for output, also by using a short-list, the array was processed and sent to the printer in a column-wise order. The *format of the display* was a row with three entries, however, and therefore, element A(3,1) was placed as the third output in the first row. This could easily lead the user to believe that the third displayed value in row 1 is element A(1,3). This is not the case, as the next line of output shows.

If it were desired to assign values from the data file to the elements of the array A in a row-wise order or to dispatch its internally stored element values to the printer also by rows, the program could be changed to appear as

```
      PROGRAM SHORT
      REAL A(3,3)
      OPEN(UNIT=9,FILE='TESTER')
      REWIND 9
      READ(9,*)((A(I,J),J=1,3),I=1,3)
      PRINT 1,((A(I,J),J=1,3),I=1,3)
    1 FORMAT(3(5X,F4.1))
      PRINT 2,A(3,1),A(1,3)
    2 FORMAT(//2(5X,F4.1))
```

Sample Problem 6.8

The operating temperature distribution in and on the surface of a furnace wall is shown in Figure 6.11. Values of temperature are stored in the 5 × 9 array TEMP. Write a FORTRAN program that searches the interior points of the wall for the value of the highest temperature. The array TEMP should be read in row-wise from a data file using an implied-DO list.

Analysis: The data is stored in such a manner that, for example, when X = 1 and Y = 3, the column index is 2, and the row index is 4. TEMP(I, J) = 600 for the (I, J) values (4, 3), (5, 3), (4, 4), (5, 4), (4, 5), (5, 5), (4, 6), (5, 6), (4, 7), and (5, 7), but these points are not considered inside the wall, and are omitted from the search. Similarly, the outer boundaries, where either I equals 1 or J equals 1 or 9, are neglected.

Solution: The algorithm and problem solution are shown in Figures 6.12a and 6.12b.

Sample Problem 6.9

For the given data file from Sample Problem 6.8, write a FORTRAN program that will count the number of array elements having the temperature value 461.

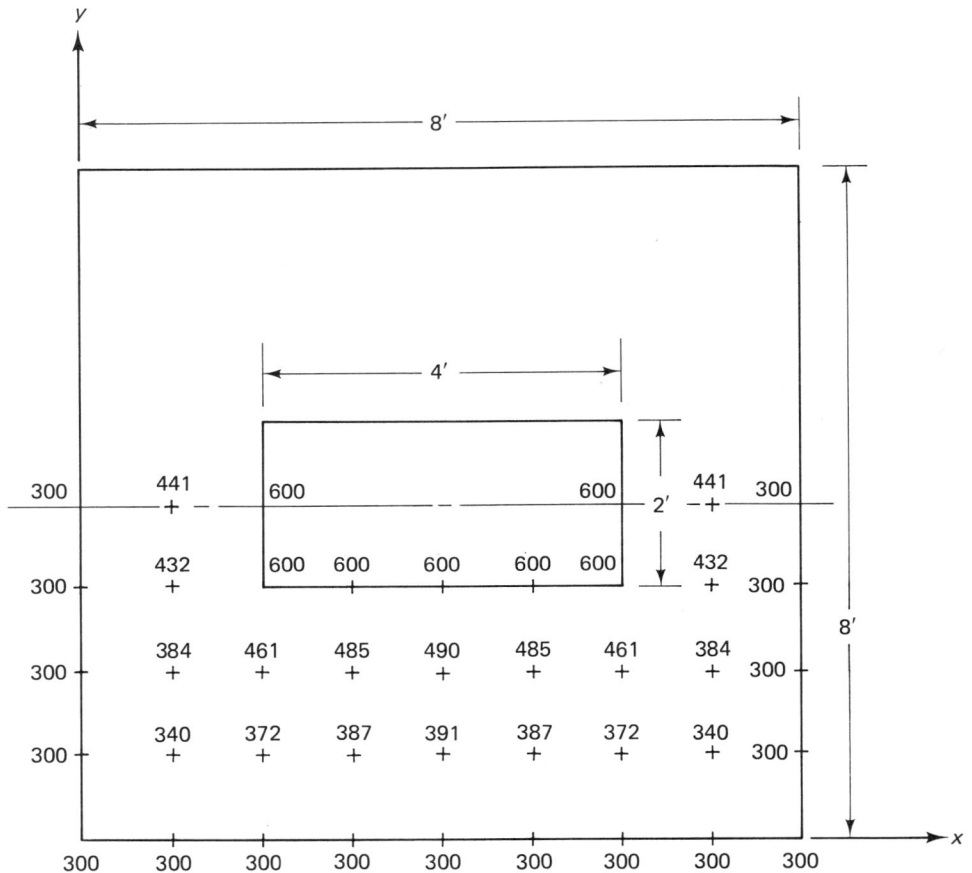

Figure 6.11

Solution: The algorithm and problem solution are shown in Figures 6.13a and 6.13b. If the temperature values at the top part of the furnace are the same as those at the bottom, how would you modify this program to search the complete furnace for a specified value?

Sample Problem 6.10

Many difficult equations that govern the behavior of engineering systems can be approximated by a set of linear simultaneous equations. Several numerical means for solving a set of linear simultaneous equations on the computer rely on having the largest element of each column of an array, specifically the coefficient array, lie on the principal diagonal. The principal diagonal element of a square array has its row number equal to its column number. (See Sample Problem 8.6, and note the discussion of partial pivoting given in CP17 of Chapter 8. Also see problem CP9 in this chapter.) To introduce the subject here, write a FORTRAN program that will search the first column

■ **Algorithm for Sample Problem 6.8.** This program finds the largest element of the 5 × 9 array FURNACE that contains the temperature distribution inside a furnace wall.

1. Read TEMP.
2. Set TMAX=TEMP(2, 2).
3. Do, with index I ranging from 2 to 3, the following:
   a. Do, with index J ranging from 2 to 8, the following:
      i. If TEMP(I, J)>TMAX then
         (a) Set TMAX=TEMP(I, J)
4. Do, with index I ranging from 4 to 5, the following:
   a. Do, with index J ranging from 2 to 8 in increments of 6, the following:
      i. If TEMP(I, J)>TMAX then
         (a) Set TMAX=TEMP(I, J)
5. Print TMAX.

Figure 6.12a

```
            PROGRAM SEARCH2

      ****************************************************************
      *                                                              *
      *    THIS PROGRAM FINDS THE LARGEST ELEMENT OF THE 5X9 ARRAY   *
      *    FURNACE THAT CONTAINS THE TEMPERATURE DISTRIBUTION INSIDE A *
      *    FURNACE WALL.                                             *
      *                                                              *
      ****************************************************************
            REAL TEMP(5,9),TMAX
            INTEGER I,J

            OPEN(UNIT=9,FILE='FURNACE')
            REWIND 9

            READ(9,*) ((TEMP(I,J),J=1,9),I=1,5)

            TMAX=TEMP(2,2)

            DO 1 I=2,3

               DO 2 J=2,8

                  IF(TEMP(I,J).GT.TMAX)THEN
                     TMAX=TEMP(I,J)

                  END IF

      2         CONTINUE
      1 CONTINUE

            DO 3 I=4,5

               DO 4 J=2,8,6
```

Figure 6.12b

```
                    IF(TEMP(I,J).GT.TMAX)THEN
                        TMAX=TEMP(I,J)

                    END IF

4        CONTINUE
3 CONTINUE

  PRINT*,' THE LARGEST TEMPERATURE INSIDE THE WALL IS ',TMAX

  END


File Furnace

300,300,300,300,300,300,300,300,300
300,340,372,387,391,387,372,340,300
300,384,461,485,490,485,461,384,300
300,432,600,600,600,600,600,432,300
300,441,600,600,600,600,600,441,300

RUN

  THE LARGEST TEMPERATURE INSIDE THE WALL IS 490.
```

Figure 6.12b   (cont.)

Algorithm for Sample Problem 6.9. This program counts the number of elements of the 5×9 array FURNACE that have the value 461.

1. Read TEMP.
2. Set ICOUNT=0.
3. Do, with index I ranging from 1 to 5, the following:
   a. Do, with index J ranging from 1 to 9, the following:
      i. If $|A(I, J)-461.0| < 1.0E-8$ then
      (a) Calculate
         $$I COUNT = I COUNT + 1$$
4. Print ICOUNT.

Figure 6.13a

```
          PROGRAM COUNT

  ****************************************************************
  *                                                              *
  *     THIS PROGRAM COUNTS THE NUMBER OF ELEMENTS OF THE 5X9    *
  *     ARRAY FURNACE THAT HAVE THE VALUE 461.                   *
  *                                                              *
  ****************************************************************

          REAL TEMP(5,9)
          INTEGER ICOUNT,I,J

          OPEN(UNIT=9,FILE='FURNACE')
          REWIND 9
```

Figure 6.13b

```
      READ(9,*) ((TEMP(I,J),J=1,9),I=1,5)
      ICOUNT=0

      DO 1 I=1,5

         DO 2 J=1,9

            IF(ABS(TEMP(I,J)-461.).LT.1.E-8)THEN
               ICOUNT=ICOUNT+1

            END IF

2        CONTINUE
1 CONTINUE

      PRINT*,' THE NUMBER OF POINTS WHERE THE TEMPERATURE ',
     + 'IS 461. IS ',ICOUNT

      END


File Furnace

300,300,300,300,300,300,300,300,300
300,340,372,387,391,387,372,340,300
300,384,461,485,490,485,461,384,300
300,432,600,600,600,600,600,432,300
300,441,600,600,600,600,600,441,300

RUN

   THE NUMBER OF POINTS WHERE THE TEMPERATURE IS 461. IS 2
```

Figure 6.13b   (cont.)

---

of the 6 × 6 array A for the position of its largest element. Exchange the value found at this position with the value found on the main diagonal, that is, at index position (1, 1). Read in your array values in a row-wise order from a data file. The new array should be printed at the terminal using an implied-DO list.

Solution: The algorithm and problem solution are shown in Figures 6.14a and 6.14b.

Sample Problem 6.11

An automobile manufacturer produces four different automobile models. The average fuel economy for each model car, that is, liters of gasoline used for 10 kilometers driven, is a very important figure for the manufacturer. So is the overall average fuel economy of the entire fleet. Write a program that will read test values of fuel economy from a data file and then store them in an array that is 10 × 4. Data for each model should be placed in one separate column of the array. The data file to be read has values already separated into four appropriate columns as shown. Each column contains data for a separate model.

```
.9,1.2,.5,.7
.8,1.1,.56,.71
.85,1.15,.55,.71
.91,1.21,.53,.72
.93,1.19,.52,.73
.94,1.18,.51,.72
.98,1.16,.57,.715
```

■   **Algorithm for Sample Problem 6.10.** This program finds the largest element in the first column of a 6 × 6 array A and places it on the diagonal through an element interchange with A(1,1).

1. Read A.
2. Set AMAX=A(1,1),IMAX=1.
3. Do, with index I ranging from 2 to 6, the following:
   a. If A(I,1)>AMAX then
      i. Set IMAX=I
      ii. Set AMAX=A(I,1)
4. If IMAX≠1 then
   a. Set ATEMP=A(1,1)
   b. Set A(1,1)=A(IMAX,1)
   c. Set A(IMAX,1)=ATEMP
5. Print A.

Figure 6.14a

```
      PROGRAM SWAP
************************************************************************
*                                                                    *
*      THIS PROGRAM FINDS THE LARGEST ELEMENT IN THE FIRST           *
*      COLUMN OF A 6X6 ARRAY A AND PLACES IT ON THE DIAGONAL          *
*      THROUGH AN ELEMENT INTERCHANGE WITH A(1,1).                   *
*                                                                    *
************************************************************************
      REAL A(6,6),AMAX,ATEMP
      INTEGER IMAX,I

      OPEN(UNIT=9,FILE='MATRIX')
      REWIND 9

      READ(9,*) ((A(I,J),J=1,6),I=1,6)

      AMAX=A(1,1)
      IMAX=1

      DO 1 I=2,6

          IF(A(I,1).GT.AMAX)THEN
            AMAX=A(I,1)
              IMAX=I

          END IF

    1 CONTINUE

      IF(IMAX.NE.1)THEN
          ATEMP=A(1,1)
          A(1,1)=A(IMAX,1)
          A(IMAX,1)=ATEMP
```

Figure 6.14b

```
      END IF

      PRINT 3,  ((A(I,J),J=1,6),I=1,6)
    3 FORMAT(6(5X,F7.2))

      END
```

File Matrix

```
  10.9
  17.5
  18.9
  14.0
  11.0
  16.1
  14.1
  11.0
  19.2
 -13.2
  20.3
 -11.0
  10.3
  12.3
   4.2
   3.4
  15.6
  14.3
11.0
   5.6
  14.1
  18.0
   7.8
   7.8
  17.8
  22.3
  -4.3
   9.1
  18.1
  18.4
  16.2
   6.1
   2.0
  -5.2
   7.
  21.
```

RUN

17.80	17.50	18.90	14.00	11.00	16.10
14.10	11.00	19.20	-13.20	20.30	-11.00
10.30	12.30	4.20	3.40	15.60	14.30
11.00	5.60	14.10	18.00	7.80	7.80
10.90	22.30	-4.30	9.10	18.10	18.40
16.20	6.10	2.00	-5.20	7.00	21.00

Figure 6.14b   (cont.)

```
  .84,1.5,.58,.70
  .91,1.3,.59,.69
  .93,1.1,.65,.62
```

Calculate the average fuel economy for each model and save them in a one-dimensional array. Calculate the overall fleet average. Print all your answers at the terminal.

Analysis: The average fuel economy for each model is given by

$$AV_j = \frac{\sum_{i=1}^{10} A_{i,j}}{10}$$

The total fleet economy is

$$AV_{\text{tot}} = \frac{\sum_{j=1}^{4} AV_j}{4}$$

Solution: The algorithm and problem solution appear in Figures 6.15a and 6.15b.

Algorithm for Sample Problem 6.11. This program computes the average fuel consumption of four different model cars and stores them in array AV. The overall average for the entire fleet, AVTOT, is printed at the terminal. Data values are stored in array A. Ten values are stored per car.

1. Read A.
2. Set AVTOT=0.
3. Do, with index J ranging from 1 to 4, the following:
   a. Set SUM=0
   b. Do, with index I ranging from 1 to 10, the following:
      i. Calculate
         SUM=SUM+A (I , J)
   c. Calculate
      AV (J) =SUM/1 0 . 0
      AVTOT=AVTOT+AV (J) / 4 . 0
4. Print AV.
5. Print AVTOT.

Figure 6.15a

```
        PROGRAM AUTO

****************************************************************
*                                                              *
*       THIS PROGRAM COMPUTES THE AVERAGE FUEL CONSUMPTION     *
*       OF FOUR DIFFERENT MODEL CARS AND STORES THEM IN        *
*       ARRAY AV. THE OVERALL AVERAGE FOR THE ENTIRE FLEET,    *
*       AVTOT, IS PRINTED AT THE TERMINAL. DATA VALUES ARE     *
*       STORED IN ARRAY A. TEN VALUES ARE STORED PER CAR.      *
*                                                              *
****************************************************************

        REAL A(10,4),AV(4),AVTOT,SUM
        INTEGER J,I
```

Figure 6.15b

```
OPEN(UNIT=9,FILE='AUTOS')
REWIND 9

READ(9,*) ((A(I,J),J=1,4),I=1,10)

AVTOT=0.0

DO 1 J=1,4
    SUM=0.0

        DO 2 I=1,10
            SUM=SUM+A(I,J)

2       CONTINUE

        AV(J)=SUM/10.0
        AVTOT=AVTOT+AV(J)/4.

1 CONTINUE

DO 3 I=1,4
    PRINT 4,I,AV(I)

3 CONTINUE

PRINT 5,AVTOT
4 FORMAT(5X,'THE AVERAGE FOR MODEL ',I1,' IS ',F5.3)
5 FORMAT(//,5X,'THE OVERALL FLEET AVERAGE IS ',F5.3)

END
```

```
File Autos

.9,1.2,.5,.7
.8,1.1,.56,.71
.85,1.15,.55,.71
.91,1.21,.53,.72
.93,1.19,.52,.73
.94,1.18,.51,.72
.98,1.16,.57,.715
.84,1.5,.58,.70
.91,1.3,.59,.69
.93,1.1,.60,.62

RUN

    THE AVERAGE FOR MODEL 1 IS   .899
    THE AVERAGE FOR MODEL 2 IS  1.209
    THE AVERAGE FOR MODEL 3 IS   .551
    THE AVERAGE FOR MODEL 4 IS   .702


    THE OVERALL FLEET AVERAGE IS   .840
```

Figure 6.15b   (cont.)

6.4 THREE-DIMENSIONAL ARRAYS

An array having three subscripts is called a **three-dimensional array.** The array can be thought of as containing two-dimensional groups of information in different planes or levels. A three-dimensional array is very much analogous to a function of three variables, that is, $f(x, y, z)$. In an array, integer indices are substituted for the independent continuous variables x, y, and z. A three-dimensional element might appear as

```
A(I,J,K)
```

The first index appearing in the parentheses is the row number, the second index is the column number, and the third index is the plane number of the element. A DIMENSION statement for a three-dimensional array would need to specify the range of all three indices.

DIMENSION array1$(ll_r : ul_r, ll_c : ul_c, ll_p : ul_p), \ldots$

The array could also be dimensioned with a type statement. On input or output of data, if the short-list form is used, information will be entered in or displayed in a column-wise order for different levels. This means that the first index changes the fastest (the most number of times), the second index varies less rapidly, and the third index changes the least.

Implied-DO lists can also be used with input/output statements for three-dimensional arrays. For example, if A is a $10 \times 5 \times 7$ array, the following statements might typically appear in a user's program:

```
REAL A(10,5,7)
READ *,(((A(I,J,K),I=1,10),J=1,5),K=1,7)
PRINT 5,(((A(I,J,K),K=7,1,-1),I=1,10,J=1,5)
```

The rules for processing three-dimensional arrays follow closely those previously described for one- and two-dimensional arrays.

It is also possible to use an array with even more than three dimensions in FORTRAN; FORTRAN allows for as many as seven dimensions for a single array. These arrays are not often used, however, because the programmer seldom needs to relate that many different variables to one another.

*6.5 MATRICES

In mathematics, a two-dimensional array containing an ordered set of numeric quantities and having M rows and N columns is called an $M \times N$ **matrix.** A one-dimensional array with a row of n entries is called a $1 \times N$ **row matrix.** Similarly, a one-dimensional array with a column of m entries is called an $M \times 1$ **column matrix.** In engineering, matrices are very important and play a dominant role in modern numerical analysis performed on the computer. Several matrix operations commonly encountered in engineering work are of sufficient interest and are elementary enough to discuss here. Some further definitions will be given first, though, before proceeding.

A **square matrix** has the same number of rows as it has columns. If all elements except for those lying on the principal diagonal of a square matrix are 0, the matrix is called a **diagonal matrix.** If, in addition, these diagonal terms all have the value 1, the matrix is called an **identity or unit matrix.** A matrix with all 0 values is called a **null matrix.**

When the off-diagonal terms of a matrix [**A**] (square brackets, [ ], denote a two-dimensional matrix; braces, { }, denote a column or row matrix) are equal about the diagonal — that is, the elements $A(1, 3) = A(3, 1)$, $A(3, 5) = A(5, 3)$, and so on — the matrix is called a **symmetric matrix.** The resulting matrix derived from the matrix [**A**] after the indices for the matrix have been interchanged is called the *transpose* of [**A**], or $[\mathbf{A}]^t$. The transpose of an $M \times N$ matrix then becomes an $N \times M$ matrix. The transpose of a column matrix is a row matrix, and vice versa. For a symmetric matrix, the transpose of the matrix equals the original matrix. See Figure 6.16 for examples.

The operations of matrix addition or subtraction are mathematically defined by

$$\{\mathbf{C}\} = \{\mathbf{A}\} \pm \{\mathbf{B}\}$$

or

$$C(I) = A(I) \pm B(I) \quad \text{for } I = 1 \text{ to } N \tag{6.1}$$

where {**A**}, {**B**}, and {**C**} are $N \times 1$ column matrices or $1 \times N$ row matrices. For the $M \times N$ matrices [**A**] and [**B**], the resultant matrix [**C**] is given by

$$[\mathbf{C}] = [\mathbf{A}] \pm [\mathbf{B}]$$

or

$$C(I, J) = A(I, J) \pm B(I, J) \quad \text{for } I = 1 \text{ to } M \quad \text{and} \quad J = 1 \text{ to } N \tag{6.2}$$

$$\begin{bmatrix} 5 & 4 & 2 \\ 4 & 3 & 1 \\ 2 & 1 & 6 \end{bmatrix} \qquad \begin{bmatrix} 2 & 0 & 0 & 0 \\ 0 & 2 & 0 & 0 \\ 0 & 0 & 2 & 0 \\ 0 & 0 & 0 & 2 \end{bmatrix} \qquad \begin{bmatrix} 1 & 0 & 0 & 0 \\ 0 & 1 & 0 & 0 \\ 0 & 0 & 1 & 0 \\ 0 & 0 & 0 & 1 \end{bmatrix}$$

3 X 3 symmetric matrix   4 X 4 diagonal matrix   4 X 4 identity matrix

$$\{1 \quad 2 \quad 3 \quad 4\}^t = \begin{Bmatrix} 1 \\ 2 \\ 3 \\ 4 \end{Bmatrix} \qquad \qquad \begin{bmatrix} 4 & 3 \\ 2 & 1 \\ 4 & 6 \end{bmatrix}^t = \begin{bmatrix} 4 & 2 & 4 \\ 3 & 1 & 6 \end{bmatrix}$$

Transpose of a row matrix is a column matrix   Transpose of a 3 X 2 matrix

Figure 6.16

Matrix multiplication for the $M \times L$ matrix [A] and the $L \times N$ matrix [B] is represented by

$$[C] = [A][B] \tag{6.3}$$

where [C] is the resultant $M \times N$ matrix. Mathematically, the elements of [C] can be expressed by

$$C(I, J) = \sum_{K=1}^{L} A(I, K)B(K, J) \quad \text{for } I = 1 \text{ to } M \quad \text{and} \quad J = 1 \text{ to } N \tag{6.4}$$

Note that [A][B] does not equal [B][A] — that is, matrix multiplication is not commutative. Note also that the total number of columns of [A] must be equal to the total number of rows of [B]. For the $N \times 1$ column matrices {A} and {B}, the product of {A} and {B} is

$$C = \{A\}^t\{B\} \tag{6.5}$$

where C is a scalar (has magnitude only). Using equation (6.4),

$$C = C(1, 1) = \sum_{K=1}^{N} A^t(1, K)B(K, 1) \tag{6.6}$$

For the $1 \times N$ row matrices {A} and {B}, the product of {A} and {B}^{t} is

$$C = \{A\}\{B\}^t \tag{6.7}$$

A scalar can be regarded as a matrix having only one element. It is sometimes called a *zero-order tensor*. A vector can be represented by a row or a column matrix. Each element of the matrix, having only one index, represents a component value of the vector. A vector is known mathematically as a *first-order tensor*. A two-dimensional array or matrix is classified as a *second-order tensor*.

Sample Problem 6.12

Write a FORTRAN program that will add two 5×4 matrices [A] and [B] together. The matrices are to be entered row-wise from a file called MATMATH. Matrix [B] follows matrix [A] in the data file. The resultant matrix, [C], should be displayed at the terminal using four entries (columns) to a line (row).

Solution: The algorithm and problem solution are shown in Figures 6.17a and 6.17b.

Sample Problem 6.13

Write a FORTRAN program that will multiply together two matrices [A] and [B]. Matrix [A] is 5×4 and matrix [B] is 4×5. The 5×5 resultant matrix, [C], should be displayed at the terminal using five entries per line. The matrices are to be entered row-wise from a file called MATMATH (see Sample Problem 6.12). Matrix [B] follows matrix [A] in the data file.

Solution: The algorithm and problem solution are shown in Figures 6.18a and 6.18b.

Algorithm for Sample Problem 6.12. This program adds together two 5×4 matrices [**A**] and [**B**] and stores the result in matrix [**C**].

1. Read A,B.
2. Do, with index I ranging from 1 to 5, the following:
   a. Do, with index J ranging from 1 to 4, the following:
      i. Calculate
         C(I , J)=A(I , J)+B(I , J)
3. Print C.

Figure 6.17a

```
              PROGRAM MATADD

     ****************************************************************
     *                                                              *
     *      THIS PROGRAM ADDS TOGETHER THE TWO 5X4 MATRICES [A]      *
     *      AND [B] AND STORES THE RESULT IN MATRIX [C].             *
     *                                                              *
     ****************************************************************
              REAL A(5,4),B(5,4),C(5,4)
              INTEGER I,J

              OPEN(UNIT=9,FILE='MATMATH')
              REWIND 9

              READ(9,*) ((A(I,J),J=1,4),I=1,5)
              READ(9,*) ((B(I,J),J=1,4),I=1,5)

              DO 1 I=1,5

                  DO 2 J=1,4
                      C(I,J)=A(I,J)+B(I,J)

         2        CONTINUE

         1 CONTINUE

              PRINT 3, ((C(I,J),J=1,4),I=1,5)
         3 FORMAT(4(5X,F7.2))

              END

     File Matmath

     3,7,4,5
     2,0,3,4
     5,4,1,2
     5,6,0,1
     3,3,1,4
     2,4,3,5
     1,1,5,0
     3,3,0,2
     1,2,3,4
     2,2,3,3
```

Figure 6.17b

RUN

5.00	11.00	7.00	10.00
3.00	1.00	8.00	4.00
8.00	7.00	1.00	4.00
6.00	8.00	3.00	5.00
5.00	5.00	4.00	7.00

Figure 6.17b   (cont.)

Algorithm for Sample Problem 6.13. This program premultiplies a 4×5 matrix [**B**] by a 5×4 matrix [**A**]. The results are stored in the 5×5 matrix [**C**].

1. Read A,B.
2. Do, with index I ranging from 1 to 5, the following:
   a. Do, with index J ranging from 1 to 5, the following:
      i. Set C(I, J)=0
      ii. Do, with index J ranging from 1 to 5, the following:
         (a) Calculate
            C(I,J)=C(I,J)+A(I,K)*B(K,J)
3. Print C.

Figure 6.18a

```
        PROGRAM MATMUL

****************************************************************
*                                                              *
*       THIS PROGRAM PREMULTIPLIES A 4X5 MATRIX [B] BY A 5X4   *
*       MATRIX [A]. THE RESULTS ARE STORED IN 5X5 MATRIX [C].  *
*                                                              *
****************************************************************

        REAL A(5,4),B(4,5),C(5,5)
        INTEGER I,J

        OPEN(UNIT=9,FILE='MATMATH')
        REWIND 9

        READ(9,*) ((A(I,J),J=1,4),I=1,5)
        READ(9,*) ((B(I,J),J=1,5),I=1,4)

        DO 1 I=1,5

            DO 2 J=1,5
                C(I,J)=0.

                DO 3 K=1,4
                    C(I,J)=C(I,J)+A(I,K)*B(K,J)

3           CONTINUE

2       CONTINUE
```

Figure 6.18b

```
1 CONTINUE

  PRINT 4, ((C(I,J),J=1,5),I=1,5)
4 FORMAT(5(3X,F9.2))

  END
```

File Matmath

```
3,7,4,5
2,0,3,4
5,4,1,2
5,6,0,1
3,3,1,4
2,4,3,5
1,1,5,0
3,3,0,2
1,2,3,4
2,2,3,3
```

RUN

```
        33.00        65.00        23.00        59.00        51.00
```

THE AVERAGE FOR MODEL 4 IS .702

THE OVERALL FLEET AVERAGE IS .840

```
20.00        22.00        17.00        28.00        23.00
22.00        46.00        20.00        45.00        26.00
20.00        52.00        17.00        46.00        26.00
25.00        37.00        18.00        38.00        27.00
```

Figure 6.18b (cont.)

Sample Problem 6.14

In engineering analyses of structures, the relationship between applied forces and re-sulting structural displacements can be represented by the matrix equation

$$\{Y\} = [F]\{P\}$$

where $\{Y\}$ and $\{P\}$ are column vectors of displacements and applied forces, respec-tively, and $[F]$ is called the flexibility matrix. For instance, element $F(I,J)$ represents the displacement at point I of the structure resulting from a unit load applied at point J. The total displacement at point I would then be computed by

$$Y(I) = F(I, 1) \times P(1) + F(I, 2) \times P(2) \times \ldots \times F(I, N) \times P(N)$$

where N is the total number of points, or nodes, studied in the structure. For the canti-levered beam shown in Figure 6.19, write a FORTRAN program that will solve for the deflection at points 1, 2, and 3 resulting from the given applied loads. The flexibility

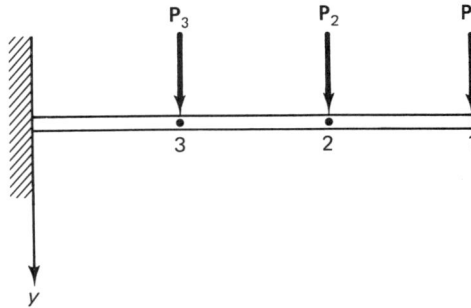

Figure 6.19

matrix may be taken as

$$[\mathbf{F}] = \begin{bmatrix} 45.0 & 23.3 & 6.7 \\ 23.3 & 13.3 & 4.2 \\ 6.7 & 4.2 & 1.7 \end{bmatrix} 10^{-2} \quad mm/kN$$

and $P(1)$ and $P(2) = 10.0$, and $P(3) = 22.0$ kilonewtons.

Analysis: The deflections may be found from matrix multiplication:

$$Y(I) = \sum_{J=1}^{3} F(I,J) \times P(J) \quad I = 1, 2, 3$$

Solution: The computer program is shown in Figure 6.20.

```
      PROGRAM FLEXMAT

*********************************************************************
*                                                                 *
*     THIS PROGRAM FINDS THE DEFLECTION AT NODE POINTS ALONG A    *
*     BEAM BY PREMULTIPLYNG THE GIVEN LOAD MATRIX {P} BY THE      *
*     BEAM'S FLEXIBILITY MATRIX [F]. PROGRAM DATA RESIDES IN      *
*     FILE BEAM.                                                  *
*                                                                 *
*********************************************************************
      REAL F(3,3),P(3),Y(3)
      INTEGER I,J

      OPEN(UNIT=9,FILE='BEAM')
      REWIND 9

      READ(9,*) ((F(I,J),J=1,3),I=1,3)
      READ(9,*) (P(I),I=1,3)

      DO 1 I=1,3
         Y(I)=0.

         DO 2 J=1,3
            Y(I)=Y(I)+F(I,J)*P(J)

 2       CONTINUE
```

Figure 6.20

```
   1 CONTINUE

     PRINT *,' THE DEFLECTIONS ARE:'
     PRINT 4, (I,Y(I),I=1,3)
   4 FORMAT(5X,'Y(',I1,') = ',E9.3,' MM')

     END
```

File Beam

```
45.E-2,23.3E-2,6.7E-2
23.3E-2,13.3E-2,4.2E-2
6.7E-2,4.2E-2,1.7E-2
10.,10.,22.
```

RUN

```
THE DEFLECTIONS ARE:
   Y(1) =  .830E+01 MM
   Y(2) =  .458E+01 MM
   Y(3) =  .146E+01 MM
```

Figure 6.20 (cont.)

REVIEW

For handling very large amounts of associated data, FORTRAN provides a means of storing values under one collective or group name. The group is called an array, and it uses integer indices, or subscripts, to distinguish the array's components, or elements. One-dimensional arrays use only a single subscript, written in parentheses, to identify array elements. To inform the FORTRAN compiler of how much memory space is needed for the array, a DIMENSION or explicit type statement is used. (The size of an array cannot be specified in both a DIMENSION and a type statement.) The syntaxes for these statements are

DIMENSION array$_1$(ll : ul),array$_2$(ll : ul), . . .

type array$_1$(ll : ul),array$_2$(ll : ul), . . .

where ll and ul represent the maximum lower and upper limits for the range of the array's subscripts, respectively. The upper limit must be greater than or equal to the lower limit. In addition, every element of an array must be of the same type.

DIMENSION and type statements are specification statements and must appear at the very beginning of a program before any executable statement is listed. More than one DIMENSION or type statement can be used in a program. When the lower limit of a DIMENSION or type statement is omitted, a default value of 1 is assigned. The DIMENSION or type statement then has the form

DIMENSION array$_1$(ul),array$_2$(ul), . . .

type array$_1$(ul),array$_2$(ul), . . .

For input or output of data, listing of the array elements must accompany the DATA, READ, PRINT, or WRITE statements. Elements can be listed individually or in a group. When the entire group is to be processed, the short-list technique can be employed, in which only the array's name is used. The compiler then assumes that all group elements are to be listed in serial order. When specific control of the elements' processing order is required or when only a partial group of elements is to be processed, a DO loop or an implied-DO list can be used. The latter method is preferred. A PRINT statement having an implied-DO list might appear as

```
PRINT *,(A(I),I=1,N)
```

The compiler interprets the input/output statement having an implied-DO list or a short list as one executable statement having multiple elements listed in series. For an explicit DO loop, the input/output operation is performed only once every time the loop is traversed. Assuming that the READ list consists solely of a single array element, only one data entry will be made during each pass through the loop. The data file must then be adjusted accordingly so that only one of the array's elements appears per record line. For the same reason, only one of the array's elements can be written per output line when the array is displayed.

Two-dimensional arrays are arrays that use two integer subscripts or indices, enclosed within parentheses, to identify their elements. Two-dimensional arrays can be thought of as a group of columns or rows. The first index found within the parentheses defines an element's row number; the second index defines its column number. The DIMENSION statement for a two-dimensional array appears as

DIMENSION $\text{array}_1(ll_r:ul_r,ll_c:ul_c), \ldots$

where r stands for row and c for column. When using the lower limit default value of 1, the DIMENSION statement appears as

DIMENSION $\text{array}_1(ul_r,ul_c),\text{array}_2(ul_r,ul_c), \ldots$

Type declarations can also be used to dimension the array. When processing two-dimensional arrays, the arrays are usually treated in row-wise or column-wise order using nested DO loops. The index associated with the innermost DO loop varies the most.

Input and output operations for two-dimensional arrays are similar to those for one-dimensional arrays. When using the short-list technique, it is important to remember that arrays are filled or displayed in a column-wise order, starting with column 1 in the array and proceeding consecutively to the right. When using implied-DO lists, two indices must be employed. The first index, which appears immediately after the innermost right parenthesis, is filled first (varies the fastest). It acts like an inner nested DO loop.

KEY TERMS

array
array element
array name
column matrix
column-wise
diagonal matrix
DIMENSION statement
dimensioning
identity or unit matrix
implied-DO list
matrix

null matrix
one-dimensional array
row matrix
row-wise
short-list technique
square matrix
subscript
subscripted variable
symmetric matrix
two- and three-dimensional arrays

REVIEW QUESTIONS

1. In the following problems, the arrays are to be initialized as indicated. Show what values *each* element of the array is assigned.

(a)
```
    REAL A(5)
    DO 7 I=5,1,-1
        A(I)=2.0*(6-I)
  7 CONTINUE
```

(b)
```
    REAL A(3,3)
    DO 7 I=1,3
        DO 8 J=1,3
            A(I,J)=(I+J)
  8     CONTINUE
  7 CONTINUE
```

(c)
```
    REAL A(0:5)
    DO 7 I=1,6
        A(I-1)=SIN(0.2*I)
  7 CONTINUE
```

(e)
```
    INTEGER K(2,2)
    DATA J,K/1,2,4,6,8/
```

(d)
```
    INTEGER K(3,3)
    DO 7 I=1,3
    M=2
    DO 8 J=1,3
        K(I,J)=M+I*J
  8     CONTINUE
  7 CONTINUE
```

2. In the following problems, assume that the data file was entered as shown. Determine and show the values assigned to each element of the indicated arrays.

```
13.524.614.3
10.4 6.2 7.8
12.713.8-9.4
11.4 1.9-7.2
10.9
13.3
21.4
19.4
17.3
```

(a)
```
    REAL B(3)
    READ(7,1) B
  1 FORMAT(F4.1)
```

(b)
```
    REAL B(3)
    READ(7,1)(B(I),I=1,3)
  1 FORMAT(3F4.1)
```

(c)   REAL B(3)
       DO 8 I=1,3
          READ(7,1) B(I)
     8 CONTINUE
     1 FORMAT(F4.1)
(e)   REAL B(3,3)
       READ(7,1)((B(I,J),J=1,3),I=1,3)
     1 FORMAT(3F4.1)

(d)   REAL B(3,3)
       READ(7,1) B
     1 FORMAT(3F4.1)

3. Assuming that the data file in Problem 2 has been changed so that a blank column appears between entries, determine and show the values that each element of the indicated arrays will be assigned.

```
13.5 24.6 14.3
10.4  6.2  7.8
12.7 13.8 -9.4
11.4  1.9 -7.2
10.9
13.3
21.4
19.4
17.3
```

(a) REAL B(3)
    READ(7,*) B
(c) REAL B(3,3)
    READ(7,*)((B(I,J),J=1,3),I=1,3)

(b)   REAL B(3)
       DO 7 I=1,3
          READ(7,*) B(I)
     7 CONTINUE

4. Assuming that the elements of the one-dimensional array B have the values B(1) = 3.2, B(2) = 9.4, and B(3) = 6.8, show how the output will appear with the following commands:

(a) PRINT *,B

(b) PRINT *,(B(I),I=1,3)

(c)   DO 7 I=1,3
       PRINT *,B(I)
   7 CONTINUE

5. Assuming that the array B is two-dimensional, indicate whether the printed output will appear in rows or in a single column.

(a) PRINT *,B

(b) PRINT *,((B(I,J),I=1,3),J=1,3)

(c)   DO 7 J=1,3
       DO 8 I=1,3
          PRINT *,B(I,J)
   8   CONTINUE
   7 CONTINUE

In the following questions, fill in the missing information that is needed in the program blocks shown to accomplish the indicated tasks.

6. Assign a value of 0 to each element of array A.

   (a)   REAL A(0:10)   **(b)**   REAL A(10,10)

      DO _ _ _ I =_ _ _ _ _ _ _ _ _   DO _ _ _ I =_ _ _ _ _ _ _ _ _

           A(_ _)=_ _ _ _ _ _   DO _ _ _ J _ _ _ _ _ _

     7 CONTINUE   A(_ _ , _ _)_ _ _ _ _

                                  8   CONTINUE

                                  7 CONTINUE

7. Copy the values in array A to array B in sequential order.

   (a)   REAL A(0:10),B(11)   **(b)**   REAL A(10),B(5,2)
                                DO 7 I=1,5

      DO _ _ _ I =_ _ _ _ _ _ _ _ _ _ _ _   B(I,1)=A(_ _ _)

           B(_ _)_ _ _ _ _ _ _ _ _ _ _ _   B(I,2)=A(_+_)

     7 CONTINUE   7 CONTINUE

8. Count how many times a specific value, VAL, is found in the array A.

   (a)   REAL A(-5:5)
          J=0

      DO _ _ I =_ _ _ _ _ _ _ _ _ _

          IF(_ _ _ _ _ _ _ _ _ _ _)_ _ _ _ _ _ _ _ _

     5 CONTINUE
   (b)   REAL A(4,2)
          K=0

      DO _ _ I =_ _ _ _ _ _ _ _

          DO _ _ J=_ _ _ _ _ _ _

          IF(_ _ _ _ _ _ _ _)_ _ _ _ _

     5   CONTINUE
     6 CONTINUE

9. Search for a specific value, VAL, in a two-dimensional array and print out its index location.

        REAL A(10,10)

        DO _ _ _ I =_ _ _ _ _ _ _ _ _ _

```
      DO _ _  J=_ _ _ _ _ _ _ _ _

          IF  (_ _ _ _ _ _ _ _ _ _ _)  THEN

              I1=_ _ _ _ _ _ _ _

              J1=_ _ _ _ _ _ _ _

              PRINT *,_ _ _ _ _ _ _ _
          END IF
   5      CONTINUE
   6  CONTINUE
```

10. Search for the largest absolute value, AMAX, in a two-dimensional array and print the value.

```
      REAL  A(10,10)

      AMAX=_ _ _ _ _ _ _ _ _ _ _ _

      DO _ _ _  I=_ _ _ _ _ _ _ _ _

          DO _ _ _  J=_ _ _ _ _ _ _

              IF  (_ _ _ _ _ _ _ _ _ _ _)  _ _ _ _ _ _ _ _ _
   7      CONTINUE
   8  CONTINUE

      PRINT *,_ _ _ _ _ _ _ _ _ _
```

11. Sort in descending order the elements of each *column* of a two-dimensional array.

```
      REAL  A(10,10)
      DO 7  I=1,10
          DO 8  J=9,1,-1

              DO 9  K=_ _ _ _ _ _ _

                  IF  (A(_ _,_ _)._ _ .A(_ _,_ _))  THEN

                      T=_ _ _ _ _ _ _ _

                      A(K,I)=_ _ _ _ _ _ _ _ _ _

                          A(K+1,I)=_ _ _ _ _ _ _ _ _
                  END IF
   9          CONTINUE
   8      CONTINUE
   7  CONTINUE
```

12. Sort in descending order the elements of each *row* of a two-dimensional array.

```
REAL  A(10,10)
DO  7  I=1,10
    DO  8  J=9,1,-1

        DO  9  K=_ _ _ _ _ _ _

            IF  (A(_ _,_ _)._ _.A(_ _,_ _))  THEN

                T=_ _ _ _ _ _ _ _

                A(I,K)=_ _ _ _ _ _ _ _ _

                A(I,K+1)=_ _ _ _ _ _ _ _
            END  IF
9           CONTINUE
8       CONTINUE
7  CONTINUE
```

13. Find the average value of all the elements in a three-dimensional array.

```
REAL  A(7,3,5)

SUM=_ _ _ _ _ _ _ _ _

NTOTAL=_ _ _ _ _ _ _ _ _ _

DO  _ _  I=_ _ _ _ _ _ _

    DO  _ _  J=_ _ _ _ _ _ _

        DO  _ _  K=_ _ _ _ _ _ _

            _ _ _ _ _ _ _ _ _ _ _ _ _

            _ _ _ _ _ _ _ _ _ _ _ _ _
7           CONTINUE
8       CONTINUE
9  CONTINUE

AV=_ _ _ _ _ _ _ _ _ _ _ _ _
PRINT  *,AV
```

14. Write a FORTRAN program block that will sort the one-dimensional array A(20) in descending order.

15. Write a FORTRAN program block that will perform a binary search of the array A(0:20) to find a specified value, after which it will print out its index location. The file is assumed to be sorted in descending order with no duplicate entries present. Print an error message if the value sought cannot be found.

16. Write a FORTRAN program block that will transpose (interchange rows and columns, so that A(I, J) = A(J, I), etc.) the two-dimensional array A(4, 4). Use the variable name T for temporary storage.

COMPUTER PROJECTS

CP1. Rework Sample Problem 5.5 using two arrays for the *x* and *y* values.

CP2. Rework Sample Problem 5.7 using an array for the *y* values. Assume NDATA to be known beforehand so that the array for *y* can be properly dimensioned.

CP3. Rework Sample Problem 5.10 using an array for the *y* values. Assume NDATA to be known beforehand so that the array for *y* can be properly dimensioned.

CP4. For the file shown, write a FORTRAN program using a binary search algorithm to locate the position in the array (index) of an element having a specified value. The value searched for should be entered at the keyboard, but the file must be read in from external memory. If found, the index should be printed on the monitor. Provision must also be made for the case when the value searched for does not exist in the file. In that event, an appropriate message should appear. Test your program with the values 487.9 and 421.8. (Refer to Sample Problem 6.6.)

```
DATA FILE:
550.7
527.8
520.6
510.3
495.0
487.9
480.6
477.9
471.2
462.8
455.5
441.6
432.9
423.7
417.4
405.8
388.2
376.7
356.1
344.8
```

CP5. Write a FORTRAN program that will use a bubble-sort algorithm to rearrange an arbitrary-sized, one-dimensional array in descending order. The new sorted array should be printed at the terminal. Test your program on the following data file, which is to be read in from external memory. (See Sample Problem 6.7.)

```
DATA FILE:
119.1
328.7
```

```
258.9
567.3
111.4
236.7
258.9
239.4
478.8
455.6
498.7
388.4
525.8
174.6
348.2
278.3
190.7
594.4
498.7
239.9
398.5
```

CP6. Write a program to evaluate the function $z = x \ln A$ for values of x from 0 to 10 in increments of 0.1. Take A as 2.4. The resulting z values should be stored in a one-dimensional array.

CP7. Write a program to evaluate $z = e^x \cos y \sin x$ for values of x ranging from 0 to 1 in increments of 0.1, and values of y ranging from 0 to 0.4 in increments of 0.2. The resulting z values should be stored in a two-dimensional array that has eleven rows and three columns that correspond to the different values x and y assume. For example, element $Z(4, 2)$ would then correspond to the functional value found when $x = 0.3$ and $y = 0.2$.

CP8. Repeat CP7 but this time store your values in a one-dimensional array having thirty-three elements. The first eleven elements should be reserved for the values of z found when y is 0. When $y = 0.2$, the second eleven elements are to be used, and so on. Element $Z(13)$, for example, then corresponds to the functional value $z(0.1, 0.2)$.

CP9. Write a FORTRAN program that will read a two-dimensional square array A and search for the largest element in column 1. Once this element is found, the row that contains it should be interchanged with the first row of the array so that the largest element will now appear in position $A(1, 1)$. With a suitable test case, demonstrate that your program works.

CP10. Write a FORTRAN program that will read a two-dimensional square array A and search for the largest element in each column. Once each element is found, its value should be interchanged with the value currently located in that column's main diagonal position. With a suitable test case, demonstrate that your program works.

CP11. Write a FORTRAN program that will read values of automobile fuel economy for four different car models from a data file and store them in an array that is $5 \times 4 \times 2$. Data for each car model (five values per year) should be placed in one separate column of the array and each year's results should lie in a separate level or plane. For your data file use the following information:

MODEL

Car 1	Car 2	Car 3	Car 4	
0.9	1.2	0.5	0.7	
0.8	1.1	0.56	0.71	
0.85	1.15	0.55	0.71	1985
0.91	1.21	0.53	0.72	
0.93	1.19	0.52	0.73	
0.94	1.18	0.51	0.72	
0.98	1.16	0.57	0.715	
0.84	1.5	0.58	0.70	1986
0.91	1.3	0.59	0.69	
0.93	1.1	0.60	0.62	

Element $A(2, 3, 2)$ of your array, the second test result on model 3 in 1986, should be 0.57. Calculate the fuel-economy average of each model for each year. Calculate the overall fuel average of the fleet for each year. Print all your answers at the terminal.

CP12. Write a program that will take two polynomials of the form

$$A_0 x^n + A_1 x^{n-1} + \ldots + A_n$$

and

$$B_0 x^m + B_1 x^{m-1} + \ldots + B_m$$

and multiply them together. The polynomial coefficients are to be read in and stored in arrays A and B, respectively. Other values to be entered are n, m, and x. As a test case use

$$x^4 + 3x^3 + 2.1x^2 + 5.0x + 8.0$$

and

$$3.1x^3 + 2.0x^2 + 7.1x + 18.1$$

Evaluate the resultant product polynomial at $x = 4.1$.

CP13. Write a FORTRAN program that will linearly interpolate or extrapolate between the data points given in the accompanying table. Use arrays to store the x and y values. What does y equal when x equals 0.439? When x equals 1.134, what is y?

x	y
0.0	0.0
0.1	0.11052
0.2	0.24428
0.3	0.40496
0.4	0.59673
0.5	0.82436
0.6	1.09327
0.7	1.40963
0.8	1.78043
0.9	2.21364
1.0	2.71828

CP14. Repeat CP12 of Chapter 4. Use arrays this time to store the values of B and H. Test your results for the case when $B = 79$ kilolines per square inch.

CP15. The location of the center of mass for a system of particles is given by

$$X_{\text{c.m.}} = \frac{\sum_{i=1}^{n} x_i M_i}{\sum_{i=1}^{n} M_i}, \qquad Y_{\text{c.m.}} = \frac{\sum_{i=1}^{n} y_i M_i}{\sum_{i=1}^{n} M_i}, \qquad Z_{\text{c.m.}} = \frac{\sum_{i=1}^{n} z_i M_i}{\sum_{i=1}^{n} M_i}$$

Write a FORTRAN program that will compute the center of mass for a system of particles where the mass and location for each particle are provided (see the accompanying table). In this program, use the two-dimensional array POSIT to store the x, y, and z coordinates of each particle, and the one-dimensional array MASS to store the mass data. Array MASS is real.

Particle	x (cm)	y (cm)	z (cm)	M (kg)
1	2.3	−1.1	2.5	1.11
2	−0.3	2.1	1.3	0.98
3	1.7	3.0	−1.2	0.39
4	1.1	2.1	3.7	0.50
5	2.3	−0.84	3.4	0.45

Answers: $X_{\text{c.m.}} = 1.314$, $Y_{\text{c.m.}} = 0.781$, $Z_{\text{c.m.}} = 2.029$

CP16. From the data given in CP15, write a computer program that will solve for the system's *mass moments of inertia* referenced to the center of mass. The formulas for the mass moment of inertia are

$$I_{xx} = \sum_{i=1}^{n} [(y_i - Y_{\text{c.m.}})^2 + (z_i - Z_{\text{c.m.}})^2] M_i$$

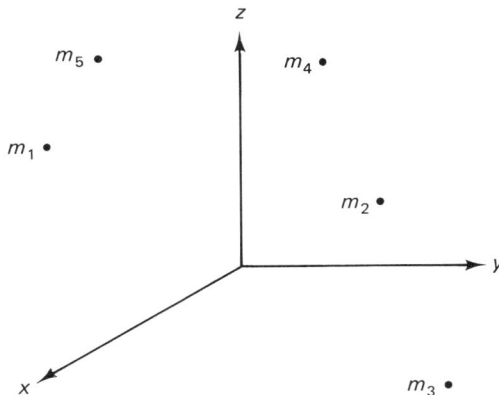

CP6.15

$$I_{yy} = \sum_{i=1}^{n} [(x_i - X_{c.m.})^2 + (z_i - Z_{c.m.})^2]M_i$$

$$I_{zz} = \sum_{i=1}^{n} [(x_i - X_{c.m.})^2 + (y_i + Y_{c.m.})^2]M_i$$

Use the arrays POSIT and MASS.

CP17. Continue CP16 by finding the system's *mass products of inertia*. Mass products of inertia referenced to the origin are defined by

$$I_{xy} = I_{yx} = -\sum_{i=1}^{n} x_i y_i M_i$$

$$I_{xz} = I_{zx} = -\sum_{i=1}^{n} x_i z_i M_i$$

$$I_{yz} = I_{zy} = -\sum_{i=1}^{n} y_i z_i M_i$$

CP18. Shown in the figure is a thin plate acted on by the planar forces given in the accompanying table. Write a FORTRAN program that will sum the force components in the x and y directions, F_x and F_y, respectively. (See Chapter 2, CP14.) Calculate the resultant of the total force. Use two one-dimensional arrays, FORCE and THETA.

	F (newtons)	θ (degrees)
1	1000	10
2	1350	-15
3	2130	130
4	1500	210
5	3100	310
6	450	180

CP19. From the data provided in CP18, write a computer program that will solve for the total moment about point O. The moment can be found from the expression

$$M_O = \sum_{i=1}^{n} [x_i(F_i)_y - y_i(F_i)_x]$$

The forces are located at

Force	x (cm)	y (cm)
1	3.3	2.1
2	1.3	-1.3
3	-1.6	4.1
4	-1.6	-4.1
5	0	-3.4
6	-2.5	-3.3

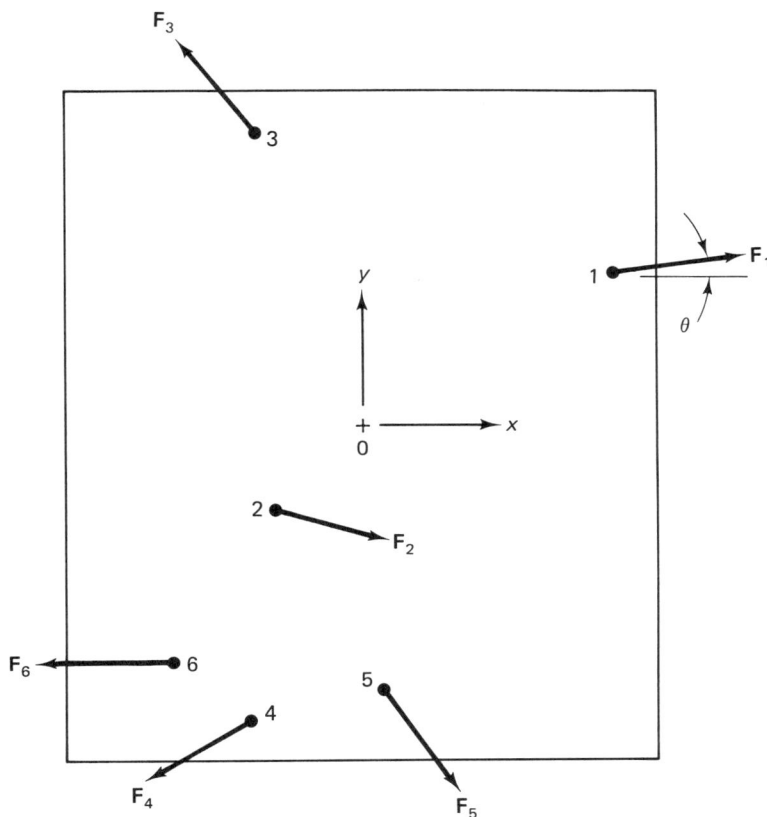

CP6.18

Use arrays FORCE, THETA, and POSIT. POSIT is a two-dimensional array containing the x, y force positions.

*CP20. Let the $m + 1$ mth-degree polynomials, $P_k(x)$, be defined by

$$P_k(x) = \frac{(x - x_0) \times (x - x_1) \times \ldots \times (x - x_m)}{(x - x_k)}$$

where $k = 0, 1, 2, 3, \ldots, m$. For example, if $m = 2$,

$$P_0(x) = (x - x_1) \times (x - x_2) = x^2 - (x_1 + x_2) \times x + x_1 \times x_2$$
$$P_1(x) = (x - x_0) \times (x - x_2) = x^2 - (x_0 + x_2) \times x + x_0 \times x_2$$
$$P_2(x) = (x - x_0) \times (x - x_1) = x^2 - (x_0 + x_1) \times x + x_0 \times x_1$$

Write a program that will evaluate all the aforementioned $m + 1$ mth-degree polynomials for a specified x value. The answers should be stored in array P. Display your results at the terminal. Take as a test case $m = 2$, $x = 0.13$, $x_0 = 0.0$, $x_1 = 0.1$, and $x_2 = 0.2$.

*CP21. The Lagrange interpolation formula shown uses $m + 1$ mth-degree polynomials, $P_k(x)$, where $k = 0, 1, 2, \ldots, m$, to interpolate or extrapolate for values between or outside the given data points (x_k, y_k) of a table (see CP13 and CP20).

$$y = \sum_{k=0}^{m} \frac{y_k \times P_k(x)}{P_k(x_k)}$$

Use the table given in CP13 and a tenth-degree Lagrange interpolation formula to find the values of y at $x = 0.439$ and $x = 1.134$.

CP22. In the design of concrete members, the water-to-cement ratio used determines the eventual compressive strength that the member attains. For example, the accompanying table describes various mixes of cement and the compressive strength achieved after twenty-eight days. Write a program that will read the values for this table from an external file using arrays (each value in the table is an array member) and then duplicate the table at the terminal. The program should also be capable of linearly interpolating for strength values in the table when water-to-cement ratios are entered at the keyboard. Test your program with the value 5.3. Your output should appear as shown.

```
      I         I         I         I         I         I
1234567890123456789012345678901234567890123456789012345678901 2345
```

CONCRETE DESIGN

WATER–CEMENT RATIO GALLONS PER SACK OF CEMENT	COMPRESSIVE STRENGTH AFTER 28 DAYS CURING PSI
8.00	1500
7.50	2000
6.75	2500
6.00	3000
5.00	3750
4.00	5000

THE GIVEN WATER–CEMENT RATIO IS X.XX.

ITS ASSOCIATED STRENGTH VALUE IS XXXX.X.

*CP23. When a group of fasteners is loaded eccentrically (that is, the external load does not pass through the group's center), the total load experienced by each bolt or rivet of the group will consist of two load components. The first component is called the direct shear load, and is brought about by each bolt or rivet's equal participation in resisting the external load. (A shear load is any load that acts parallel to the member's resisting cross-sectional area.) The second load component is due to the resultant moment (see CP19) about the group's center, which is a result of the load's eccentricity. This loading is called the torsional shear load. If the fasteners have equal diameters, the total shear stress in any fastener i may therefore be found with the aid of the following formulas:

Shear load V in fastener i with N fasteners present is

$$(V_i)_x = \frac{F_x}{N} - \frac{T(y_i - y_C)}{J}$$

$$(V_i)_y = \frac{F_y}{N} + \frac{T(x_i - x_C)}{J}$$

The torque, T, in terms of the external load components, is

$$T = (x_L - x_C)F_y - (y_L - y_C)F_x$$

where (x_L, y_L) is the load application point. In these expressions, J is the unit area polar moment of inertia

$$J = \sum_{i=1}^{N} R_i^2$$

Also,

$$R_i^2 = (x_i - x_C)^2 + (y_i - y_C)^2$$

and

$$x_C = \frac{\sum_{i=1}^{N} x_i}{N}, \qquad y_C = \frac{\sum_{i=1}^{N} y_i}{N}$$

Total shear load is given by

$$V_i = [(V_i)_x^2 + (V_i)_y^2]$$

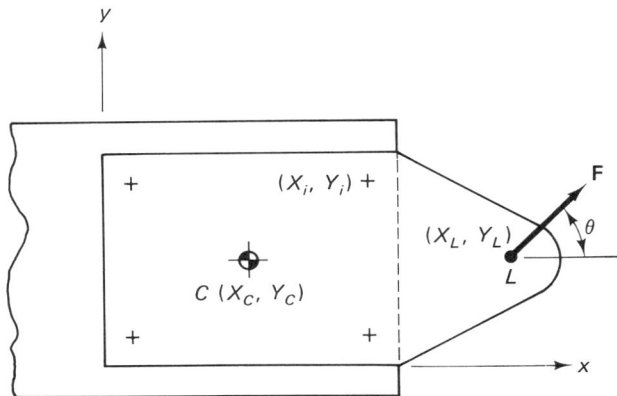

CP6.23

and for single shear, the shear stress, τ_i, is

$$\tau_i = \frac{V_i}{A}$$

where A is the cross-sectional area of the fastener.

Write a FORTRAN program that will solve for all the shear stresses present in an arbitrary-sized fastener group. Test your results when $N = 4$, $F = 0.2$ meganewton (MN), $A = 0.05$ meter, $x_L = 0.1$ meter, $y_L = 0.25$ meter, $\theta = 30$ degrees, and

Bolt	x	y (meters)
1	0	0
2	0.05	0
3	0	0.05
4	0.05	0.05

Shear stress is given in kilopascals (kPa). A kilopascal is equal to 1 kilonewton/m^2. How would you modify your program to include other external forces?

*CP24. A transformation of force components to a new coordinate system is effected by the following equation:

$$F_{\text{new}}(I) = \sum_{J=1}^{3} l(I, J) \times F(J) \qquad (I = 1, 2, 3)$$

where $F_{\text{new}}(I)$ are the new components, $F(J)$ are the old ones, and $l(I, J)$ are elements of the *real* direction cosine [**l**] matrix describing the rotation. Note that I and J equal to 1 denotes the new and old x directions, respectively, I equal to 2 and J equal to 3 stands for the new y and old z directions, and so on. Write a FORTRAN program that will read the x, y, and z components of a force vector from one data file, read the associated direc-

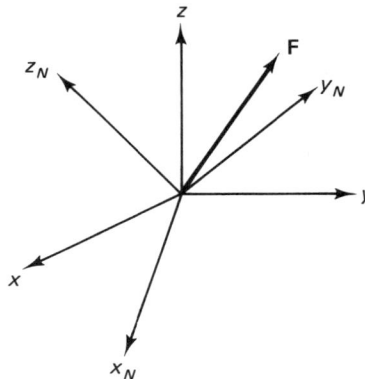

CP6.24

tion cosine matrix from another data file, and then calculate the new force components using a DO loop. Answers should be printed at the terminal in the following format:

```
        |        |        |        |        |        |
123456789012345678901234567890123456789012345678901234567890123 45
                TRANSFORMATION  OF  FORCE  COMPONENTS
    NEW  FORCE  COMPONENTS,  N              OLD  FORCE  COMPONENTS,  N
    ----------------------                  ----------------------
        FX  =  XX.XXXEXXX                       FX  =   -.121E+03
        FY  =  XX.XXXEXXX                       FY  =    .893E+03
        FZ  =  XX.XXXEXXX                       FZ  =    .513E+04
```

The direction cosine matrix is given as

$$[\mathbf{l}] = \begin{bmatrix} .0580 & .9665 & .2500 \\ -.8995 & -.0580 & .4330 \\ .4330 & -.2500 & .8660 \end{bmatrix}$$

*CP25. Transformation of plane stress components resulting from coordinate rotation is given by

$$ST_{\mathrm{N}}(I,J) = \sum_{K=1}^{z} \sum_{M=1}^{z} l(I,K) \times ST(K,M) \times l(J,M)$$

for I and J values ranging from 1 to 2. As in CP24, $[\mathbf{l}]$ represents the real direction cosine matrix, I and J the new x and y directions, and K and M the old x- and y-coordinates: an integer value of 1 stands for x; an integer value of 2 stands for y. Matrix $[\mathbf{ST}]_{\mathrm{N}}$ is the new plane stress matrix. Write a FORTRAN program that will read the elements of the plane stress matrix $[\mathbf{ST}]$ from one data file, read the associated direction cosine matrix from another data file, and then calculate the new transformed stress compo-

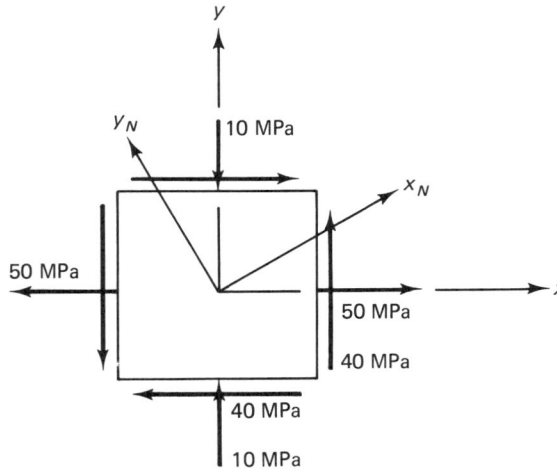

CP6.25

nents using nested DO loop structure. Answers should be printed at the terminal in the following format:

```
      |           |           |           |           |           |
1234567890123456789012345678901234567890123456789012345678901234 5
             TRANSFORMATION  OF  PLANE  STRESS
        OLD  STRESS  COMPONENTS,  MPA        NEW  STRESS  COMPONENTS,  MPA
        ------------------------            ------------------------------

               50.0  40.0                            F5.1  F5.1
        ST  =                               STN  =
               40.0-10.0                             F5.1  F5.1
```

The direction cosine matrix is as follows

$$[\mathbf{l}] = \begin{bmatrix} .8944 & .4472 \\ -.4472 & .8944 \end{bmatrix}$$

*CP26. Transformation of mass moments of inertia components (see CP16 and CP17) resulting from a coordinate rotation is given by

$$IN_N(I,J) = \sum_{K=1}^{3} \sum_{M=1}^{3} l(I,K) \times IN(K,M) \times l(J,M)$$

for I, J, and K values ranging from 1 to 3. As in CP25, $[\mathbf{l}]$ represents the real direction cosine matrix, I and J the new x-, y-, or z directions, and K and M the old x-, y-, or z-coordinates. $[\mathbf{IN}]_N$ is the new real mass moment of inertia matrix. Write a FORTRAN program that will read the elements of the mass moments of inertia matrix $[\mathbf{IN}]$ from one data file, read the associated direction cosine matrix from another data file, and then calculate the new transformed mass moments of inertia components using nested DO loop structure. Answers should be printed at the terminal in the following format:

```
      |           |           |           |           |           |
1234567890123456789012345678901234567890123456789012345678901234 5
             TRANSFORMATION  OF  INERTIA  COMPONENTS
        OLD  INERTIA  COMPONENTS            NEW  INERTIA  COMPONENTS
              KG-M*M                             KG-M*M
        ------------------------            ------------------------------
             450.   -60.   100.                 F6.1  F6.1  F6.1

        IN  =  -60.   500.    7.            INN  =  F6.1  F6.1  F6.1

             100.    7.   550.                   F6.1  F6.1  F6.1
```

The direction cosine matrix is as follows

$$[\mathbf{l}] = \begin{bmatrix} .0580 & .9665 & .2500 \\ -.8995 & -.0580 & .4330 \\ .4330 & -.2500 & .8660 \end{bmatrix}$$

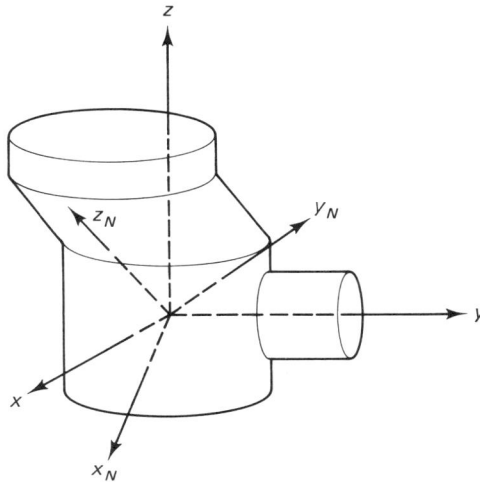

CP6.26

*CP27. Write a FORTRAN program that will read in two two-dimensional matrices, [A] and [B]. Matrix [A] will always be square, but [B] might be rectangular. Once the values have been entered, the program should then calculate the value of

$$[C] = [A]^t [B]$$

where $[A]^t$ denotes the transpose of [A]. Your program should include a built-in check to see that the number of columns of [A] equals the number of rows of [B] present. If a mismatch is present, an error message should be printed. Test your results on the following case:

$$[A] = \begin{bmatrix} 1 & 4 & 6 \\ 3 & 2 & 1 \\ 4 & 6 & 2 \end{bmatrix} \quad \text{and} \quad [B] = \begin{bmatrix} 3 & 1 \\ 1 & 0 \\ 2 & 4 \end{bmatrix}$$

*CP28. One way that a determinant can be evaluated is through the use of the summation formula

$$|\mathbf{B}| = \sum_{I=1}^{3} \sum_{J=1}^{3} \sum_{K=1}^{3} \delta(I, J, K) \times B(1, I) \times B(2, J) \times B(3, K)$$

where $\delta(I, J, K)$, the mathematical Levi-Civita density function, takes on the values

$$\delta(I, J, K) = 0 \quad \text{for } I = J, \text{ or } I = K, \text{ or } J = K$$

$$= +1 \quad \text{for an even permutation of } IJK, \text{ i.e., } 123, 231, \text{ or } 312$$

$$= -1 \quad \text{for an odd permutation, i.e., } 132, 213, \text{ or } 321$$

Write a computer program that will evaluate a 3×3 determinant using the foregoing formula. Test your program on the matrix

$$[\mathbf{B}] = \begin{bmatrix} 10 & 2 & 4 \\ 1 & 8 & 5 \\ 0 & 4 & 6 \end{bmatrix}$$

CHAPTER 7

Functions and Subroutines

The *X-29* research aircraft. The forward-swept–wing aircraft was designed to be aerodynamically unstable by the use of canards to enhance its maneuverability. The aircraft must use an on-board digital computer to continuously stabilize and control the plane's flight. *(Photo courtesy of NASA Ames-Dryden Flight Research Facility.)*

7.1 INTRODUCTION

In many programs there are groups of calculations that are performed repeatedly, for example, $\sin(x)$, e^x, and $n!$. To simplify program preparation, FORTRAN provides for a means of writing such groups as **subprograms.** The subprograms, compiled separately, can be called whenever needed by the **main program,** or even from another subprogram. Data is passed to the subprogram and later retrieved by the **calling program** after it is processed. Each time the calling program accesses the subprogram, the subprogram's statements are performed no differently than if they had been a permanent part of the calling program. By using these **program modules,** which are only coded once, instead of actually typing and inserting the identical code every time the same calculations are needed, a great amount of programming effort and time is saved.

There are other equally important reasons why a programmer should try to employ program modules or modular design whenever possible. For large programs, for example, it proves much easier to develop and debug separate smaller units than the entire program at once. Through the use of subprograms, modules can be written, tested, and corrected separately, even independently by co-workers, and then appended to the calling program. In engineering work, this is a very typical practice. Successful programs are modified, improved, and expanded upon continuously, often by different programmers. By using subprograms, the proven program parts never need be tampered with.

Furthermore, while a module can be used repeatedly within the same main program and any of its subprograms, it can also be employed at any time with entirely different programs. This is possible because subprogram modules are typically stored in separately named files. They lie dormant in the user's library until the operating system addresses and temporarily appends them, through a linking process, to a main calling program. (The exact details of how the linking is done depends on the local operating system.) As each subprogram module is a unique file, any module can be accessed by any program. It is possible to save subprograms in the same file as the main program, but this is seen as being ineffective, and it would preclude having the subprograms appended to other main programs.

The overall effect of using modules is to create simpler, smaller program units that are easier to read and understand. In addition, it allows the programmer to use a top-down, prioritized approach to algorithm development and program structure. With top-down design a programming task can be broken down into smaller problems, each being treated separately. At the highest level, the major or top tasks that a program must perform are outlined and developed. At the next level down, the necessary subtasks are defined and allotted to subprograms. Then at the next lower level, these subprograms might have to call upon other subprograms to obtain information before they can respond. Program development continues in this manner until no further clarification or resolution is necessary. Once the program is totally outlined, source code is written (see Section 1.4). It might be mentioned that coding is also usually completed in a top-down manner; that is, the calling programs are written before the subprograms. When proceeding downward in this manner, empty subprograms called **stubs** are introduced to substitute for the yet undeveloped subprogram code for testing purposes. (Calling programs written solely to test subroutines are called **drivers**.) Coding then continues from top to bottom until all the lowest-level units or modules are completed.

FORTRAN provides for two structural forms of subprograms: **functions** and **subroutines.** Function subprograms return only one value to the calling program, regardless of the length of the subprogram, whereas subroutines may return as many values as the program requires. The reader is also reminded of the intrinsic functions that are available in FORTRAN (see Section 2.6).

7.2 STATEMENT FUNCTIONS

The simplest FORTRAN function is only one statement long and therefore is called a **statement function.** Statement functions are not really a separate subprogram module, since they are included and compiled with the calling program. The statement function can be used to replace any single statement computation that appears repetitively in a program. Statement functions are defined just once: in the beginning of the program before the first executable statement and after any specification statements. They are called out by writing the function name and its arguments. The statement function, which is nonexecutable, has the syntax

name(argument list)=expression

Since the value of the expression is assigned to the function name (an address location), the name given to the function must be a legal FORTRAN name; that is, it can only consist of up to six letters or digits, and the first character must be a letter. This first letter is used for implicit typing. The **formal arguments** of the statement function are variable names, or **dummy arguments.** Their purpose is to inform the compiler at the very beginning of the program of the number, order (placement), and type of arguments to expect during program execution when the **actual arguments** are specified.

Argument types may be mixed. The function expression can contain constants, variable names, the formal arguments, mathematical operators, intrinsic functions, other previously defined statement-function names, and external function subprogram names. Arrays are not allowed. Some typical statement functions follow.

```
RADIUS(X,Y)=SQRT(X**2+Y**2)

HSTAG(P,V)=P/RHO+V**2/(2.0*G)

FUNK(Z,N)=(Z+3.2)**N
```

To explicitly type a function name, a type statement is used. For example,

```
REAL INERT
INERT(B,H)=B*H**3/12.0
```

When the statement function is called, only its name and the actual argument list is given. The actual argument list must agree in number and type with the formal argument list. The order of the actual argument list must match the given order defined in the statement function. Multiple calls to the same function using different arguments is common programming practice. In the following example, remember that the actual arguments must be assigned *before* the call to the statement function. This is not shown.

```
PROGRAM ONE
RADIUS(X,Y)=SQRT(X**2+Y**2)
  .

  .

  .
Z=RADIUS(X,Y)
  .

  .
T=3.0*RADIUS(R,S)
  .

  .
Z1=SQRT(RADIUS(1.2,ZETA))
  .

  .
END
```

In the next example two statement functions are used.

```
PROGRAM TWO
REAL INERT
INERT(B,H)=B*H**3/12.0
FUNK(Z,N)=(Z+3.2)**N-5.0*Z+A
A=31.2
  .

  .

  .
R=4.0*FUNK(2.4,3)
I=5
S=3.2
R1=FUNK(S,I)
PRINT *,INERT(2.1,1.2)
  .

  .

  .
END
```

Perhaps the most common error that a beginning programmer makes is placing the statement function after an executable statement. The compiler does not like that at all, and usually responds with a long list of error messages. The compiler may immediately spot the problem, or it may assume that the function name is really an array; therefore, the program is missing a DIMENSION statement as well as improperly including noninteger values for the element indices. It may even report both types of errors at once. The beginner is well advised to place the statement function properly in the program.

Sample Problem 7.1

Rework the solution to Sample Problem 3.1. This time use a statement function instead of an assignment statement to obtain your answer. Vary the temperature from 0 to 212 degrees Fahrenheit, in 1-degree-Fahrenheit increments, using DO loop structure.

Solution: See Figures 7.1a and 7.1b.

Algorithm for Sample Problem 7.1. This program converts Fahrenheit temperature to Celsius for temperatures ranging from 0 to 212 degrees in increments of one degree. A statement function CELS is used.

1. Define CELS
   CELS(FAHR)=(FAHR-32.0)*5.0/9.0
2. Do, with index FAHR ranging from 0.0 to 212.0, the following:
   a. Print FAHR,CELS(FAHR)

Figure 7.1a

```
        PROGRAM TEMP1

   ******************************************************************
   *                                                                *
   *   THIS PROGRAM CONVERTS FAHRENHEIT TEMPERATURE TO CELSIUS FOR   *
   *   TEMPERATURES RANGING FROM 0 TO 212 IN INCREMENTS OF ONE       *
   *   DEGREE. A STATEMENT FUNCTION CELS IS USED.                    *
   *                                                                *
   ******************************************************************

        REAL CELS,FAHR

        CELS (FAHR) = (FAHR-32.)*5./9.

        DO 1 FAHR =0.,212.

           PRINT *,' THE TEMPERATURE IN F DEGREES IS = ',FAHR
           PRINT 2,CELS(FAHR)

      1 CONTINUE

      2 FORMAT(' THE TEMPERATURE IN C DEGREES IS = ',F6.2)

        END

   RUN

     THE TEMPERATURE IN F DEGREES IS = 0.
     THE TEMPERATURE IN C DEGREES IS = -17.78
     THE TEMPERATURE IN F DEGREES IS = 1.
     THE TEMPERATURE IN C DEGREES IS = -17.22
     THE TEMPERATURE IN F DEGREES IS = 2.
     THE TEMPERATURE IN C DEGREES IS = -16.67
     THE TEMPERATURE IN F DEGREES IS = 3.
     THE TEMPERATURE IN C DEGREES IS = -16.11
                  .
                  .
                  .
```

Figure 7.1b

7.3 FUNCTION SUBPROGRAMS

In contrast with the one-line statement function, a complete function subprogram is an independent module with its own name and its own END statement. It is compiled separately from the main program and can contain many lines of code. When stored in a file separate from the main program, it can be appended (linked) to other subprogram units and the main program in any order—first, second, last. Since subprograms are compiled separately, they may repeat statement numbers and variable names that may also be found in the main program without producing program error; however, this is not recommended practice, since it can be confusing to other users.

A function subprogram is identified by the nonexecutable FUNCTION **statement.** The FUNCTION statement has the syntax

FUNCTION name(argument list)

The FUNCTION statement must be the very first statement of the subprogram. Since a value is assigned to the function name, the function name must follow FORTRAN syntax. Type is determined implicitly, using the first letter of the function's name. If explicit typing is desired, one can use the alternative FUNCTION statement

type FUNCTION name(argument list)

For example,

REAL FUNCTION INERT(X,Y)

INTEGER FUNCTION SAIL(V,W,X)

If the function is explicitly typed, however, the calling program must also have an explicit type statement bearing the function's name. For the foregoing two functions, the type statements would appear in the main program as

```
PROGRAM TYPE
REAL INERT
INTEGER SAIL
```

The function's formal argument list gives information to the compiler regarding the number, order, and type of arguments to expect. Argument types may be mixed and arrays are allowed. When using an array, the array must be dimensioned *both* in the subprogram and in the calling program. The actual array name may be different from its corresponding dummy array name. Only the actual array name is dimensioned in the calling program. The dimensions of the arrays in the subprogram and in the calling program do not have to be the same. It is recommended, however, that these dimensions be identical. (See Section 7.4.) It is also possible for a function subprogram to have no formal argument list. In that case, the FUNCTION statement appears as

FUNCTION name( )

where the parentheses must be included.

The statements immediately following the FUNCTION statement are the nonexecutable specification statements that dimension and type the subprogram's dummy arguments and local variables. These local variables are independent of the main program's variables. They may even bear the same name as variables in the calling program without inducing error. The other nonexecutable statements needed for subprogram processing—for example, DATA and local statement functions—are placed immediately following the subprogram's specification statements.

Within the function itself, the function name must be assigned a value, that is, it must appear to the left of an equal sign in some assignment statement. The single value that the function subprogram returns is borne by the function's name. A RETURN **statement** transfers control immediately back to the original calling statement in the main program or subprogram. Several RETURN statements and return paths are permissible in a subprogram. Finally, the very last statement of a subprogram must be an END statement. It also acts as a RETURN statement.

Sample Problem 7.2

Write a general function subprogram that will find the maximum absolute element value in a one-dimensional array having twelve elements. The array elements are to be passed from the calling program to the subprogram.

Solution: The algorithm and problem solution are shown in Figures 7.2a and 7.2b. Note that intrinsic functions can be used in subprograms. Note also that the function name is assigned different values at several points in the program. It is typed explicitly in the FUNCTION statement.

When calling the function from the main program or from another subprogram, the function's name is given along with the actual arguments. The computer accomplishes data "transfer" between the two independent program untis by assigning the

---

Algorithm for Sample Problem 7.2. This function subprogram finds the maximum absolute value present in a one-dimensional array having twelve elements.

1. Set MAXA=ABS(X(1)).
2. Do, with index I ranging from 2 to 12, the following:
   a. If ABS(X(I))>MAXA then
      i. Set MAXA=ABS(X(I))
3. Return to the calling program.

Figure 7.2a

```
      REAL FUNCTION MAXA (X)

***************************************************************
*                                                             *
*     THIS FUNCTION SUBPROGRAM FINDS THE MAXIMUM ABSOLUTE VALUE  *
*     PRESENT IN A ONE-DIMENSIONAL ARRAY HAVING 12 ELEMENTS.     *
*                                                             *
***************************************************************

      REAL X(12)

      MAXA=ABS(X(1))

      DO 1 I=2,12

          IF(ABS(X(I)).GT.MAXA) THEN
              MAXA=ABS(X(I))

          END IF

  1 CONTINUE

      RETURN

      END
```

Figure 7.2b

same address in memory to the formal arguments of the subprogram and their counterparts from the actual argument list. The one-to-one correspondence between formal and actual arguments is established by the order in which the variable names appear on their respective lists. Since information is shared between the program units via common memory addresses, it is important that the actual and formal arguments match in number, order, and type. In addition, if the subfunction has an array in the formal argument list, or if it is explicitly typed, associated specification statements must also be included for the function and the associated array in the calling program.

Function subprograms can be called repeatedly from the main program or from another subprogram. To call for a function program with a missing argument list, one writes the function's name followed by empty parentheses. For example,

```
Y=3.0*ZETA( )
```

Sample Problem 7.3

Modify the four-bar linkage problem of Chapter 6 (Sample Problem 6.3). This time write a main program that uses the function subprogram MAXA from Sample Problem 7.2 to find the maximum output angle PHI.

Solution: See the solution given in Figure 7.4. Notice how the function name is explicitly typed in the calling program as well. Also, observe how the function is called by using the function name and the actual arguments. Only the actual arguments are dimensioned or typed in the calling program.

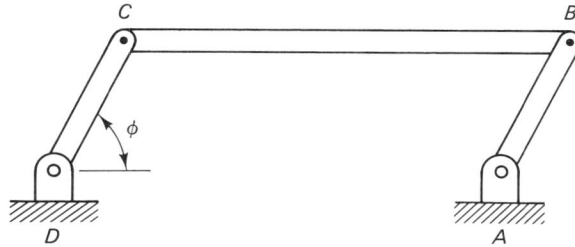

Figure 7.3

```
          PROGRAM CALLER
    ****************************************************************
    *                                                              *
    *   THE PROGRAM CALLS OUT THE FUNCTION MAXA FROM SAMPLE PROBLEM *
    *   7.2 TO FIND THE MAXIMUM OUTPUT ANGLE PHI.                   *
    *                                                              *
    ****************************************************************

          REAL PHI(12),MAXA

          OPEN(UNIT=7,FILE='PHILE')
          REWIND 7

          READ(7,*) (PHI(I),I=1,12)
          PRINT *,' THE LARGEST OUTPUT ANGLE IN THE ARRAY HAS THE',
        + ' VALUE ',MAXA(PHI)

          END

File Phile

24.4
38.8
61.6
86.3
110.0
129.0
131.7
108.3
78.9
55.3
37.8
26.1


RUN

    THE LARGEST OUTPUT ANGLE IN THE ARRAY HAS THE VALUE 131.7
```

Figure 7.4

Sample Problem 7.4

Write a program that will develop a table of sin *x* for the *x*-values ranging from 0 to 1 in increments of 0.1 radians. Use a Taylor's series expansion to approximate sin *x*, keeping only the first four terms of the expansion. See Sample Problems 5.2 and 5.3 for the analysis.

Solution: The algorithm and problem solution are shown in Figures 7.5a and 7.5b. Notice how statement address numbers in the calling program and subprogram are independent from each other (statement address number 1). Also note that the variable name N is independently used in each program and has a different value in the main program than in the two subprograms.

Algorithm for Sample Problem 7.4. This program develops the values of SIN(X) using a Taylor's series expansion with only the first four terms. A table is drawn up for the values of SIN(X) with X ranging from 0 to 1 in increments of 0.1. Results are compared to the SIN(X) value found using the intrinsic FORTRAN sine function.

1. Set N=4.
2. Print headings.
3. Do, with index X ranging from 0 to 1 in increments of 0.1, the following:
   a. Set SINE=0.
   b. Do, with index I ranging from 1 to N in increments of 1, the following:
      i. Calculate
         SINE=SINE+SERIES(X,I)
   c. Set Y=SIN(X)
   d. Print X,SINE,Y.

Subalgorithm SERIES(XTERM,J). The function subprogram SERIES(XTERM,J) calculates terms of the Taylor's series expansion for SIN(X).

1. Calculate
   N=2*J−1
   SERIES=((−1)**(J+1)*XTERM**N)/NFACT(N)
2. Return to the calling program.

Subalgorithm NFACT(N). The function subprogram NFACT(N) calculates N factorial.

1. Set NFACT=1.0.
2. Do, for index I ranging from 2 to N, the following:
   a. Calculate
      NFACT=NFACT*I
3. Return to the calling program.

Figure 7.5a

```
        PROGRAM SINX1

****************************************************************
*                                                              *
*  THIS PROGRAM DEVELOPS THE VALUES OF SIN(X) USING A TAYLOR'S  *
*  SERIES EXPANSION WITH ONLY THE FIRST FOUR TERMS. A TABLE IS  *
*  DRAWN UP FOR THE VALUES OF SIN(X) WITH X RANGING FROM 0 TO   *
*  1 IN INCREMENTS OF 0.1. RESULTS ARE COMPARED TO THE SIN(X)   *
*  VALUE FOUND USING THE INTRINSIC FORTRAN SINE FUNCTION.       *
*                                                              *
****************************************************************

        REAL X,SINE,SERIES,Y
        INTEGER I,N

        N=4

        PRINT 1
        PRINT 2
        PRINT 3
        PRINT 2

        DO 5 X=0.,1.,.1
            SINE=0.

            DO 6 I=1,N
                SINE=SINE+SERIES(X,I)

6           CONTINUE
            Y=SIN(X)

            PRINT 4, X,SINE,Y

5 CONTINUE

1 FORMAT(T27,'TABLE FOR SIN(X)',//)
2 FORMAT(65('-'))
3 FORMAT(T14,'X',TR18,'SINX',TR15,'SIN(X)')
4 FORMAT(T6,E14.7,2(TR6,E14.7))

        END


        REAL FUNCTION SERIES(XTERM,J)

        REAL NFACT,XTERM
        INTEGER N,J

        N=2*J-1
        SERIES=((-1)**(J+1)*XTERM**N)/NFACT(N)

        RETURN
        END


        REAL FUNCTION NFACT(N)

        INTEGER I,N

        NFACT=1.

        DO 1 I=2,N
            NFACT=NFACT*I
```

Figure 7.5b

```
        1 CONTINUE

          RETURN
          END


    RUN

                        TABLE FOR SIN(X)
```

```
    ----------------------------------------------------------------
              X                    SINX              SIN(X)
    ----------------------------------------------------------------
          .0000000E+00         .0000000E+00        .0000000E+00
          .1000000E+00         .9983342E-01        .9983342E-01
          .2000000E+00         .1986693E+00        .1986693E+00
          .3000000E+00         .2955202E+00        .2955202E+00
          .4000000E+00         .3894183E+00        .3894183E+00
          .5000000E+00         .4794255E+00        .4794255E+00
          .6000000E+00         .5646424E+00        .5646425E+00
          .7000000E+00         .6442176E+00        .6442177E+00
          .8000000E+00         .7173557E+00        .7173561E+00
          .9000000E+00         .7833258E+00        .7833269E+00
          .1000000E+01         .8414683E+00        .8414710E+00
```

Figure 7.5b   (cont.)

If it is desired to save a local variable when leaving a subprogram (for instance if one has a counter within the subprogram to keep track of how many times the function has been called), one can use the SAVE **statement.** (Note that DATA statements are only operable during compilation time, so that even their information will be missing the next time the subprogram is entered.) The SAVE statement is nonexecutable and has the syntax

SAVE variable list

It appears only in the subprogram. If the variable list is missing, all local subprogram variables will be saved. For example,

```
REAL FUNCTION INERT(X,Y)
INTEGER S,U,I
SAVE LOW
DATA LOW/3/
   .
   .
   .
LOW=LOW+1
   .
   .
   .
RETURN
END
```

Note that the SAVE statement, must appear before the DATA statement. When the function subprogram is called out again, the *last* value of LOW attained in the subprogram will become its initial value upon subprogram reentry. The DATA statement is not in effect, since it only initializes variables at compilation time.

7.4 SUBPROGRAMS AND ARRAYS

Because actual and formal arguments are linked via common address locations, programming errors may arise if the dimensions of the subprogram's arrays and those of the matching calling program's arrays are not identical. For one-dimensional arrays, it is important to check that enough subprogram array storage space has been allocated so that access to the calling program's data is not curtailed.

To avoid this problem, and to give the subprogram generality, **adjustable dimensions** may be employed for the subprogram's arrays. With adjustable dimensioning, the subprogram's array sizes are specified in a type or DIMENSION statement using integer variables rather than integer constants. The integer variable names are then included in the formal argument list. (The main program's arrays must still have their dimensions given by integer constants.) The desired sizes of the subprogram's arrays are later established during program execution by assigning integer values to corresponding actual arguments, that is, the arguments corresponding to the variable names that represent the subprogram's array sizes. *The subprogram consequently fits any calling program without need of formally changing dimension statements and then recompiling.* To show how this is accomplished, note how the following main program calls out the function VARY. The function VARY includes the dummy array X, which has an adjustable dimension L. During program execution, L is given the value 20.

```
PROGRAM MAIN
REAL A(20),VARY
   .
   .
   .
N=20
Y=VARY(A,N)
   .
   .
   .
END

REAL FUNCTION VARY(X,L)
REAL X(L)
   .
   .
   .
RETURN
END
```

The same procedure holds for multidimensional arrays. For example, if array X of FUNCTION VARY were two-dimensional, the following revisions would be needed:

```
PROGRAM MAIN1
REAL A(20,10),VARY
  .
  .
  .
N=20
M=10
Y=VARY(A,N,M)
  .
  .
END

REAL FUNCTION VARY(X,L,K)
REAL X(L,K)
  .
  .
RETURN
END
```

The calling program could also have specified N and M directly as 20 and 10, respectively, in the argument list.

The multidimensional dummy array may have dimensions that are not equal to the dimensions of the actual array. For multidimensional arrays, having the dummy and actual arrays differ in size could lead to serious programming errors. This is because multidimensional arrays are stored sequentially in the computer in a *column-wise* manner. A 3 × 2 array B, for instance, is actually stored in the computer in six consecutive positions (BS):

```
BS(1)=B(1,1)
BS(2)=B(2,1)
BS(3)=B(3,1)
BS(4)=B(1,2)
BS(5)=B(2,2)
BS(6)=B(3,2)
```

Note, then, what happens when an attempt is made to pass along a 2 × 2 sub-array of B to function ZERO in the next example.

```
PROGRAM EXAMPLE
REAL B(3,2),ZERO
  .
  .
  .
```

```
Y=ZERO(B,2,2)
   .
   .
   .
END

REAL FUNCTION ZERO(X,N,M)
REAL X(N,M)
   .
   .
   .
RETURN
END
```

Because the dummy array is dimensioned to receive a 2×2 array, it assigns storage space as follows:

```
XS(1)=X(1,1)
XS(2)=X(2,1)
XS(3)=X(1,2)
XS(4)=X(2,2)
```

When the information is passed (matched), XS expects and receives the first four elements of BS. However the first four elements of BS are not $B(1,1)$, $B(1,2)$, $B(2,1)$, and $B(2,2)$. They are $B(1,1)$, $B(2,1)$, $B(3,1)$, and $B(1,2)$. The values ultimately assigned to the X elements are

```
X(1,1)=B(1,1)
X(2,1)=B(2,1)
X(1,2)=B(3,1)
X(2,2)=B(1,2)
```

which was not intended at all. This type of hidden error is not picked up by the computer. The only way to avoid it is to always pass the complete array of the main program to the subprogram and to size the subprogram's array accordingly.

Sample Problem 7.5

Write a general function subprogram that finds the largest element value present in any two-dimensional array. Write a calling program to test the function using the furnace data given in Sample Problem 6.8.

Solution: The algorithm and problem solution are given in Figures 7.7a and 7.7b. Note how the variable dimensions M and N are also assigned to the function's DO loops. The maximum value of TEMP is returned to the calling program.

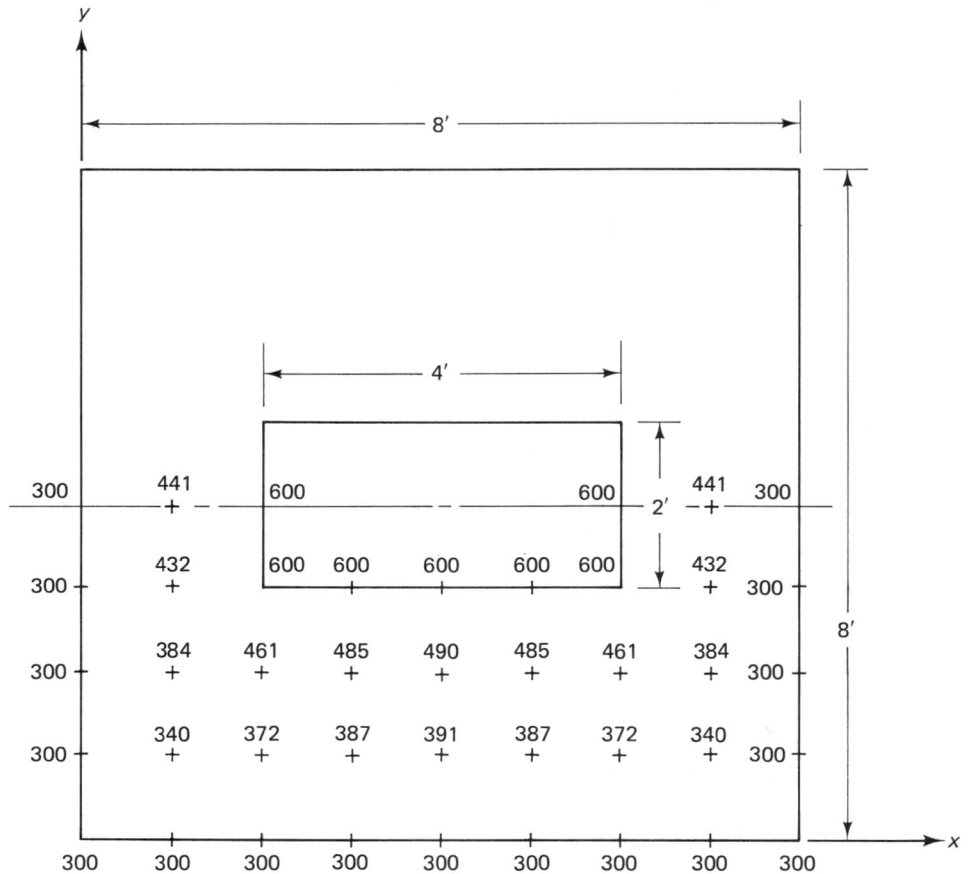

Figure 7.6

■ **Subalgorithm XMAX(X,M,N) of Sample Problem 7.5.** This function sub-program finds the largest element present in an M × N two-dimensional array X.

1. Set XMAX=X(1,1).
2. Set IMAX=JMAX=1.
3. Do, with index I ranging from 1 to M, the following:
   a. Do, with index J ranging from 1 to N, the following:
      i. If X(I,J)>XMAX then
         (a) Set XMAX=X(I,J)
4. Return to the calling program.

Figure 7.7a

```
       REAL FUNCTION XMAX(X,M,N)

*****************************************************************
*                                                               *
*   THIS FUNCTION SUBPROGRAM FINDS THE LARGEST ELEMENT PRESENT   *
*   IN AN M X N TWO-DIMENSIONAL ARRAY X.                        *
*                                                               *
*****************************************************************

       REAL X(M,N)
       INTEGER I,M,N

       XMAX=X(1,1)

       DO 1 I=1,M

           DO 2 J=1,N

               IF(X(I,J).GT.XMAX) THEN
                   XMAX=X(I,J)

               END IF

2          CONTINUE

1   CONTINUE

       RETURN
       END


       PROGRAM MAINS

       REAL TEMP(5,9),XMAX
       INTEGER I,J

       OPEN(UNIT=9,FILE='FURNACE')
       REWIND 9

       READ (9,*) ((TEMP(I,J),J=1,9),I=1,5)
C
C   TO AVOID SEARCHING THE OUTSIDE WALL TEMPERATURES, OR THE
C   TEMPERATURES INSIDE THE FURNACE ITSELF, THESE ARRAY VALUES
C   ARE SET EQUAL TO ZERO DURING THE SEARCH.
C
       DO 8 J=1,9,8

           DO 9 I=1,5

               TEMP(I,J)=0.

9          CONTINUE

8   CONTINUE

       REAL FUNCTION NFACT(N)

       INTEGER I,N

       NFACT=1.

       IF(N.GE.2) THEN

           DO 1 I=2,N
               NFACT=NFACT*I
```

Figure 7.7b

```
  1      CONTINUE

         END IF

         RETURN
         END
```

```
RUN
```

```
                        TABLE FOR SIN(X)
```

```
-------------------------------------------------------------
        X                    SINX                SIN(X)
-------------------------------------------------------------
   .0000000E+00          .0000000E+00          .0000000E+00
   .1000000E+00          .9983342E-01          .9983342E-01
   .2000000E+00          .1986693E+00          .1986693E+00
   .3000000E+00          .2955202E+00          .2955202E+00
   .4000000E+00          .3894183E+00          .3894183E+00
   .5000000E+00          .4794255E+00          .4794255E+00
   .6000000E+00          .5646424E+00          .5646425E+00
   .7000000E+00          .6442176E+00          .6442177E+00
   .8000000E+00          .7173557E+00          .7173561E+00
   .9000000E+00          .7833258E+00          .7833269E+00
   .1000000E+01          .8414683E+00          .8414710E+00
```

```
         DO 10 J=3,7

             DO 12 I=4,5
                 TEMP(I,J)=0.

  12         CONTINUE

  10     CONTINUE

         DO 22 J=2,8
             TEMP(1,J)=0.

  22     CONTINUE

         PRINT *,' THE LARGEST TEMPERATURE INSIDE THE WALL IS ',
       + XMAX(TEMP,5,9)

         END
```

```
RUN
```

```
THE LARGEST TEMPERATURE INSIDE THE WALL IS 490.
```

Figure 7.7b   (cont.)

7.5 SUBROUTINES

Subroutines are more useful than function subprograms because they are designed
to return more than one output value. Values are "passed" and "returned" through
the subroutine's argument list instead of using the subroutine's name (Since corre-

sponding formal and actual arguments share the same address location in memory, when a formal argument is changed in the subprogram, its corresponding actual argument immediately assumes and retains this new value. The subroutine's actual argument list is therefore capable of receiving and carrying information *back* to the calling program.) The subroutine's name is not of great importance here, since it is not used to represent a value. Though it must be a legal FORTRAN name, it has no type associated with it.

Since a subroutine is compiled separately, the very first statement of a subroutine must identify it to the compiler. The SUBROUTINE **statement** performs this task.

SUBROUTINE name(argument list)

The name of the subroutine does not appear again within the subprogram, and as mentioned, is not assigned any value. The formal argument list of the subprogram informs the compiler about the number, order, and type of arguments to expect. In addition to variables, the argument list may include names of dummy arrays. As with function subprograms, adjustable dimensioning is commonly used for array specifications. The actual dimensions of the array are then carried by the subprogram's arguments to the subroutine. When there are no arguments the SUBROUTINE statement has the reduced form

SUBROUTINE name

in which the argument list parentheses are not needed.

Since the subroutine is independent of the calling program, its statement numbers and local-variable names may be identical to those used in the calling program without error occurring. The subroutine must have an END statement, which also serves as a RETURN. Additional RETURN statements and paths are permitted. Explicit typing of the formal arguments and the local variables is performed in the beginning of the subroutine before any executable statement.

To reference a subroutine, one uses a CALL **statement.** A CALL statement has the form

CALL subroutine name(argument list)

The actual arguments must match the subroutine's formal arguments in number, order, and type. Actual arguments may be constants, variables, variable expressions, arrays, subscripted array elements, or the names of other functions or subroutine procedures. In the latter case, an EXTERNAL or an INTRINSIC **statement** must be included in the calling program (see Section 7.7).

Since the actual arguments have a dual purpose — passing input information and returning output results — some of the actual arguments may not at first be defined when the subroutine is called. These arguments (output arguments) only receive their values within the subroutine. Similarly, the arguments that are passed as input might be assigned new values within the subroutine and upon program return would be found permanently changed, since they now represent output variables.

Actual arguments and array names are typed and dimensioned in the calling program. If no arguments are present, the CALL statement takes the form

CALL subroutine name

in which no parentheses are needed. Subroutines without argument lists are commonly used for input and output operations as well as for program management.

Sample Problem 7.6

Write a subroutine that will take the position of an arbitrary point given in polar coordinates (r, θ) and convert it to the Cartesian coordinates (x, y) (see the analysis for Sample Problem 3.5). The angle θ is assumed to be given in radians. Also, write a driver program to verify the subroutine.

Solution: The algorithm and problem solution are shown in Figures 7.8a and 7.8b. The subroutine is listed first in the figure for clarity's sake.

*Sample Problem 7.7

A first-order ordinary differential equation having the form

$$\frac{dy}{dx} = f(y, x) \tag{7.1}$$

with the initial condition $y = y_0$ when x equals x_0, can be readily solved on the computer using numerical approximation. From calculus, the Taylor's series expansion for y is given by

$$y = y_0 + \left(\frac{dy}{dx}\right)_{x=x_0} \Delta x + \left(\frac{d^2y}{dx^2}\right)_{x=x_0} \frac{\Delta x^2}{2!} + \cdots \tag{7.2}$$

Algorithm for Sample Problem 7.6

1. Enter R,THETA (in degrees).
2. Calculate
   THETA=THETA/57.29578
3. Call PLCART(R,THETA,X,Y).
4. Print X,Y.

Subalgorithm PLCART(R,THETA,X,Y). This subroutine converts the position of a point given in polar coordinates R and THETA to Cartesian X- and Y-coordinates. THETA must be in radians.

1. Calculate
   X=R*COS(THETA)
   Y=R*SIN(THETA)
2. Return to the calling program.

Figure 7.8a

```
          SUBROUTINE PLCART(R,THETA,X,Y)

*********************************************************************
*                                                                 *
*    THIS SUBROUTINE CONVERTS THE POSITION OF A POINT GIVEN IN     *
*    POLAR COORDINATES R AND THETA TO CARTESIAN X-Y COORDINATES.   *
*    THETA MUST BE IN RADIANS.                                     *
*                                                                 *
*********************************************************************

          REAL X,R,THETA,Y

          X=R*COS(THETA)
          Y=R*SIN(THETA)

          RETURN
          END



          PROGRAM TEST

          REAL R,THETA,X,Y

          PRINT*,' ENTER R AND THEN THETA. THETA MUST BE IN DEGREES.'
          READ*,R,THETA

          THETA=THETA/57.29578

          CALL PLCART(R,THETA,X,Y)

          PRINT 1,X,Y

        1 FORMAT(' THE X AND Y POSITION OF THE GIVEN POINT IS (',
         + E11.4,',',E11.4,')')

          END

   RUN

     ENTER R AND THEN THETA. THETA MUST BE IN DEGREES.
   ? 2.0,45.0
     THE X AND Y POSITION OF THE GIVEN POINT IS (   .1414E+01,   .1414E+01)
```

Figure 7.8b

A linear approximation to the solution for y at $x = x_0 + \Delta x$ can then be written as

$$y = y_0 + \left(\frac{dy}{dx}\right)_{x=x_0} \Delta x \tag{7.3}$$

where the higher-order terms have been discarded.

Euler's method employs equation (7.3) successively, using each new y and x value found as initial values (x_0, y_0) for the next cycle of calculations. When using the computer, the solution to the differential equation is never found in functional form but is instead represented by a series of numerical values of y at succeeding x values.

The problem with this simple but powerful method is that there is numerical error introduced when discarding the higher-order terms. This error is called truncation error and increases as the step size Δx increases. The truncation error here is of the order $(\Delta x)^2$. If the step size is decreased too much, on the other hand, the computer itself will introduce significant roundoff error, since more calculations will be needed to cover the same range of *x* values. Still, many engineering problems are amenable to this method, and it is easy to program. The problem here is to write a general subroutine that uses Euler's method to solve a first-order differential equation. Show what a typical calling program might look like to solve the equation

$$\frac{dy}{dx} = y^3x + y$$

Solution: The algorithm and problem solution are shown in Figures 7.9a and 7.9b. Notice the FORTRAN WHILE structure used in the main program. The calculation proceeds until X is greater than XF decreased by DX/10. This is done to prevent roundoff problems that could lead to an extra value of X and Y being calculated. So that the main program and all its modules are as general as possible, a separate function subprogram DERIV is used to define the differential equation. Alternatively, a statement function in the main program could be used.

Algorithm for Sample Problem 7.7.   This main program is used as the driver for subroutine EULER.

1. Enter X0,Y0,DX,XF.
2. Print column headings.
3. Print X0,Y0.
4. Set X=X0.
5. While X<(XF−DX/10.0) then
   a. Call EULER(X0,Y0,X,Y,DX)
   b. Print X,Y
   c. Set X0=X,Y0=Y

Subalgorithm EULER(X0,Y0,X,Y,DX).   This subroutine uses Euler's method to solve a first-order differential equation. The differential equation is specified in the function subprogram DERIV.

1. Calculate
   ```
   YD=DERIV(X0,Y0)
   Y=Y0+YD*DX
   X=X+DX
   ```
2. Return to the calling program.

Subalgorithm DERIV(X0,Y0).   This function subprogram calculates the derivative value: the value of the given function at X0 and Y0.

1. Calculate
   ```
   DERIV=Y0**3*X0+Y0
   ```
2. Return to the calling program.

Figure 7.9a

```
      SUBROUTINE EULER(X0,Y0,X,Y,DX)

****************************************************************
*                                                              *
*  THIS SUBROUTINE USES EULER'S METHOD TO SOLVE A FIRST-ORDER  *
*  DIFFERENTIAL EQUATION. THE DIFFERENTIAL EQUATION IS         *
*  SPECIFIED IN THE FUNCTION SUBPROGRAM 'DERIV'.               *
*                                                              *
****************************************************************

      REAL YD,DERIV,X,X0,DX,Y0,Y

      YD=DERIV(X0,Y0)
      Y=Y0+YD*DX
      X=X0+DX

      RETURN
      END



      PROGRAM DIFFEQ

****************************************************************
*                                                              *
*     THIS MAIN PROGRAM IS USED AS THE DRIVER FOR SUBROUTINE   *
*     EULER.                                                    *
*                                                              *
****************************************************************

      REAL X0,Y0,DX,XF,X,Y

      PRINT*,'ENTER THE INITIAL VALUES OF X AND Y, IN THAT ORDER'
      READ*,X0,Y0
      PRINT*,'ENTER THE VALUE OF DX, THE INTEGRATION STEP SIZE.'
      READ*,DX
      PRINT*,'ENTER THE FINAL VALUE OF X, THE INDEPENDENT ',
     +  'VARIABLE.'
      READ*,XF

      PRINT 1
    1 FORMAT(T20,'X VALUES',TR24,'Y VALUES'/)
      PRINT 2,X0,Y0
    2 FORMAT(T14,E14.7,TR18,E14.7)

      X=X0
C
C TO PREVENT AN EXTRA VALUE BEING PRINTED DUE TO ROUNDOFF ERROR,
C XF IS REDUCED BY DX/10.
C

C
C BEGIN WHILE BLOCK STRUCTURE
C
    3 IF(X.LT.XF-DX/10.) THEN

      CALL EULER(X0,Y0,X,Y,DX)

      PRINT 2,X,Y

      X0=X
      Y0=Y
      GO TO 3
```

Figure 7.9b

```
      END IF

      END



      REAL FUNCTION DERIV(X0,Y0)

      REAL X0,Y0

      DERIV=Y0**3*X0+Y0

      RETURN
      END


   RUN

    ENTER THE INITIAL VALUES OF X AND Y, IN THAT ORDER
   ? 0.0,1.0
    ENTER THE VALUE OF DX, THE INTEGRATION STEP SIZE.
   ? 0.01
    ENTER THE FINAL VALUE OF X, THE INDEPENDENT VARIABLE.
   ? 0.2
                      X VALUES                    Y VALUES

                   .0000000E+00                 .1000000E+01
                   .1000000E-01                 .1010000E+01
                   .2000000E-01                 .1020203E+01
                   .3000000E-01                 .1030617E+01
                   .4000000E-01                 .1041252E+01
                   .5000000E-01                 .1052116E+01
                   .6000000E-01                 .1063220E+01
                   .7000000E-01                 .1074573E+01
                   .8000000E-01                 .1086187E+01
                   .9000000E-01                 .1098074E+01
                   .1000000E+00                 .1110247E+01
                   .1100000E+00                 .1122718E+01
                   .1200000E+00                 .1135502E+01
                   .1300000E+00                 .1148613E+01
                   .1400000E+00                 .1162070E+01
                   .1500000E+00                 .1175887E+01
                   .1600000E+00                 .1190085E+01
                   .1700000E+00                 .1204683E+01
                   .1800000E+00                 .1219702E+01
                   .1900000E+00                 .1235165E+01
                   .2000000E+00                 .1251097E+01
```

Figure 7.9b[1]   (cont.)

The subroutine EULER is called every time the FORTRAN WHILE loop is traversed. Each time through the loop, it receives the new values of X0, Y0, and DERIV or YD. The constant value of DX is also passed. It returns to the main program the values of X and Y, which are then printed out before X is tested again. The subroutine is listed first again for clarity's sake, but the order in which the main program and subprogram appear has no practical significance.

*Sample Problem 7.8

Another very popular and effective numerical procedure that can be used to solve a first-order ordinary differential equation is the Runge-Kutta fourth-order method. The Runge-Kutta fourth-order method has a truncation error on the order of dx^5. Typical engineering calculations have $dx = 0.001$, and as a result, the truncation error becomes almost insignificant. For the first-order differential equation

$$\frac{dy}{dx} = f(x, y) \tag{7.1}$$

with $y = y_0$ at $x = x_0$, the Runge-Kutta solution has the form

$$y = y_0 + \frac{K_1 + 2.0(K_2 + K_3) + K_4}{6.0} \tag{7.4}$$

where

$$K_1 = dx \times f(x_0, y_0)$$

$$K_2 = dx \times f\left(x_0 + \frac{dx}{2.0}, y_0 + \frac{K_1}{2.0}\right)$$

$$K_3 = dx \times f\left(x_0 + \frac{dx}{2.0}, y_0 + \frac{K_2}{2.0}\right) \tag{7.5}$$

$$K_4 = dx \times f(x_0 + dx, y_0 + K_3)$$

Write a subroutine that implements the Runge-Kutta fourth-order method to solve the differential equation of Sample Problem 7.7. Include in your solution a calling program for the subroutine.

Solution: The algorithm and problem solution are shown in Figures 7.10a and 7.10b. Again, the subroutine is listed first only for clarity's sake. Note how in the subprogram the variables K1, K2, K3, and K4 must be declared as real. Similarly, note how the function subprogram YD is called out four different times using modified arguments. The function uses the dummy arguments S and T so as not to confuse the user with the ambiguous variable names X and Y. The actual values for the variables X and Y are only defined at the end of the step interval. Again XF is adjusted to forestall round-off difficulties.

Algorithm for Sample Problem 7.8. This main program is used as the driver for subroutine RK4TH.

1. Enter X0,Y0,DX,XF.
2. Print column headings.
3. Print X0,Y0.
4. Set X=X0.
5. While X<(XF−DX/10.0) then
   a. Call RK4TH(X0,Y0,X,Y,DX)
   b. Print X,Y
   c. Set X0=X,Y0=Y

Figure 7.10a

Subalgorithm RK4TH(X0,Y0,X,Y,DX). This subroutine uses the Runge-Kutta fourth-order method to solve a first-order differential equation. The differential equation is specified in the function subprogram YD.

1. Calculate
   ```
   K1=DX*YD(X0,Y0)
   K2=DX*YD(X0+DX/2.0,Y0+K1/2.0)
   K3=DX*YD(X0+DX/2.0,Y0+K2/2.0)
   K4=DX*YD(X0+DX,Y0+K3)
   Y=Y0+(K1+2.0*(K2+K3)+K4)/6.0
   X=X0+DX
   ```
2. Return to the calling program.

Subalgorithm YD(S,T). This function subprogram calculates the derivative value: the value of the given function at X0 and Y0.

1. Calculate
   ```
   YD=T**3*S+T
   ```
2. Return to the calling program.

Figure 7.10a   (cont.)

```
         SUBROUTINE RK4TH(X0,Y0,X,Y,DX)

*********************************************************************
*                                                                  *
*     THIS SUBROUTINE USES THE RUNGE-KUTTA FOURTH-ORDER METHOD TO  *
*     SOLVE A FIRST-ORDER DIFFERENTIAL EQUATION. THE DIFFERENTIAL  *
*     EQUATION IS SPECIFIED IN THE FUNCTION SUBPROGRAM 'YD'.        *
*                                                                  *
*********************************************************************

         REAL K1,K2,K3,K4,DX,YD,Y0,X

         K1=DX*YD(X0,Y0)
         K2=DX*YD(X0+DX/2.,Y0+K1/2.)
         K3=DX*YD(X0+DX/2.,Y0+K2/2.)
         K4=DX*YD(X0+DX,Y0+K3)

         Y=Y0+(K1+2.*(K2+K3)+K4)/6.
         X=X0+DX

         RETURN
         END


         PROGRAM DIFFEQ

*********************************************************************
*                                                                  *
*     THIS MAIN PROGRAM IS USED AS THE DRIVER FOR SUBROUTINE       *
*     RK4TH.                                                        *
*                                                                  *
*********************************************************************
```

Figure 7.10b

```
      REAL X0,Y0,DX,XF,X,Y

      PRINT*,'ENTER THE INITIAL VALUES OF X AND Y, IN THAT ORDER'
      READ*,X0,Y0
      PRINT*,'ENTER THE VALUE OF DX, THE INTEGRATION STEP SIZE.'
      READ*,DX
      PRINT*,'ENTER THE FINAL VALUE OF X, THE INDEPENDENT '
     + 'VARIABLE.'
      READ*,XF

      PRINT 1
    1 FORMAT(T20,'X VALUES',TR24,'Y VALUES'/)
      PRINT 2,X0,Y0
    2 FORMAT(T14,E14.7,TR18,E14.7)

      X=X0
C
C DX/10. IS SUBTRACTED FROM XF TO AVOID POSSIBLE EXTRA OUTPUT
C CAUSED BY ROUNDOFF ERROR.
C
C
C BEGIN WHILE BLOCK CONSTRUCTION
C
    3 IF(X.LE.XF-DX/10.) THEN

          CALL RK4TH(X0,Y0,X,Y,DX)

          PRINT 2,X,Y

          X0=X
          Y0=Y
          GO TO 3

      END IF

      END


      REAL FUNCTION YD(S,T)

      REAL S,T

      YD=T**3*S+T

      RETURN
      END


RUN

 ENTER THE INITIAL VALUES OF X AND Y, IN THAT ORDER
? 0.0,1.0
 ENTER THE VALUE OF DX, THE INTEGRATION STEP SIZE.
? 0.01
 ENTER THE FINAL VALUE OF X, THE INDEPENDENT VARIABLE.
? 0.2
                    X VALUES                        Y VALUES

                  .0000000E+00                    .1000000E+01
                  .1000000E-01                    .1010101E+01
                  .2000000E-01                    .1020411E+01
                  .3000000E-01                    .1030938E+01
                  .4000000E-01                    .1041690E+01
```

Figure 7.10b   (cont.)

```
.5000000E-01                    .1052679E+01
.6000000E-01                    .1063914E+01
.7000000E-01                    .1075406E+01
.8000000E-01                    .1087167E+01
.9000000E-01                    .1099210E+01
.1000000E+00                    .1111547E+01
.1100000E+00                    .1124192E+01
.1200000E+00                    .1137161E+01
.1300000E+00                    .1150470E+01
.1400000E+00                    .1164136E+01
.1500000E+00                    .1178177E+01
.1600000E+00                    .1192612E+01
.1700000E+00                    .1207464E+01
.1800000E+00                    .1222755E+01
.1900000E+00                    .1238509E+01
.2000000E+00                    .1254754E+01
```

Figure 7.10b   (cont.)

7.6 COMMON BLOCKS OF STORAGE

It is also possible in FORTRAN to pass numeric values from a subprogram to a calling program, and vice versa, without having specific subprogram arguments carry the information. The alternative method is to create **common blocks** of shared storage area in memory for the main program and subprograms. Any information stored in a common location is immediately made available for direct use in each of the program units. As a consequence, common blocks present a means of reducing a lengthy and expanding subprogram argument list. They also offer the advantage that when a variable in a common block is changed in one subprogram unit, the new value is automatically made available to all other units sharing the same common storage. Updating is then performed without having to formally call the program units to pass the information via the argument list. For larger programs, with many subprograms, this is an important advantage to the programmer. It is not at all unusual in engineering programs to see a full page at the beginning of a program devoted solely to defining shared common storage for all the associated subroutines.

Common blocks are specified in memory by the COMMON **statement.** There are two types of common blocks. The **blank common block** and the **named common block.** To specify a blank common block one uses the syntax

COMMON list

where list is a list of variables and array names separated by commas. Any type of variable may be included in the list.

The COMMON statement is nonexecutable and must appear before any executable statement in all the subprograms or main program that are to share information. The variable names selected for the same memory location in each program unit may be different, but they must agree in number, type, and order. It is usually not good practice, though, to use different variable names for the same address in different program units, as this leads to difficulty in following the source code. Variable names on the list may not be used as subprogram or main program arguments. Arrays are also allowed on the list, but adjustable dimensioning is not permitted.

To dimension the array in each program unit, either a DIMENSION or type statement is used, or the dimensions for the array are specified along with the array name in the COMMON statement. An array's dimensions cannot be specified more than once. Multidimensional arrays are stored in a column-wise manner. An example of how blank common blocks are used follows.

```
PROGRAM MAIN
REAL X,Y,REGN,Z(10,10)
COMMON   X,Y,Z
  .
  .
  .
END

SUBROUTINE ARC(R,S)
REAL X,Y,R,S,Z(10,10)
COMMON X,Y,Z
  .
  .
  .
RETURN
END

REAL FUNCTION REGN(I)
REAL X,Y,Z(10,10)
INTEGER I
COMMON X,Y,Z
  .
  .
  .
RETURN
END
```

Information concerning variables X, Y, and any Z are shared directly between the main program and the subroutine and function subprograms. Not all subprograms may need to share information. In that case, the COMMON statement would only need to appear in those program units that required direct access.

There is only one blank common storage area allowed per program. If the program's variable list becomes too large, *named* common storage can be used. Named common storage is specified by the following COMMON statement:

$$\text{COMMON }/name_1/list_1/name_2/list_2 \ldots$$

or

$$\text{COMMON }/name_1/list_1$$

and

$$\text{COMMON }/name_2/list_2$$

Any number of named common storage blocks may be established. The chief advantage of using named common storage over blank common storage is that variable lists can be selectively shared between different subprograms. For example:

```
PROGRAM MAIN
REAL Z(10,10),X,Y,T,U
COMMON /ABLE/X,Y,Z
COMMON /BAKER/T,U
     .
     .
     .
END

SUBROUTINE ARC(R,S)
REAL Z(10,10),X,Y,R,S,A,B
COMMON /ABLE/X,Y,Z
COMMON /CHARLS/A,B
     .
     .
     .
RETURN
END

SUBROUTINE BETA(D)
REAL D,T,U,A,B
COMMON /BAKER/T,U
COMMON /CHARLS/A,B
     .
     .
     .
RETURN
END
```

The program MAIN shares information concerning variables X, Y, and Z with subroutine ARC, and information about variables T and U with subroutine BETA. Subroutines ARC and BETA share information about variables A and B. A variable name can only appear on one list. Variables appearing on matching lists must agree in number, type, and order, as before.

Variables that are named in common block lists cannot be initialized using DATA statements. A special FORTRAN subprogram containing no executable statements, called a **block data subprogram,** is employed for initialization purposes. A block data subprogram is specified by the BLOCK DATA **statement.** The statement has the form

BLOCK DATA

or

BLOCK DATA name

Only one block data subprogram may be unnamed in a program. A typical block data program might appear as follows:

```
BLOCK DATA ONE
REAL Z(10),X,Y,T,U
COMMON /ABLE/X,Y,Z
COMMON /BAKER/T,U
DATA X,Y,Z,T,U/3.2,5.7,10*2.0,2.3,6.1/
END
```

In this example, the variable lists in the two named common blocks ABLE and BAKER are initialized. Note that the last statement of the block data subprogram must be an END statement. Block data subprograms can contain IMPLICIT, PARAMETER, DIMENSION, type, COMMON, SAVE, EQUIVALENCE or DATA statements.

For the variable list associated with blank common storage, a special block data subprogram name is used. The BLOCK DATA statement then appears as

```
BLOCK DATA BLKDAT
```

The reader is cautioned that not all compilers allow for data initialization of blank common variables.

Sample Problem 7.9

The piston shown in Figure 7.11 moves to the right, after compressed air is suddenly admitted to the cylinder through the valve opening, driving an effective dynamic load of 450 pounds. Write a FORTRAN program, using subprogram modules, that solves for the displacement, velocity, and acceleration of the piston for times ranging from 0.0 to 0.8 seconds in 0.004-second increments (200 time steps). Assume that the piston starts from rest. The following data is known:

A_1 – area of orifice, 0.144 in.^{2}

g – gravity constant, 32.2 ft/sec^2

γ – specific heat ratio, 1.4

c_p – specific heat of air at constant pressure, 0.24 Btu/lbm-Rankine

c_v – specific heat of air at constant volume, 0.171 Btu/lbm-Rankine

T_0 – temperature inside the cylinder initially, 70.0° F

p_0 – atmospheric pressure, 14.7 lb/in.^{2}

V_0 – volume inside cylinder initially, 100.0 in.^{3}

R – gas constant, 53.35 ft-lb/Btu

J – conversion factor, 778.0 ft-lb/Btu

A – piston cross-sectional area, 56.0 in.^{2}

W – effective dynamic weight of piston plus load, 450.0 lb

C – valve orifice coefficient, 0.2 lbm/sec

Analysis: The purpose of this problem is to demonstrate, in a meaningful way, how subroutines are commonly used in engineering practice. When reviewing this problem, one should concentrate on the total structure of the solution and study how the subpro-

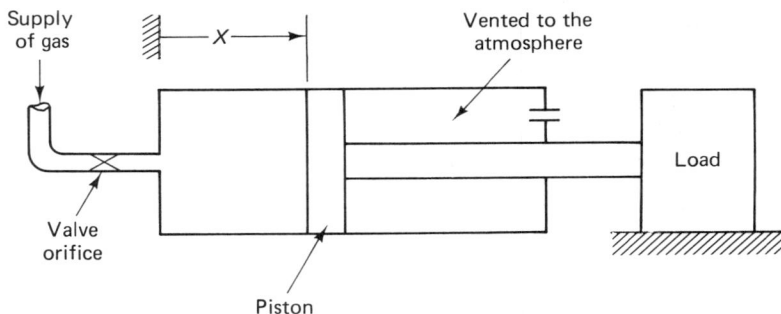

Figure 7.11

grams, with their named common blocks, interplay with one another. The governing physical relationships are of secondary importance to this goal. The behavior of a piston as it moves under the pressure of an expanding gas is a very difficult problem to solve analytically. The reader should bear this in mind when the following advanced material is presented.

To construct a model for the gas-expansion process, the analytic problem will be conveniently divided up into four parts: the flow description, the thermodynamics of the gas in the cylinder, the dynamics of the piston, and the numerical integration of the derived rate equations. By dividing the tasks, a subroutine can be written to describe and treat each part separately. A driver program will be used to monitor the solution and provide initial data, and a final subroutine will be added to handle output operations. (See Figure 7.12.) The following formulas are from advanced course work.

Flow Description

The flow of gas into the cylinder through the valve orifice is governed by the equation

$$\frac{dm}{dt} = C\left[\left(\frac{p}{p_1}\right)^{2.0/\gamma} - \left(\frac{p}{p_1}\right)^{(\gamma+1.0)/\gamma}\right]^{1/2}$$

where dm/dt (lbm/sec) is the rate of mass flowing into the cylinder, C (lbm/sec) is the valve orifice coefficient, p (lb/ft^2) is the internal pressure, p_1 (lb/ft^2) is the supply pressure, and γ is the specific heat ratio.

Thermodynamics of the Gas

There are several equations that govern the thermodynamic model. First, the gas must obey the perfect gas law:

$$p = \frac{mRT}{V}$$

where p is pressure (lb/ft^2), m is the mass (lbm), R is the gas constant (ft-lb/lbm-° R), T is the temperature on the Rankine scale, and V (ft^3) is the volume. The volume, in turn, can be found from

$$V = V_0 + A \times x$$

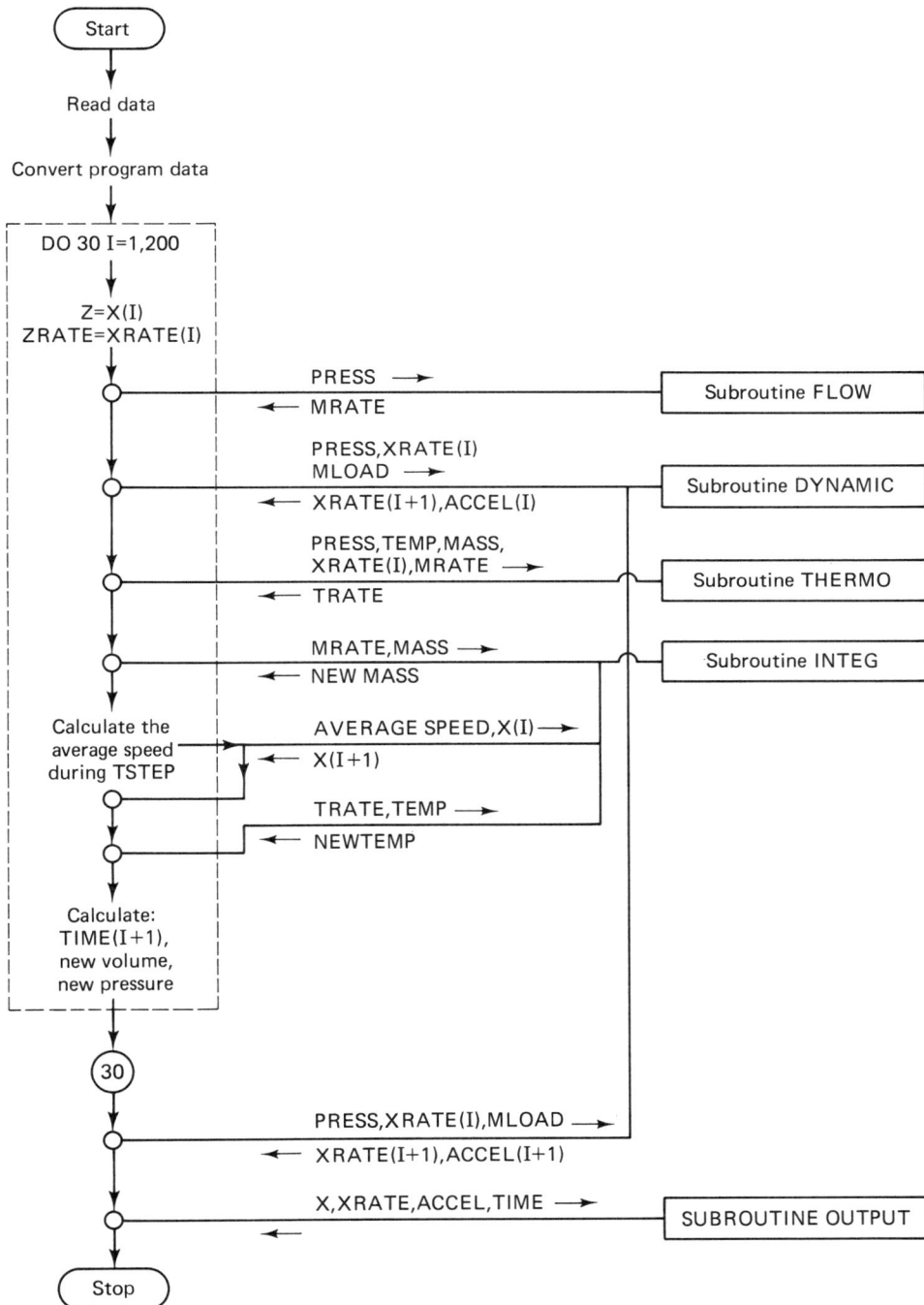

Figure 7.12

where V_0 (ft^3) is the initial volume, A (ft^2) is the cross-sectional area of the piston, and x (ft) is the piston displacement. The rate of temperature change is then given by

$$\frac{dT}{dt} = \frac{\left\{ -\dfrac{pA}{J}\dfrac{dx}{dt} + \dfrac{dm}{dt}(c_p \cdot T_1 - c_v \cdot T) \right\}}{m \cdot c_v}$$

dx/dt (ft/sec) is the velocity of the piston, J (778.0 ft-lb/Btu) is a thermodynamic conversion factor, T_1 is the supply temperature of the incoming gas in degrees Rankine, and c_v and c_p are the specific heats of the gas at constant volume and constant pressure, respectively.

The Dynamics of the Piston

The acceleration of the piston is expressed by

$$a = \frac{A(p - p_0)}{W/g}$$

W (lb) is the effective dynamic load on the piston, g (ft/sec^2) is the gravity constant, and p_0 (lb/ft^2) is the value of atmospheric pressure. Since the acceleration will be assumed constant during each time step of the study (0.004 sec), the speed of the piston will vary uniformly at an average rate:

$$v_{AV} = a \times \frac{t_{step}}{2.0}$$

where t_{step} (sec) is the time step. The final velocity, or rate of change of position at the end of each time step, is given by

$$v = v_{IN} + a \times t_{step}$$

where v_{IN} (ft/sec) is the initial velocity of the piston at the beginning of each particular time step, and v (ft/sec) is the velocity at the end of the step. (v will become the initial velocity for the next succeeding pass through the loop.)

Numerical Integration

With all the rates except velocity (for which we will use the average value, v_{AV}) assumed constant during an integration time step, the new values of each variable—temperature, mass, and displacement—can be found from

$$\text{final value} = \text{initial value} + \text{rate} \times t_{step}$$

Solution: The solution is shown in Figure 7.13. It is very common in large engineering companies to divide a problem like this up into subproblems, each with its separate subprogram. In this way, different departments can be assigned specific parts of the problem with which they are more familiar by reason of training and experience. It also allows for easier program expansion and maintenance. For example, in the subroutine FLOW, a suggestion was made for later expansion and improvement of the subprogram by calculating the valve-orifice coefficient directly instead of assuming a value. Similarly, there is room in the subroutine OUTPUT to incorporate code, or call out another established subroutine, to plot the results. It was for this reason that data was stored in an array and printed later collectively instead of being immediately displayed after being

```
            PROGRAM PISTON
  *************************************************************
  *                                                          *
  * THIS PROGRAM SOLVES FOR THE DISPLACEMENT, VELOCITY, AND  *
  * ACCELERATION OF A PISTON IN A GAS-DRIVEN MOTOR. THE      *
  * GAS, UNDER PRESSURE AND ELEVATED TEMPERATURE, IS         *
  * SUDDENLY ADMITTED TO THE CYLINDER WHEREAFTER IT EXPANDS  *
  * AND MOVES THE PISTON AND ITS ATTACHED LOAD. THE MAIN     *
  * PROGRAM MANAGES THE PROGRAM SOLUTION WHICH IS DEVELOPED  *
  * IN SEVERAL SUBROUTINES.                                  *
  *                                                          *
  *     A1 - AREA OF ORIFICE, .144 IN**2                     *
  *     G  - GRAVITY CONSTANT, 32.2 FT/S**2                  *
  *     GAMMA - SPECIFIC HEAT RATIO, 1.4                     *
  *     CP - SPECIFIC HEAT OF AIR AT CONSTANT                *
  *          PRESSURE, 0.24 BTU/LBM-RANKINE                  *
  *     CV - SPECIFIC HEAT OF AIR AT CONSTANT                *
  *          VOLUME, 0.171 BTU/LBM-RANKINE                   *
  *     T0 - TEMPERATURE INSIDE THE CYLINDER INITIALLY,      *
  *          70. FAHRENHEIT                                  *
  *     P0 - ATMOSPHERIC PRESSURE, 14.7 PSI                  *
  *     V0 - VOLUME INSIDE CYLINDER INITIALLY, 100 IN**3     *
  *     M0 - INITIAL MASS OF AIR INSIDE CYLINDER, LBM        *
  *     R  - GAS CONSTANT, 53.35 FT-LB/LBM-RANKINE           *
  *     J  - CONVERSION FACTOR, 778 FT-LB/BTU                *
  *     A  - PISTON CROSS-SECTIONAL AREA, 56 IN**2           *
  *     WEIGHT - WEIGHT OF PISTON PLUS LOAD, 450 LB          *
  *     P1 - PRESSURE OF SUPPLY AIR, PSI                     *
  *     T1 - TEMPERATURE OF SUPPLY AIR, FAHRENHEIT           *
  *     TFINAL - FINAL TIME, S                               *
  *     X - DISPLACEMENT OF THE PISTON, FT                   *
  *     XRATE - VELOCITY OF THE PISTON, FT/S                 *
  *     ACCEL - ACCELERATION OF THE PISTON, FT/S**2          *
  *     TSTEP - INTEGRATION TIME STEP, S                     *
  *     MASS - MASS OF THE GAS INSIDE THE CYLINDER, LBM      *
  *     TEMP - TEMP OF THE GAS INSIDE THE CYLINDER, RANKINE  *
  *     PRESS - PRESSURE OF THE GAS INSIDE THE CYLINDER, PSF *
  *     FLOWC - NOZZLE COEFFICIENT, LBM/S                    *
  *     VOL - VOLUME OF AIR INSIDE THE CYLINDER.             *
  *     MRATE - MASS FLOW RATE, LBM/S                        *
  *     TRATE - TEMPERATURE RATE CHANGE, RANKINE             *
  *                                                          *
  *************************************************************


            REAL X(201),XRATE(201),ACCEL(201),TIME(201)
            REAL MASS,M0,MRATE,MASSN,MLOAD,J
            REAL A1,GAMMA,P1,R,T1,G,TSTEP,A,P0,CP,CV,WEIGHT,T0,V0
            REAL TFINAL,PRESS,TEMP,Z,ZRATE,ZRATEN,ZN,VOL,ZTEMP,
           + TRATE,ZACCEL
            INTEGER I

            COMMON /ALPHA/A1,GAMMA,P1,R,T1
            COMMON/BETA/G
            COMMON /LAMBDA/MLOAD,TSTEP
            COMMON /DELTA/A,P0
            COMMON /OMEGA/CP,CV,J

            DATA A1,G,GAMMA,CP,CV,T0,P0,V0,R,J,A,WEIGHT/.144,32.2,
           + 1.4,.24,.171,70.,14.7,100.,53.35,778,56.,450./

            PRINT*,' ENTER THE VALUE OF SUPPLY PRESURE IN PSI'
            READ*,P1
```

Figure 7.13

```
      PRINT*,' ENTER THE VALUE OF SUPPLY SOURCE',
     + ' TEMPERATURE IN DEGREES FAHRENHEIT'
      READ*,T1
      PRINT*,' ENTER THE FINAL TIME FOR YOUR ANALYSIS'
      READ*,TFINAL
C
C CONVERTING INCH UNITS TO FEET, FAHRENHEIT TO RANKINE, AND
C PSI TO PSF (POUNDS PER SQUARE FOOT).
C
      P0=P0*144.
      V0=V0/1728.
      A1=A1/144.
      A=A/144.
      P1=P1*144.
      T1=T1+459.67
      T0=T0+459.67
C
C ESTABLISHING THE INITIAL CONDITIONS AND CALCULATING
C PROBLEM CONSTANTS. TSTEP IS THE TIME STEP FOR THE
C SOLUTION.
C
      TSTEP=TFINAL/200.
      X(1)=0.
      XRATE(1)=0.
      TIME(1)=0.
      PRESS=P0
      TEMP=T0
      MLOAD=WEIGHT/G
      M0=P0*V0/(R*T0)
      MASS=M0
C
C THE SOLUTION IS NOW BEGUN.
C
      DO 30 I=1,200

         Z=X(I)
         ZRATE=XRATE(I)

         CALL FLOW(PRESS,MRATE)

         CALL DYNAMIC(PRESS,ZRATE,ZRATEN,ZACCEL)

         XRATE(I+1)=ZRATEN
         ACCEL(I)=ZACCEL

         CALL THERMO(PRESS,TEMP,TRATE,ZRATE,MASS,MRATE)

         CALL INTEG(MRATE,MASSN,MASS)

         MASS=MASSN
C
C NOTICE THAT SINCE THE ACCELERATION IS ASSUMED CONSTANT
C OVER THE TIME STEP, THE RATE CHANGES UNIFORMLY. AN
C AVERAGE VALUE IS USED.
C
         ZRATE=(ZRATE+ZRATEN)/2.

         CALL INTEG(ZRATE,ZN,Z)

         X(I+1)=ZN

         CALL INTEG(TRATE,ZTEMP,TEMP)
```

Figure 7.13   (cont.)

```
            TEMP=ZTEMP
            TIME(I+1)=TSTEP*(I)
            VOL=V0+A*X(I+1)
            PRESS=MASS*R*TEMP/VOL

      30 CONTINUE
C
C   THE LAST VALUE OF ACCEL IS NOW CALCULATED. NOTE THAT
C   AFTER LEAVING THE LOOP, I EQUALS 201.
C
            CALL DYNAMIC(PRESS,ZRATE,ZRATEN,ZACCEL)

            ACCEL(I)=ZACCEL

            CALL OUTPUT(X,XRATE,ACCEL,TIME)

            END



            SUBROUTINE FLOW(PRESS,MRATE)

      *************************************************************
      *                                                           *
      *   THIS SUBROUTINE CALCULATES THE MASS RATE OF FLOW,       *
      *   MRATE. PROGRAM INPUT IS CYLINDER GAS PRESSURE AND       *
      *   TEMPERATURE. OTHER CONSTANTS ARE OBTAINED FROM THE      *
      *   CALLING PROGRAM VIA NAMED COMMON BLOCKS.                *
      *                                                           *
      *************************************************************

            REAL MRATE,A1,GAMMA,P1,R,T1,G,FLOWC,COEFF,FLOWT,PRESS
            COMMON /ALPHA/A1,GAMMA,P1,R,T1
            COMMON/BETA/G

            FLOWC=0.2
C
C   IN A MORE THOROUGH ANALYSIS,THE VALUE OF THE NOZZLE
C   COEFFICIENT CAN BE OBTAINED BY THE FOLLOWING EXPRESSION.
C   NOTE HOW THE SUBROUTINE CAN BE LATER EXPANDED.
C
C   FLOWC=A1*P1*SQRT(2.*G*GAMMA/((GAMMA-1)*R*T1)
C
            COEFF=(GAMMA+1)/GAMMA
            FLOWT=FLOWC*((PRESS/P1)**(2./GAMMA)-(PRESS/P1)
          + **COEFF)
            MRATE=FLOWT**.5

            RETURN
            END



            SUBROUTINE DYNAMIC(PRESS,ZRATE,ZRATEN,ZACCEL)

      *************************************************************
      *                                                           *
      *   THIS SUBROUTINE CALCULATES THE PISTON ACCELERATION AND  *
      *   NEW PISTON VELOCITY, OR RATE. PROGRAM INPUT IS GAS      *
      *   PRESSURE AND PISTON VELOCITY (ZRATE). OTHER INPUT COMES *
      *   FROM THE CALLING PROGRAM VIA NAMED COMMON BLOCKS.       *
      *                                                           *
      *************************************************************

            REAL MLOAD,ZACCEL,A,PRESS,P0,ZRATEN,ZRATE,TSTEP
            COMMON /LAMBDA/MLOAD,TSTEP
            COMMON /DELTA/A,P0
```

Figure 7.13   (cont.)

```
      ZACCEL=A*(PRESS-P0)/MLOAD
      ZRATEN=ZRATE+ZACCEL*TSTEP

      RETURN
      END


      SUBROUTINE THERMO(PRESS,TEMP,TRATE,ZRATE,MASS,MRATE)

**************************************************************
*                                                          *
*    THIS SUBROUTINE CALCULATES THE RATE OF TEMPERATURE    *
*    RISE INSIDE THE CYLINDER AND THE GAS PRESSURE. PROGRAM *
*    INPUT IS GAS TEMPERATURE, PISTON VELOCITY (ZRATE),    *
*    PISTON DISPLACEMENT (Z), AND THE MASS OF CYLINDER GAS *
*    ACCUMULATED. OTHER PROGRAM INPUT IS FROM NAMED COMMON *
*    BLOCKS.                                               *
*                                                          *
**************************************************************

      REAL MASS,MRATE,TRATE1,PRESS,A,ZRATE,J,CP,CV,T1,TEMP,
     + TRATE
      COMMON /DELTA/A,P0
      COMMON /ALPHA/A1,GAMMA,P1,R,T1
      COMMON /OMEGA/CP,CV,J

      TRATE1=-PRESS*A*ZRATE/J +MRATE*(CP*T1-CV*TEMP)
      TRATE=TRATE1/(CV*MASS)

      RETURN
      END


      SUBROUTINE INTEG(RATE,ZOUT,ZIN)

**************************************************************
*                                                          *
*    THIS SUBROUTINE FINDS THE NEXT VALUE OF THE VARIABLE   *
*    ZIN ASSUMING THAT ITS RATE OF CHANGE IS CONSTANT OVER  *
*    THE TIME-STEP INTERVAL.                                *
*                                                          *
**************************************************************

      REAL ZOUT,ZIN,RATE,TSTEP
      COMMON/LAMBDA/ MLOAD,TSTEP

      ZOUT=ZIN+RATE*TSTEP

      RETURN
      END


      SUBROUTINE OUTPUT(X,XRATE,ACCEL,TIME)

**************************************************************
*                                                          *
*    THIS SUBROUTINE FORMS A TABLE FOR THE OUTPUT VALUES OF *
*    PISTON DISPLACEMENT, SPEED, AND ACCELERATION.         *
*                                                          *
**************************************************************

      REAL X(201),XRATE(201),ACCEL(201),TIME(201)
      INTEGER J
```

Figure 7.13   (cont.)

```
                PRINT 5
                PRINT 6

                DO 7 J=1,201
                    PRINT 9, TIME(J),X(J),XRATE(J),ACCEL(J)

           7 CONTINUE
C
C   A PLOT PROGRAM COULD BE ADDED HERE LATER. NOTE AGAIN THE
C   ADVANTAGE OF USING PROGRAM SUBMODULES.
C
C       CALL PLOT(X,XRATE,ACCEL,TIME)
C
           5 FORMAT(T9,'T-S',T23,'X-FT',T37,'XRATE-FT/S',T57,
           + 'ACCEL-FT/S*S')
           6 FORMAT(T9,3('-'),T23,4('-'),T37,10('-'),T57,12('-')/)
           9 FORMAT(T2,E10.3,T17,E10.3,T37,E10.3,T59,E10.3)

             RETURN

             END


     RUN

       ENTER THE VALUE OF SUPPLY PRESURE IN PSI
     ? 27.0
       ENTER THE VALUE OF SUPPLY SOURCE TEMPERATURE IN DEGREES FAHRENHEIT
     ? 100.0
       ENTER THE FINAL TIME FOR YOUR ANALYSIS
     ? 0.045
               T-S            X-FT          XRATE-FT/S        ACCEL-FT/S*S
               ---            ----          ----------        ------------

           .000E+00       .000E+00          .000E+00           .000E+00
           .225E-03       .000E+00          .000E+00           .525E+00
           .450E-03       .133E-07          .118E-03           .105E+01
           .675E-03       .665E-07          .355E-03           .158E+01
              .
              .
              .
```

Figure 7.13   (cont.)

calculated. The number of points allowed for a plot or table of results was arbitrarily set in this problem to 200. This quantity, in turn, determines what the time step, TSTEP, should be. The user enters the final time (TFINAL) desired for the run, and the step size is then calculated by

```
TSTEP=TF I NAL/200.0
```

The program begins with the user entering the supply pressure and temperature of the gas, and the final time desired for the run. (refer to Figure 7.13). Program data is then converted to standard units, and the program next enters the DO loop that will develop the problem's solution. Each pass through the loop advances the solution one time step. The loop is repeated 200 times. Note that at the beginning of each repetition, for each new I value, initial values of displacement, X(I), and velocity, XRATE(I), are assigned to the "scratch pad" variables Z and ZRATE. This is done so that the original variables, X(I) and XRATE(I), will not have to be modified during the loop's execution. Only Z and ZRATE will change.

The subroutine FLOW, having the arguments PRESS and MRATE (pressure and rate of mass flow), is then called out. Notice that not all of the defined constants in the main program are needed in FLOW. The named common blocks permit selective access to P1 and GAMMA. If the subroutine is expanded later, the values of A1, G, R, and T1 are already provided for.

Following the return of control to the main program, the subroutine DYNAMIC is called. It receives values of PRESS and ZRATE and returns values of ZACCEL (acceleration) and ZRATEN (new velocity) via the argument list. Named common blocks make the variables MLOAD, TSTEP, A, P0, and V0 accessible to DYNAMIC. It should be noted that subprograms FLOW and DYNAMIC are independent of one another. Subroutine DYNAMIC could be called before subprogram FLOW. When subprogram DYNAMIC returns control to the calling program, the output values for velocity (XRATE) and acceleration (ACCEL) are updated to the values they will assume at the next time step (the future values).

With MRATE now known from the subprogram FLOW, the program THERMO can be called out. (The subroutine FLOW must be executed before the program THERMO). Although the future value of the velocity (XRATE) is now available from the program DYNAMIC, note that it is the current value of the velocity that is used to calculate the rate of temperature change. Named common blocks provide access to the variables A1, GAMMA, P1, R, T1, CP, CV, J, A, and P0. Program input arguments are PRESS, TEMP, MRATE, MASS, and ZRATE. The subprogram output argument is TRATE.

The last part of the loop requires three successive calls to the subroutine INTEG to integrate the rate value over one time step. The rate value is assumed constant for that time step and then updated and changed in the subroutines during the next pass through the loop. This subroutine has RATE and ZIN (initial value) as input arguments, and returns ZOUT to the calling program. The named common block LAMBDA provides the value of TSTEP. The calling program first calculates the new mass in the cylinder (MASS) using INTEG. Then two more calls are made to determine the future values of displacement (X(I + 1)) and temperature (TEMP). Note that an average velocity or rate is used to get displacement, since the velocity's value is assumed to change linearly during each time step. Once these values are found, TIME is incremented to match the new values, and then the pressure (PRESS) and volume (VOL) are updated before the loop is repeated.

When emerging from the DO loop, the values of all the variables except acceleration are available for printout. The acceleration has still not been calculated at TFINAL; therefore, one last call must be made to the program DYNAMIC. After this is accomplished, the subroutine OUTPUT is called. This program is designed to print the problem's results in tabular form. It should be possible later to insert a plotting routine, or a call to a plotting subprogram.

This problem and its solution clearly illustrate the need for thorough testing of a program before it is released to co-workers. For program verification, use a supply pressure of 27 psi and an input temperature of 100° F. Use a final time of 0.8 seconds. If a final time of 1.8 seconds or larger is used, the program will not run. This is not caused by a programming error but results from the assumption that the rates of change of the variables are constant during each time interval. Not surprisingly, this approximation becomes more and more inaccurate as the time step increases, and eventually the program fails.

Choosing too large a time step causes further problems as well. The very nature of the gas admission and expansion process leads to a very rapid buildup of pressure inside the cylinder in a very short period of time. If the time step is taken too large, the sudden pressure buildup will be missed, and only its averaged effects will be felt later on in time. The rule in program development, therefore, is to verify; and after verification is complete, verify again. As can be seen here, it is sometimes not easy to anticipate program problems that might arise and thereby design an adequate test.

7.7 THE EXTERNAL AND INTRINSIC STATEMENTS

It was mentioned earlier in Section 7.5 that function or subroutine names can be used as actual arguments in a subroutine CALL statement. This allows, for example, even the choice of numerical procedures to be a variable in a subprogram. When a user-defined function or subroutine is used as an actual argument, an EXTERNAL specification statement must be given in the calling program. The nonexecutable EXTERNAL statement has the form

EXTERNAL subprogram list

and must precede all the executable statements and statement function definitions of the calling program. If instead of a user-defined function, specific names of the intrinsic functions given in Appendix B are used as actual arguments (not the generic names), an INTRINSIC specification statement must be given. It too is nonexecutable and must precede the first executable statement of the calling program. It has the form

INTRINSIC function list

Sample Problem 7.10

Write a main program that will use either Euler's method or the Runge-Kutta fourth-order method (see Sample Problems 7.7 and 7.8) to solve a first-order differential equation. Information as to which routine to use is to be provided by the user at the terminal when the program is executed. The function representing the first-order differential equation is user defined.

Solution: The algorithm and problem solution are given in Figures 7.14a, 7.14b, 7.15a and 7.15b. The main program needs an EXTERNAL statement that references subroutines EULER and RK4TH because they are actual arguments of the subroutine SOLVE. The subroutine SOLVE calls out either of these two numerical methods. To accomplish this, the subroutine name of EULER or RK4TH, an actual argument, replaces the formal argument F of SOLVE when the program is run. Which of the two subroutines is selected depends on the value of L entered by the user. (For Euler's method L equals 0; for L equal to 1, the Runge-Kutta method is used.) Note also that two WHILE blocks are included: one for Euler's method and one for the Runge-Kutta method. Once more, XF is adjusted to prevent roundoff difficulties for the displayed output.

■ **Algorithm for Sample Problem 7.10.** This program uses either Euler's method or the Runge-Kutta fourth-order method to solve a first-order equation. Information as to which procedure is to be used is provided by the operator at the terminal: "0" is for Euler's method, and "1" is for the Runge-Kutta method. The first-order equation is specified in the function subprogram YD.

1. Specify common DX.
2. Specify RK4TH and EULER as external procedure.
3. Enter X0,Y0,DX,XF.
4. Enter L.
5. Print column headings.
6. Print X0,Y0.
7. Set X=X0.
8. While X<(XF−DX/10.0) and L=0 then
   a. Call SOLVE(EULER,X0,Y0,X,Y)
   b. Print X,Y
   c. Set X0=X,Y0=Y
9. While X<(XF−DX/10.0) and L=1 then
   a. Call SOLVE(RK4TH,X0,Y0,X,Y)
   b. Print X,Y
   c. Set X0=X,Y0=Y

Figure 7.14a

```
          PROGRAM MAIN

     *****************************************************************
     *                                                               *
     *    THIS PROGRAM USES EITHER EULER'S METHOD OR THE RUNGE-KUTTA  *
     *    FOURTH-ORDER METHOD TO SOLVE A FIRST-ORDER EQUATION.        *
     *    INFORMATION AS TO WHICH PROCEDURE IS TO BE USED IS PROVIDED *
     *    BY THE OPERATOR AT THE TERMINAL: "0" IS FOR EULER'S METHOD, *
     *    AND "1" IS FOR THE RUNGE-KUTTA METHOD. THE FIRST-ORDER      *
     *    EQUATION IS SPECIFIED IN THE FUNCTION SUBPROGRAM YD.        *
     *                                                               *
     *****************************************************************

          REAL DX,X0,Y0,XF,X,Y
          INTEGER L
          EXTERNAL RK4TH,EULER
          COMMON DX

          PRINT*,'ENTER THE INITIAL VALUES OF X AND Y, IN THAT ORDER'
          READ*,X0,Y0
          PRINT*,'ENTER THE VALUE OF DX, THE INTEGRATION STEP SIZE.'
          READ*,DX
          PRINT*,'ENTER THE FINAL VALUE OF X, THE INDEPENDENT ',
         + 'VARIABLE.'
          READ*,XF
```

Figure 7.14b

```
      PRINT*,'IF YOU WANT EULER''S METHOD ENTER 0. IF YOU WANT',
    + ' THE RUNGE-KUTTA METHOD ENTER 1.'
      READ*,L
      PRINT 1
    1 FORMAT(T20,'X VALUES',TR24,'Y VALUES'/)
      PRINT 2,X0,Y0
    2 FORMAT(T14,E14.7,TR18,E14.7)

      X=X0

C
C   BEGIN WHILE BLOCK STRUCTURE.
C

    3 IF(X.LT.(XF-DX/10.).AND.L.EQ.0) THEN

          CALL SOLVE(EULER,X0,Y0,X,Y)

          PRINT 2,X,Y

          X0=X
          Y0=Y
          GO TO 3

      END IF

C
C   BEGIN WHILE BLOCK STRUCTURE.
C
    4 IF(X.LT.(XF-DX/10.).AND.L.EQ.1) THEN

          CALL SOLVE(RK4TH,X0,Y0,X,Y)

          PRINT 2,X,Y

          X0=X
          Y0=Y

          GO TO 4

      END IF

      END


RUN

 ENTER THE INITIAL VALUES OF X AND Y, IN THAT ORDER
? 0.0,1.0
 ENTER THE VALUE OF DX, THE INTEGRATION STEP SIZE.
? 0.01
 ENTER THE FINAL VALUE OF X, THE INDEPENDENT VARIABLE.
? 0.2
 IF YOU WANT EULER'S METHOD ENTER 0. IF YOU WANT THE RUNGE-KUTTA METHOD
? 1
                      X VALUES                      Y VALUES

                    .0000000E+00                  .1000000E+01
                    .1000000E-01                  .1010101E+01
                    .2000000E-01                  .1020411E+01
                    .3000000E-01                  .1030938E+01
                    .4000000E-01                  .1041690E+01
                    .5000000E-01                  .1052679E+01
                    .6000000E-01                  .1063914E+01
```

Figure 7.14b   (cont.)

```
.7000000E-01                    .1075406E+01
.8000000E-01                    .1087167E+01
.9000000E-01                    .1099210E+01
.1000000E+00                    .1111547E+01
.1100000E+00                    .1124192E+01
.1200000E+00                    .1137161E+01
.1300000E+00                    .1150470E+01
.1400000E+00                    .1164136E+01
.1500000E+00                    .1178177E+01
.1600000E+00                    .1192612E+01
.1700000E+00                    .1207464E+01
.1800000E+00                    .1222755E+01
.1900000E+00                    .1238509E+01
.2000000E+00                    .1254754E+01
```

Figure 7.14b   (cont.)

Subalgorithm SOLVE(F,X0,Y0,X,Y). This subroutine is used to manage the solution, calling out the proper numerical subroutine method F.

1. Call F(X0,Y0,X,Y).

2. Return to the calling program.

Subalgorithm RK4TH(X0,Y0,X,Y). This subroutine uses the Runge-Kutta fourth-order method to solve a first-order differential equation. The differential equation is specified in the function subprogram YD.

1. Specify common DX.

2. Calculate

```
K1=DX*YD(X0,Y0)
K2=DX*YD(X0+DX/2.0,Y0+K1/2.0)
K3=DX*YD(X0+DX/2.0,Y0+K2/2.0)
K4=DX*YD(X0+DX,Y0+K3)
Y=Y0+(K1+2.0*(K2+K3)+K4)/6.0
X=X0+DX
```

3. Return to the calling program.

Subalgorithm EULER(X0,Y0,X,Y). This subroutine uses Euler's method to solve a first-order differential equation. The differential equation is specified in the function subprogram YD.

1. Specify common DX.

2. Calculate

```
Y=Y0+YD(X0,Y0)*DX
X=X0+DX
```

3. Return to the calling program.

Figure 7.15a

■ **Subalgorithm YD(S,T).** This function subprogram calculates the derivative value: the value of the given function at X0 and Y0.

1. Calculate
   YD=T**3*S+T
2. Return to the calling program.

Figure 7.15a   (cont.)

```
      SUBROUTINE SOLVE(F,X0,Y0,X,Y)

      REAL X0,Y0,X,Y

      CALL F(X0,Y0,X,Y)

      RETURN
      END



      SUBROUTINE RK4TH(X0,Y0,X,Y)
*****************************************************************
*                                                               *
*    THIS SUBROUTINE USES THE RUNGE-KUTTA FOURTH-ORDER METHOD TO *
*    SOLVE A FIRST-ORDER DIFFERENTIAL EQUATION. THE DIFFERENTIAL *
*    EQUATION IS SPECIFIED  IN THE FUNCTION SUBPROGRAM 'YD'.     *
*                                                               *
*****************************************************************
      REAL K1,K2,K3,K4,DX,X0,X,Y0,YD
      COMMON DX

      K1=DX*YD(X0,Y0)
      K2=DX*YD(X0+DX/2.,Y0+K1/2.)
      K3=DX*YD(X0+DX/2.,Y0+K2/2.)
      K4=DX*YD(X0+DX,Y0+K3)
      Y=Y0+(K1+2.*(K2+K3)+K4)/6.
      X=X0+DX

      RETURN
      END



      SUBROUTINE EULER(X0,Y0,X,Y)
*****************************************************************
*                                                               *
*    THIS SUBROUTINE USES EULER'S METHOD TO SOLVE A FIRST-ORDER  *
*    DIFFERENTIAL EQUATION. THE DIFFERENTIAL EQUATION IS         *
*    SPECIFIED  IN THE FUNCTION SUBPROGRAM 'YD'.                 *
*                                                               *
*****************************************************************
      REAL X0,X,Y0,Y,DX,YD
      COMMON DX
```

Figure 7.15b

```
Y=Y0+YD(X0,Y0)*DX
X=X0+DX

RETURN
END
REAL FUNCTION YD(S,T)

REAL T,S

YD=T**3*S+T

RETURN
END
```

Figure 7.15b   (cont.)

REVIEW

The two types of subprograms included in FORTRAN are functions and subroutines. FORTRAN functions include the FORTRAN-supplied intrinsic functions, user-defined statement functions, and function subprograms. The arguments of functions can be constants, variables, variable expressions, arrays (except for statement functions), or other functions.

Statement functions are one statement long. They are not a separate program module but appear in the beginning of the calling program as a statement, before the program's first executable statement and after any specification statements. To specify a statement function one uses

name(argument list)=expression

The name can be any legal FORTRAN name. The formal (dummy) argument list informs the compiler of the number, order, and type of variable names that will follow. When the statement function is referenced or called, the actual arguments are given. They must match in number, order, and type with the formal arguments.

Function subprograms allow more than one statement to be included in the function definition. The function subprogram is an independent unit that is compiled separately. It has its own name, variable, statement numbers, and END statement. It is appended to the calling program either before or after it. The function subprogram is identified by a FUNCTION statement,

FUNCTION name(argument list)

or, when explicitly typed,

type FUNCTION name(argument list)

The FUNCTION statement must be the very first statement of any subprogram. Since a value is assigned to the function name, the name must follow FORTRAN syntax. Type is determined implicitly. If the function is explicitly typed, the calling program must also have an explicit type statement that refers to the function's name.

The formal argument list may include arrays. When arrays are used they must be dimensioned both in the subprogram and in the calling program. The calling program dimensions the actual array name, and the function subprogram dimensions the formal array name. It is recommended that these dimensions match. If no arguments are given, the FUNCTION statement appears as

FUNCTION name( )

in which the parentheses must be included.

Since subprograms are independent, all local variables are lost when the subprogram is exited; that is, control returns to the calling program. To retain the values of specified local variables, one uses the nonexecutable SAVE statement

SAVE variable list

which should appear in the subprogram before the first executable statement, or DATA statement. If the variable list is missing, all local variables will be saved.

To provide subprogram generality when arrays are present, one can employ adjustable dimensions. With adjustable dimensioning, the subprogram's array sizes are specified by integer variables rather than by integer constants. The ultimate desired size of the subprogram's array is later passed as an actual argument, along with the array's name.

Subroutines are more versatile than functions, since they are designed to return more than one output value. Values are passed and returned through the subroutine's arguments rather than through its name. Each subroutine argument is assigned a dual role: when the subroutine is called, it passes information into the subroutine, and later, when the subroutine is exited, it bears information out.

Subroutines are independent program units that are compiled separately. The first statement of a subroutine must be the SUBROUTINE statement:

SUBROUTINE name(argument list)

The subroutine's name is not assigned a value or type. It must be a legal FORTRAN name, however. The formal argument list may include variables and arrays. As with function subprograms, adjustable dimensioning is commonly used. If no arguments are present, the SUBROUTINE statement appears as

SUBROUTINE name

The subroutine's local variables and statement numbers are independent from the calling program. Like the function subprogram, subroutines must have a final END statement, which also serves the purpose of a RETURN. Additional RETURN statements and paths are permissible.

To reference a subroutine one uses the CALL statement,

CALL subroutine name(argument list)

or, when no arguments are present,

CALL subroutine name

When a user-defined function or subroutine is used as an argument, a non-executable EXTERNAL statement must appear in the beginning of the calling program, before the first executable statement and any statement function definitions. The EXTERNAL statement appears as

EXTERNAL subprogram list

If the function is from the FORTRAN-supplied library, the nonexecutable INTRINSIC statement is used. It has the form

INTRINSIC function list

It is also possible to pass numeric values between program units by assigning variable and array names to a common block of shared storage. Common blocks of memory are specified by the nonexecutable blank COMMON statement,

COMMON list

or the named COMMON STATEMENT,

COMMON /name$_1$/list$_1$/name$_2$/list$_2$ . . .

in which the list contains variable and array names separated by commas. COMMON statements are specification statements and must appear before the first executable statement in all the program units that are to share information. The variables on the list must agree in number, type, and order in each COMMON statement. Matching arrays on the list must have identical dimensions within each of their programs. Adjustable dimensioning is not allowed. Only one blank common storage area is allowed per program. Any number of named common storage areas are allowed, however.

Variable names that appear in common block lists cannot be initialized via DATA statements. A special FORTRAN subprogram containing no executable statements, called a block data subprogram, is employed for that purpose. A block data subprogram is specified by the BLOCK DATA statement. It has the form

BLOCK DATA

or

BLOCK DATA name

Only one unnamed block data subprogram is allowed per program.

KEY TERMS

actual argument	FUNCTION **statement**
adjustable dimensions	INTRINSIC **statement**
blank common block	**main program**

BLOCK DATA **statement**
block data subprogram
CALL **statement**
calling program
common block
COMMON **statement**
drivers
EXTERNAL **statement**
formal (dummy) argument
function

named common block
program module
RETURN **statement**
SAVE **statement**
statement function
stub
subprogram
subroutine
SUBROUT I NE **statement**

REVIEW QUESTIONS

1. Write arithmetic statement functions that will perform the following tasks (all arguments and function values are assumed to be real):

   (a) Find the volume of a cone.

   $$V = \frac{\pi r^2 h}{3}$$

   (b) Find the pressure of a gas.

   $$p_2 = p_1 \left(\frac{T_2}{T_1}\right)^{3.5}$$

   (c) Find the amount of decibels.

   $$db = 20 \log_{10}\left(\frac{V_2}{V_1}\right)$$

   (*Hint:* $\log_{10} x$ = ALOG10(X).)

2. With reference to the arithmetic statement functions of Problem 1, write a FORTRAN program assignment statement that will call these functions out using the following arguments:
   (a) h = H, r = R
   (b) T_1 = T1, T_2 = T2, p_1 = P1
   (c) V_2 = V2, V_1 = 0.5*V2

3. Assuming that the function subprogram shown is appended to your main program, what values are given to Z when the following assignment statements are used. Unless otherwise stated, take I = 3, R = 3.7, and S = 2.1.

```
      REAL  FUNCTION  SUM(N,A,B)
      Y=0.0
      DO 7  I=1,N
          DEN=1.0
          DO 8  K=1,I
              DEN=K*DEN
 8        CONTINUE
          Y=Y+(-1)**(I+1)*(2*I-1)/DEN
```

```
7  CONTINUE
   SUM=A*Y-B
   END
```

(a) Z=3.0*SUM(I,R,S)

(b) Z=4.0-SUM(4,3.0,2.0)

(c) Z=R/SUM(I,R,S)

4. Assuming that the accompanying FORTRAN subroutine is appended to your program, what values will be printed out if the following CALL statements are used? Unless otherwise stated, take $R(1, 1) = 2.0$, $R(1, 2) = 3.0$, $R(2, 1) = 4.0$, $R(2, 2) = 7.0$, $U = 2.5$, and $T = 1.0$.

```
SUBROUTINE IMSLL(A,B,C,D)
REAL A(2,2),B(2)
DEN=A(1,1)*A(2,2)-A(2,1)*A(1,2)
B(1)=(C*A(2,2)-D*A(1,2))/DEN
B(2)=(D*A(1,1)-C*A(2,1))/DEN
END
```

(a) CALL SUBROUTINE IMSLL(R,S,T,U)
    PRINT *,S

(b) CALL SUBROUTINE IMSLL(R,S,3*T,4.2)
    PRINT *,S

5. Fill in the missing part of the statements in the following subroutine, which was written to return a value of the polynomial

$$y = a_1 + a_2x + a_3x^2 + \ldots + a_{n+1}x^n$$

for *any* n, x and set of *a* values.

```
SUBROUTINE POLY(___ ,___ ,___ ,___)

REAL _____
Y=0.0

DO 3 I=1,__

   Y=_____
3 CONTINUE
   END
```

*6. Fill in the missing part of the statements in the following subroutine. The subroutine was written to find the transpose of matrix [**A**], which is $n \times m$. The transpose is returned via matrix [**B**] (output), which is $m \times n$.

```
SUBROUTINE TRANS(____ ,_____ ,_____ ,_____ )

REAL _____ ,_____

DO 7 I=__ ,___

   DO 8 J=__ ,___

      _____
```

```
8       CONTINUE
7  CONTINUE
   END
```

7. For the following program with its two attached subroutines, indicate what values will be printed at the terminal. Be specific as to what is contained in each record line, and clearly show what order the record lines will appear in.

```
PROGRAM TEST
COMMON A,B
COMMON /MALL/RED,BLUE
A=3.1
B=4.0
F1=4.2
CALL ABBA(F1,GLUE,2)
PRINT *,RED,BLUE
PRINT *,GLUE
END

SUBROUTINE ABBA(R,S,N)
COMMON S1,S2
COMMON /EDIT/U,K
K=1
U=R*S1
TRIO=U**2
CALL BETA(TRIO,BETT)
PRINT *,K,U
PRINT *,BETT
S=N*BETT
END

SUBROUTINE BETA(OVAL,ZERO)
COMMON /EDIT/Z1,INDEX
COMMON /MALL/GREEN,YELLOW
GREEN=15.0
YELLOW=7.0
INDEX=INDEX+1
ZERO=INDEX*EXP(-OVAL/40.0)
Z1=Z1/2.0
END
```

COMPUTER PROJECTS

CP1. Write a function subprogram and driver that receives the x- and y-coordinates of a point and then determines which quadrant the point lies in.

CP2. The probability (percentage of time) that n objects will arrive at a manufacturing assembly-line station in a certain time interval is given approximately by the *Poisson function:*

$$P(n) = \frac{A^n e^{-A}}{n!}$$

where A is the average number of objects arriving in the interval. Write a function subprogram that receives values of A and n from the main program and returns the probability of the event occurring. Test your program using values of n from 0 to 15 in increments of 1. Assume that the average value of A is 4.6.

CP3. Write a function subprogram and driver that will solve for the root of a given equation using the bisection method. The equation should be defined with a statement function. Start with the incremental search method and then, after the root has been isolated, converge to the result using the bisection method. (See Sample Problem 4.9.)

CP4. For manufactured parts that are shaped as right-circular cylinders, write a subroutine that receives from the calling program values of radius, height, and material density and returns values of surface area and mass. Test your subroutine with a driver.

CP5. Write a subroutine program and driver that will solve for the roots of a quadratic equation (see Sample Problem 4.6).

CP6. Write a subroutine that converts any millimeter data value from 1 to 800, passed as an argument, to feet and inches. Any remainder should be sorted into half-, quarter-, eighth-, sixteenth-, and thirty-second–inch parts, in that order. Results found should be returned to the calling program. Test your subroutine with a driver.

CP7. Write a subroutine and driver that will find the area lying under a given curve using Simpson's rule (see Sample Problem 5.10). Use an array to pass the curve's ordinates to the subroutine.

CP8. Write a subroutine that will linearly interpolate or extrapolate between data points given in a table. Use arrays to store the x and y values. Use the tabular values given for CP12, Chapter 4, to test your program.

CP9. Write two subroutines and a driver that will solve for the roots of a given equation. The equation and its derivative should be defined by statement functions. Start with the incremental search method and then, after the root has been isolated, converge to the result using the Newton-Raphson method. Both methods should be programmed as subroutines. (See Sample Problem 4.10.)

*CP10. Use Euler's method (see the given subroutine of Sample Problem 7.7) to solve the following differential equation:

$$\frac{dy}{dx} = -4y$$

with the initial condition that y equals 2.0 when x is 0. Let dx equal 0.1 in your solution and solve for y at x equal to 1.0. Compare your results with the exact solution,

$$y = 2e^{-4x}$$

*CP11. Use the Runge-Kutta fourth-order method (see Sample Problem 7.8) to solve the differential equation that is given in CP10. Use the same initial condition and dx values. Solve for y at x equal to 1.0. Compare your results with the exact solution,

$$y = 2e^{-4x}$$

*CP12. Write a subroutine and driver that will add two matrices together. Assume that the matrices are two-dimensional and are passed from the calling program. Results are sent back as a subroutine argument.

*CP13. Write a subroutine and driver that will multiply two matrices together. Assume that the matrices are two-dimensional and are passed from the calling program. Results are sent back as a subroutine argument.

CP14. Write a subroutine that will use a bubble-sort algorithm to rearrange an arbitrary-size, one-dimensional array in descending order. The newly sorted array should be printed at the terminal. Test your subprogram with a driver on the following data file, which is read by the calling program and passed to the subroutine. Results should be sent back to the calling program. (See Sample Problem 6.7.)

```
DATA FILE:
119.1
328.7
258.9
567.3
111.4
236.7
258.9
239.4
478.8
455.6
498.7
388.4
525.8
174.6
348.2
278.3
190.7
594.4
498.7
239.9
398.5
```

CP15. Write a subroutine that will receive a two-dimensional square array A from the calling program and search for the largest element in column 1. Once this element is found, the row that contains it should be interchanged with the first row of the array so that the largest element will now appear in position A(1, 1). Send the rearranged array back to the calling program. With a suitable driver, demonstrate that your program works.

CP16. Write a subroutine that will receive from the calling program a two-dimensional square array A and search for the largest element in each column. Once each element is found, its value should be interchanged with the value currently located in that column's main diagonal position. Send the rearranged array back to the calling program. With a suitable driver, demonstrate that your program works.

CP17. An electric motor has the torque-speed characteristic curve shown in the figure. Write a subroutine that will determine the output torque delivered by the motor, T_M, given the desired motor's operating speed, n, as subroutine input. Use linear interpolation. Test your program by finding the output torque of the motor at 830 rpm. If an attached load can be described by the relationship

$$T_{load} = K \times n$$

find what K must equal to attain the desired operating speed.

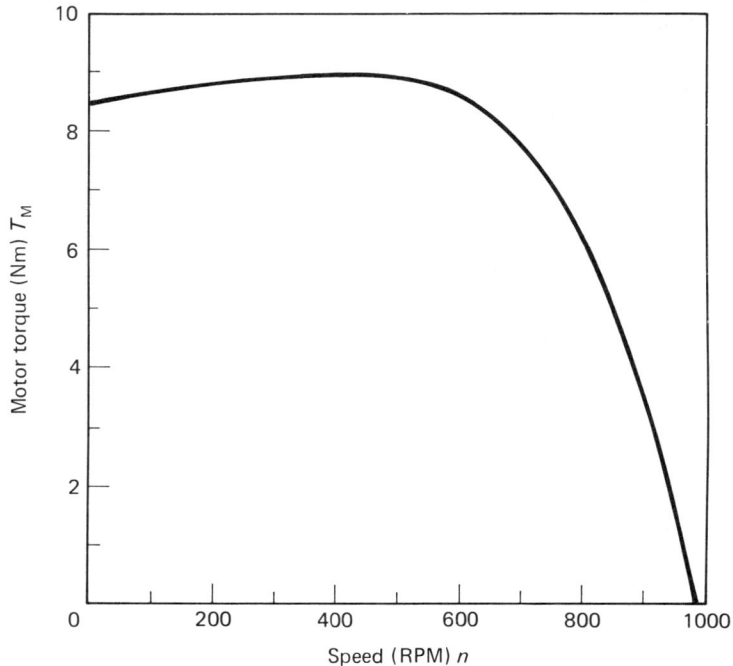

CP 7.17

CP18. The roller cam follower shown in the figure is drawn to half scale. Write a subroutine that will return the follower's displacement if the cam's angle of rotation is known. Use linear interpolation. Verify your program by finding the displacement when θ equals 70 degrees.

CP19. A water tank has a flood gate (also known as a sluice gate) at its bottom that discharges water (see the figure). The flow from the tank is given by the formula

$$Q = Kbh_s\sqrt{2gh_1}$$

where Q is the flowrate, b is the gate's width, h_s is the gate's height, h_1 is the water level in the tank, g is the acceleration of gravity, and K is the flow coefficient. The flow coefficient K changes with the height of water present in the tank. A curve showing this relationship is shown in the graph. Write a subroutine that will calculate the flow rate out of the tank given the heights h_s and h_1 and the gate width, b. Use linear interpolation. Verify your program by finding the flow when $h_s = 0.72$ m, $h_1 = 6.0$ m, and $b = 3.0$ m. Take g as 9.81 m/sec^2.

CP20. From the study of fluids in motion it is known that fluid friction in pipeline flow causes pressure (head) losses that can be calculated by the formula

$$h = f\frac{LV^2}{2Dg}$$

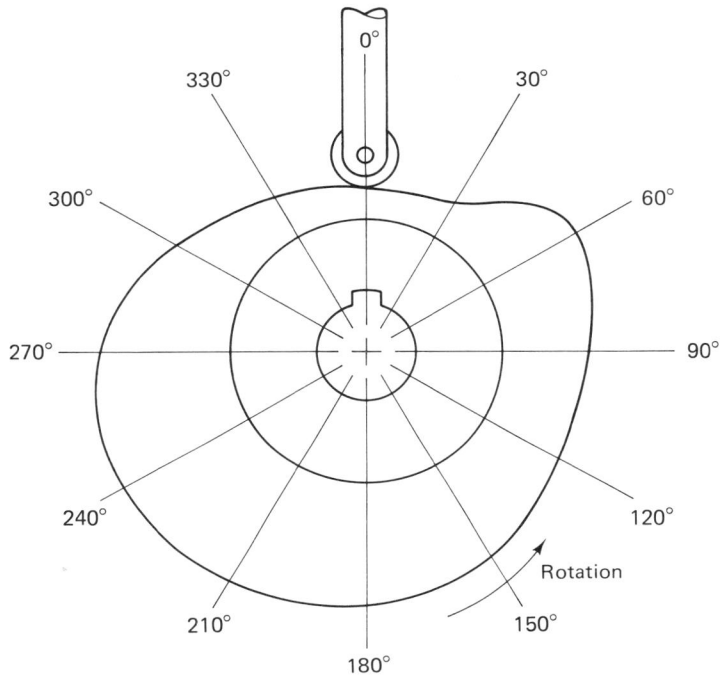

Drawn to half scale

CP 7.18

For nonturbulent flow, the constant f and N are given by

$$f = \frac{64}{N}$$

$$N = \frac{DV}{v}$$

Also,

$$V = \frac{4Q}{\pi D^2}$$

In the formulas given, h is the head loss, v is the kinematic viscosity, f is the friction factor, N is the Reynolds number, L is the effective pipe length, D is the pipe's inner diameter, V is the velocity of the flow, and $g = 32.2$ ft/sec^2. Write a function subprogram that will return the head loss when given the formula variables as input. Use

```
REAL FUNCTION HLOSS(D,V,Q,RL,VNU)
```

where VNU and RL represents v and L, respectively. Test your subprogram out by solv-

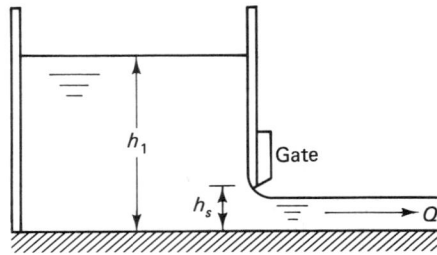

(a)

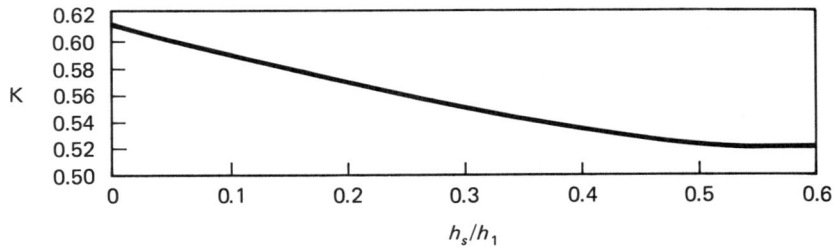

(b)

CP 7.19

ing for the *total* head loss for two pipes that are laid out in series. Print out the final and intermediate results. For both pipes, $Q = 0.005$ ft^3/sec and $v = 1.205E-5$ ft^2/sec. The first pipe has $L = 800$ ft and $D = 0.5054$ ft. The second pipe has $L = 200$ ft and $D = 0.3355$ ft. What are the units of head loss, h?

CP21. There are two types of stresses that can act on a material under load, as shown in the figure. The first, shear stress, acts parallel to an internal cross-sectional area, and it tends to cause layers of material to slide relative to one another. The second type, tensile or compressive stress, acts perpendicularly to the internal section; it is also known as normal stress. Normal stresses cause internal layers to be either stretched or crushed together. In general, a material resists normal stresses better than it does shear stresses. One theory, the maximum shear stress theory, predicts that yielding failure (nonelastic deformations) begins for a specimen under load when any of the maximum shear stresses present equals or exceeds the maximum allowable shear stress for the material. The maximum shear stresses acting can be simply found in terms of the maximum normal stresses as follows:

$$\tau_{12} = \frac{S_1 - S_2}{2.0}$$

$$\tau_{23} = \frac{S_2 - S_3}{2.0}$$

$$\tau_{13} = \frac{S_1 - S_3}{2.0}$$

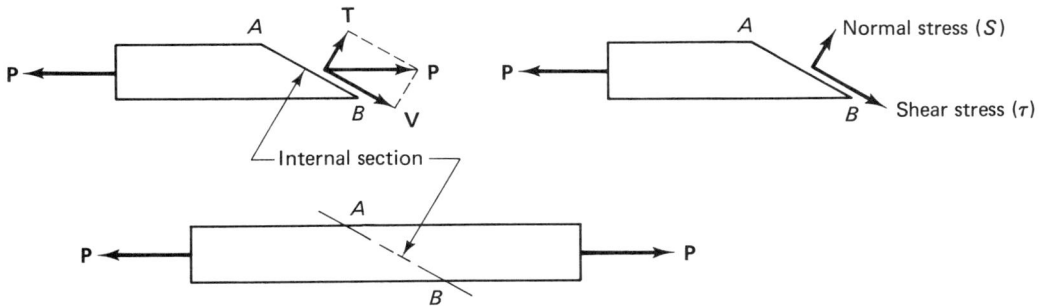

CP 7.21

Write a function subprogram that receives values of the maximum normal stresses found and the maximum allowable shear stress permissible for the material from a calling main program. The subprogram should then find the largest value of the three maximum shear stresses acting. Call this value SHEAR. It should then determine if yielding occurs. The subprogram should print out a message that "YIELDING IS PRESENT" or "YIELDING IS NOT PRESENT." Use the name

```
REAL FUNCTION SHEAR(S1,S2,S3,TALL)
```

Test your program out using values of S_1 = 10,000 psi, S_2 = 2200 psi, and S_3 = −18,600 psi. Let the material be high-strength–low-alloy ASTM-A242 steel with the maximum allowed shear stress, τ_{all}, equal to 25,000 psi. The value of SHEAR, after it is returned from the subprogram to the main program, should be printed out with an accompanying descriptive message.

CP22. A material deforms when it is acted on by forces. A measure of that deformation is the amount of strain that is present. Strain is defined as the deformation per unit length of the material. Strain measurements are usually taken with the aid of an SR-4 wire resistance strain gage that is bonded to the specimen (see the figure). When the specimen deforms, the attached wire changes length and, consequently, its electrical resistance. The relationship between strain and fractional change in resistance is given by

$$\frac{\Delta R_1}{R_1} = \varepsilon(\text{G.F.})$$

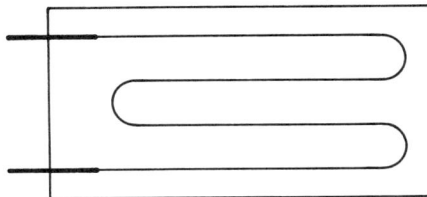

(a)

CP 7.22a

where R_1 is the resistance, ε is the strain, and G.F. is the gage factor. The small change in resistance can be measured precisely by using a Wheatstone bridge, with the strain gage acting as one leg, R_1 (see the figure). The circuit is first balanced so that no current flows through the galvanometer. After straining the specimen, the circuit is again balanced and the amount by which R_2 has been changed is read on the bridge's meter. The relationship between the changes in R_1 and R_2 is

$$\Delta R_1 = \Delta R_2 \frac{R_4}{R_3} = \Delta R_2 (R_N)$$

R_4 and R_3 are usually set to a fixed ratio R_N. Write a function subprogram that will receive $\Delta R_2, R_1, R_N$, and the gage factor as input arguments from the calling program, and return the strain value. Strain is measured in micros. One micro equals 1.0×10^{-6}. Use

        REAL FUNCTION STRAIN(DR2,R1,RN,GF)

and the given data R1 = 120 ohms, RN = .01, and GF = 2.0, passed from the main program, to test your results. The main program should print out the results found. Test your program for the cases ΔR_2 (DR2) equal to 0.8 and 1.6 ohms.

CP23. The velocity imparted to a rocket in a uniform gravity field having exhaust velocity V_E, initial velocity V_0, initial mass M_0, and final mass M is given by

$$V = V_0 + V_E \ln \frac{M_0}{M} - gt \qquad g = 9.81 \text{ m/sec}^2$$

The mass M in terms of the burning rate, b, the amount of fuel burned per second, and time t is

$$M = M_0 - bt$$

Write a FORTRAN subroutine, SUBROUTINE ROCK(B, T, V0, VE, M0, V), that will

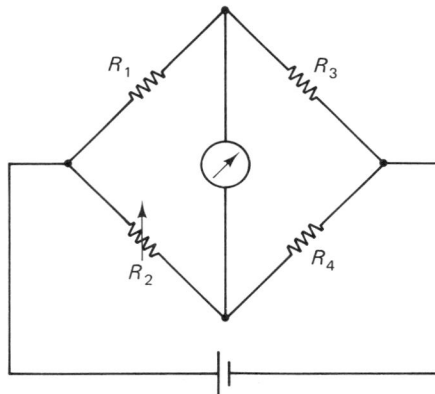

(b)

CP 7.22b

calculate and return the velocity V of the rocket given V_0, V_E, M_0, b, and t. Test your program out on the two-stage sounding rocket, which starts from rest, having the following data:

STAGE I

$$M_0 = 200 \text{ kg} \quad \text{initial mass including Stage II}$$

$$b = 14 \text{ kg/sec}$$

After ten seconds, burn out occurs, the first stage is jettisoned, and the second stage ignites. Now,

STAGE II

$$M_0 = 40 \text{ kg} \quad \text{initial mass}$$

$$b = 3.5 \text{ kg/sec}$$

V_E equals 2500 m/sec. Since the burning time here is also 10 seconds, the main program, after two calls to the subroutine, should stop when $t = 20$ seconds. Print out the velocity that is attained each second.

*CP24. The modified Euler method improves on the accuracy of the basic Euler algorithm. Because it estimates a value for the solution—in this case by Euler's method—and then improves or "corrects" this value before going on the next x-coordinate, it is classified as a predictor-corrector method. The predicted values are given by

$$P(y_{i+1}) = y_i + \left(\frac{dy}{dx}\right)_i \Delta x$$

$$P\left(\frac{dy}{dx}\right)_{i+1} = f[P(y_{i+1}), x_{i+1}]$$

where

$$\left(\frac{dy}{dx}\right)_i = f(y_i, x_i)$$

and

$$x_{i+1} = x_i + \Delta x$$

These values are then corrected to

$$y_{i+1} = y_i + \left[\left(\frac{dy}{dx}\right)_i + P\left(\frac{dy}{dx}\right)_{i+1}\right]\frac{\Delta x}{2}$$

$$\left(\frac{dy}{dx}\right)_{i+1} = f(y_{i+1}, x_{i+1})$$

The method now increments x so that

$$x_i \Leftarrow x_i + \Delta x$$

and sets

$$y_i \Leftarrow y_{i+1}$$

The same steps are repeated again until a predefined final x value is reached.

Write two subroutines that will implement the modified Euler method. Test your program with a driver using the same equation that was given in Sample Problem 7.7:

$$\frac{dy}{dx} = f(y, x) = y^3 x + y$$

Use the two subroutines

```
SUBROUTINE MODEUL(X0,Y0,Y,YD,DX)
```

and

```
SUBROUTINE DERIV(X0,Y0,YD)
```

The last subroutine evaluates the given functional representation of the derivative. Take DX = 0.01 and XF = 0.2 seconds. Initial conditions are X0 = 0.0 and Y0 = 1.0.

*CP25. The components of the product $\mathbf{C} = \mathbf{A} \times \mathbf{B}$ are given by the expression

$$C(I) = \sum_{J=1}^{3} \sum_{K=1}^{3} \delta(I, J, K) \times A(J) \times B(K) \quad I = 1, 2, 3$$

where $\delta(I, J, K)$ is the Levi-Civita density function defined in CP28 of Chapter 6. Write a FORTRAN subroutine that receives the vectors $\mathbf{A}$ and $\mathbf{B}$ from a calling program and returns the product $\mathbf{C}$ for print out. Test your results on the vectors

$$\mathbf{A} = 2\mathbf{i} + 3\mathbf{j} + 4\mathbf{k}$$

and

$$\mathbf{B} = 4\mathbf{i} + 5\mathbf{j} + 6\mathbf{k}$$

with a suitable driver. (*Hint:* Vectors are one-dimensional arrays.)

CHAPTER 8

Other Data Types

Testing for the frequency response of a structure. While measuring motion and force inputs, the instrument shown has a built-in computer that is programmed to automatically calculate and display the structure's frequency response. *(Photo courtesy of Bruel and Kjaer Instruments.)*

8.1 INTRODUCTION

In Section 2.4 we noted that FORTRAN allows for six data types: integer, real, character, logical, double precision, and complex. In the first six chapters of this book we have concentrated our attention solely on real and integer data types, since they are the predominant forms of data encountered in engineering programs. In this chapter we will complete the discussion of FORTRAN data types by investigating in detail the four other data types that are available.

8.2 CHARACTER DATA

8.2.1 Character Constants

Data that has nonnumeric symbolic form is called **character data.** The standard FORTRAN character set introduced earlier consists of the twenty-six letters from *A* to *Z*, ten numeric digits from 0 to 9, a blank, and twelve symbols: $+ - ( ) * 1 =$

, . ' $ and : (see Appendix A, Table A.2). Character data values, or constants, are called **strings.** To distinguish it from other information, a **character constant** must be enclosed within apostrophes. The *length* of a character constant is the number of symbols, blanks included, found within the apostrophes. The apostrophes are not a part of the string and are not counted. If an apostrophe is to purposely appear as one of the symbols, two consecutive apostrophes (not a quotation mark) are used. The following are typical character constants:

```
'SCIENTIFIC'

'PHASE 1'

'THE SUN''S MASS'
```

The length of the last character constant is fourteen, since the double apostrophe is interpreted as a single symbol.

8.2.2 Character Variables

A **character variable name** is a specific address location in memory where character data may be stored. The name of a character variable must consist of one to six letters or numeric digits. The first character of the name must be a letter. Character variables are declared via the CHARACTER **type statement.** The statement, which appears before the first executable statement of a program, has several useful forms. The first form is

CHARACTER*n character variable list

where n represents the length of the character strings assigned to the variable list's memory locations. The variable names on the list are separated by commas. For example,

```
CHARACTER*4 A,B,C
```

specifies that the variable names A, B, and C are of the character type and have a word size that can accommodate four characters. The statement

```
CHARACTER A,B,C
```

with n not specified, declares A, B, and C to be of the character type with a word length of one character.

The form

CHARACTER variable$_1$*n_1,variable$_2$*n_2, . . .

allows for variable names with different associated string lengths to be declared in the same CHARACTER statement. For example, the statement

```
CHARACTER A*4,D*2
```

declares A and D to be of the character type with string sizes of four and two, respectively. The statement

```
CHARACTER*4 A,D*2,B
```

establishes A, B, and D as character variables. The word length of A *and* B is four; the length of D is two. Since the length of B was not declared explicitly, it took on the string length of the character declaration for the list.

Character arrays may be specified by one of the following forms:

CHARACTER*n array name(m)
CHARACTER array name(m)*n

where n is the string length of each element and m is the array's dimension. For example,

```
CHARACTER*5 A(10)
```

```
CHARACTER*10 B(2,5)
```

```
CHARACTER*5 A(10),C(4)*3
```

A character function is declared by a FUNCTION statement. One can write, for example,

```
CHARACTER*4 FUNCTION PROOF (A,N)
```

```
CHARACTER*2 A
```

When character variables appear as the formal arguments of a subprogram, they must be declared not only in the calling program but in the subprogram as well. To allow for greater flexibility for the subprogram, the lengths of the variable strings that are arguments do not have to be specified beforehand. Instead, the following subprogram CHARACTER statement syntax may be used:

CHARACTER* (*) variable list

where (*) represents an **assumed-length specifier.** For example, returning to the function PROOF, one can write the foregoing declaration statements as

```
CHARACTER*4 FUNCTION PROOF (A,N)
```

```
CHARACTER* (*) A
```

using assumed-length specifiers. A typical calling program for all three cases might then appear as

```
PROGRAM CALL
CHARACTER*4 PROOF,C*2,Z
    .
    .
```

```
.
Z=PROOF ( C , 5 )
.

.
.
END
```

8.2.3 Character Operations

Character values can be assigned to a character variable name by using an assignment statement:

character variable name=character expression

The character variable name must appear to the left of the equal sign. A **character variable expression** is an expression that has a character value. It may contain character variable names, character constants, and character operators. For example, for the declared character variables COURSA, COURSB, and COURSC, the assignment statements

```
COURSA='PHYSICS'
COURSB=COURSC
```

assign the character string PHYSICS to character variable name COURSA, and the character value found at address COURSC to character variable COURSB. If the character variable to the left of the equal sign is assigned a string less than its declared character length, blanks are added to the right until the character word is filled (that is, it is left-justified). If the declared length of the character variable is smaller than the expression assigned to it, the extra characters are truncated on the right. For example, for the declared character variables A and B, the statements

```
CHARACTER*4  A , B
.

.
.
A='ON'
B='CLOSES'
```

will assign the values

```
A='ON_ _ '
B='CLOS'
```

where _ stands for a blank character. Note that a character variable must receive a character value. If A is a character variable, the statement

```
A=100.0
```

is not allowed. Instead, one must write

```
A='100'
```

which represents the symbol "100" but not the value.

It is possible to extract and work with a **substring** from the group of characters that are assigned to the character variable. A substring is any subset of adjacent characters. For the data word PHYSICS, all the following substrings are possible:

P   PH   PHY   PHYS   PHYSI   PHYSIC   PHYSICS

H   HY   HYS   HYSI   HYSIC   HYSICS

Y   YS   YSI   YSIC   YSICS

S   SI   SIC   SICS

I   IC   ICS

C   CS

S

Note that a substring can be one character long or it can be as long as the entire string assigned to the character variable. A substring is defined by specifying the character variable name followed by the positions of the first and last character of the substring, separated by a colon. The positions of the first and last characters and the colon are enclosed in parentheses. The character positions of the substring name may be denoted by integer constants, integer variable names, or integer variable expressions. The positions of the string are numbered from left to right, in sequential order, beginning with position 1. The last character position of the substring must be greater than or equal to the first character's numeric position. The first position's value must be positive and less than the string's length. For example, the following table shows what results if the character variable ME has the value 'MECHANICAL':

Substring	Substring value
ME(4:6)	'HAN'
ME(1:3)	'MEC'
ME(8:10)	'CAL'
ME(2:2)	'E'
ME(1:10)	'MECHANICAL'
ME(:4)	'MECH'
ME(6:)	'NICAL'

Note that when the leading position value is missing, it is assumed to be 1; when the final integer value is missing, it is assumed to be equal to the value of the variable's assigned word size. For example, if the variable ME had been declared by

CHARACTER*12 ME

ME(6:) would be equivalent to ME(6:12) and would receive the value 'NICAL _ _'.

Examples of possible substrings formed using integer variables and expressions are given by

```
ME ( I : I )

ME ( 4 : J )

ME ( M : 6 )

ME ( : K )

ME ( L : )

ME ( 2 * I : L − 2 )
```

It is also possible to assign values to a substring. The statement

```
ME ( 1 : 6 ) = ' ELECTR '
```

changes the substring from 'MECHAN' to 'ELECTR'. It subsequently changes the original string so that ME now has the value 'ELECTRICAL'.

Assignment statements can even reference other parts of the word. For example,

```
ME ( 1 : 3 ) = ME ( 4 : 6 )
```

would result in

```
ME = ' HANHAN I CAL '
```

However, it is not allowed to reference a position in the substring to the right of the equal sign if it is part of the original substring being changed. For instance,

```
ME ( 1 : 3 ) = M ( 2 : 4 )
```

would not be allowed, since it references a part of the original substring.

Since character variables are left-justified when assigned values,

```
ME ( 1 : 4 ) = ME ( 5 : 6 )
```

results in

$$ME(1:4) = \text{'AN\_\_'}$$

On the other hand,

```
ME ( 1 : 2 ) = ME ( 3 : 6 )
```

yields the truncated value

```
ME ( 1 : 2 ) = ' CH '
```

Sample Problem 8.1

An automobile manufacturer uses a thirteen-character-long identification number for the cars it produces. The tenth character position of the identification number codes information concerning the car's color. The color codes are 1 for red, 2 for blue, 3 for black, 4 for white, and 5 for brown. Write a FORTRAN program that will read an automobile's identification number and display the proper vehicle color identity on the monitor. Information should be entered at the terminal using list-directed input. Note that for list-directed input, character values are enclosed in apostrophes.

Solution: The algorithm and problem solution are given in Figures 8.1a and 8.1b. (List-directed character input/output operations will be discussed further in Section 8.2.5.) The ELSE IF logical expressions given compare the value of the tenth position of the character variable IDENT to a character constant. Remember that a character constant must be *enclosed in apostrophes*.

Character values may also be initialized with DATA statements or PARAMETER statements. Both statements must precede the first executable statement of the program, and they must follow any listed specification statements. For example, the statements

```
CHARACTER*4 A,B,C(2)
DATA A,B,C/'VEL1','VEL2',2*'  '/
```

will initialize the character variables as follows:

```
A='VEL1'
B='VEL2'
C(1)='_ _ _ _'
C(2)='_ _ _ _'
```

The variables C(1) and C(2) were only assigned one blank character in the DATA statement, but since the values assigned were less than their word size, the other

---

Algorithm for Sample Problem 8.1. This program reads a thirteen-character automobile identification number and searches the tenth position for a color code. The car's color code is printed at the terminal.

1. Enter IDENT.
2. If IDENT(10:10)=1 then
   a. Set COLOR=RED
   Else if IDENT(10:10)=2 then
   a. Set COLOR=BLUE
   Else if IDENT(10:10)=3 then
   a. Set COLOR=BLACK
   Else if IDENT(10:10)=4 then
   a. Set COLOR=WHITE
   Else if IDENT(10:10)=5 then
   a. Set COLOR=BROWN
   Else
   a. Print 'No color was found.'
3. Print COLOR.

Figure 8.1a

```
        PROGRAM CARS

*********************************************************************
*                                                                   *
*   THIS PROGRAM READS A 13 CHARACTER AUTOMOBILE IDENTIFICATION     *
*   NUMBER AND SEARCHES THE TENTH POSITION FOR A COLOR CODE.        *
*   THE CAR'S COLOR CODE IS PRINTED AT THE TERMINAL.                *
*                                                                   *
*********************************************************************

        CHARACTER*13 IDENT,COLOR*5

        PRINT*,'ENTER THE 13 CHARACTER AUTOMOBILE IDENTIFICATION',
      + ' NUMBER ENCLOSED IN'
        PRINT*,'APOSTROPHES. FOR EXAMPLE,''2476543210123''.'
        READ*,IDENT

        IF(IDENT(10:10).EQ.'1') THEN
            COLOR='RED'

        ELSE IF(IDENT(10:10).EQ.'2') THEN
            COLOR='BLUE'

        ELSE IF(IDENT(10:10).EQ.'3') THEN
            COLOR='BLACK'

        ELSE IF(IDENT(10:10).EQ.'4') THEN
            COLOR='WHITE'

        ELSE IF(IDENT(10:10).EQ.'5') THEN
            COLOR='BROWN'

        ELSE
            PRINT*,'THE COLOR WAS NOT FOUND.'
            STOP

        END IF

        PRINT*,'THE COLOR IS ',COLOR

        END


    RUN

    ENTER THE 13 CHARACTER AUTOMOBILE IDENTIFICATION NUMBER ENCLOSED IN
    APOSTROPHES. FOR EXAMPLE,'2476543210123'.
    ? '1234567895123'
    THE COLOR IS BROWN
```

Figure 8.1b

empty positions of the word were filled with blanks from the right (that is, left-justified). If a data value is too large, it will be truncated to fit the character variable's word size.

A typical PARAMETER statement used to define character data parameters appears in the following statements:

```
PROGRAM ONCE
CHARACTER*18 TITLE
PARAMETER(TITLE='RESISTANCE IN OHMS')
```

To offer greater generality in its usage, the length of the character variable that appears in a PARAMETER statement may instead be given by an assumed-length specifier in the type declaration, as follows:

```
PROGRAM ONCE
CHARACTER* (*) TITLE
PARAMETER(TITLE='INDUCTANCE IN HENRYS')
```

When working with character data, it is also possible to link strings together to form a larger string. The linking process is called **concatenation.** The FORTRAN operator that denotes this operation is called the **concatenation operator** and consists of two successive slashes, //. For example, the two strings 'MECH' and 'ANICAL' can be linked to form 'MECHANICAL' as follows:

```
'MECH'//'ANICAL'
```

The statements

```
CHARACTER*4 Y,Z*10
Y='MECH'
Z=Y//'ANICAL'
```

will assign the character string 'MECHANICAL' to the character variable Z.

If R, S, M, U, MO, and YEAR are character variables, the following expressions may be formed using concatenation operators:

```
R//S
M(1:3)//U(1:5)
'THE FIRST OF '// MO //', '// YEAR
```

As another example, suppose that

```
Z='MECHANICAL'
```

Then

```
Z(:2)//Z(9:)='MEAL'
```

and

```
Z(1:2)//Z(6:7)//Z(9:)='MENIAL'
```

It is sometimes desired to compare character values so that they may be placed in order. The character order, or **collating sequence,** established in FORTRAN is as follows: capital letters are ordered from A to Z, digits are ordered from 0 to 9, and blanks precede any character. When letters are compared with numbers, or special symbols are compared with each other or with letters and numbers, no commonly accepted ordering sequences exist. In those cases, the collating sequence employed depends on the specific numeric value encoded to each character by the computer being used. It is therefore not good programming practice to attempt to compare numbers with letters or symbols, since the results will vary from computer to computer. For example, from Table A.2 (Appendix A) we see that numbers precede letters in ASCII code. A computer using EBCDIC code would place numbers after letters. In the sequencing order for both codes, a blank precedes all characters.

To compare character strings, the matching character positions in the strings are compared with one another from left to right. If the strings are of unequal length, blanks are first added to the right of the shorter string by the computer until both strings have equal size before comparisons are made. The following table shows how character strings are compared in a logical expression using relational operators, along with each expression's results.

Logical expression	Resulting value
'A' .LT. 'B'	.TRUE.
'A' .GT. 'AB'	.FALSE.
'A' .GT. 'A1'	.FALSE.
'A1' .GT. 'A0'	.TRUE.
'ABLE' .LT. 'ABACUS'	.FALSE.
'305' .LT. '319'	.TRUE.
'305' .NE. '309'	.TRUE.
'SPEEDONE' .LE. 'SPEED TWO'	.FALSE.
'SPEEDONE' .LE. 'SPEEDTWO'	.TRUE.
'FIVE' .GE. 'FOUR'	.FALSE.
'305' .GT. 303	incorrect comparison

The reader is reminded that if a string is *equal* to another string, it must be identical, character position for character position, including blank positions.

Sample Problem 8.2

Write a FORTRAN program that will read the last names of twenty-five students from a file and place them in alphabetical order. The names in the file are assumed to lie in a single column, left-justified with no leading blanks. Each entry begins with an apostrophe in column 1 and ends with an apostrophe in column 13. This means that no entry can have more than eleven characters. Typical entries are as follows:

```
'BAKER       '
'RIDDERHOLM '
'FREDERICKSO'
'VAN DERHOLM'
```

Use list-directed input and output. Answers should be displayed at the terminal.

Solution: The algorithm and problem solution are shown in Figures 8.2a and 8.2b. Notice that a bubble sort is performed.

Algorithm for Sample Problem 8.2. This program takes a list of twenty-five last names from a file and places them in alphabetic order. The names are assumed to be in the first thirteen columns of each record line. List-directed input should be used, since the names in the file are enclosed within parentheses. The apostrophes lie in columns 1 and 13.

1. Read NAMES.
2. Do, with index K ranging from 24 to 1 in decrements of 1, the following:
    a. Do, with index I ranging from 1 to K, the following:
        i. If NAMES(I+1)<NAMES(I) then
            (a) Set TEMP=NAMES(I)
            (b) Set NAMES(I)=NAMES(I+1)
            (c) Set NAMES(I+1)=TEMP
3. Do, with index I ranging from 1 to 25, the following:
    a. Print NAMES(I)

Figure 8.2a

```
        PROGRAM SORT

******************************************************************
*                                                                *
*    THIS PROGRAM TAKES A LIST OF 25 LAST NAMES FROM A FILE AND   *
*    PLACES THEM IN ALPHABETIC ORDER. THE NAMES ARE ASSUMED TO    *
*    BE IN THE FIRST 13 COLUMNS OF EACH RECORD LINE. LIST-        *
*    DIRECTED INPUT SHOULD BE USED SINCE THE NAMES IN THE FILE    *
*    ARE ENCLOSED WITHIN APOSTROPHES. THE APOSTROPHES LIE IN      *
*    COLUMNS 1 AND 13.                                            *
*                                                                *
******************************************************************

        CHARACTER*11 NAMES (25), TEMP
        INTEGER I,K

        OPEN (UNIT=9,FILE='LASTNA')
        REWIND 9

        READ (9,*)(NAMES(I),I=1,25)
```

Figure 8.2b

```
      DO 2 K=24,1,-1

          DO 1 I=1,K

              IF (NAMES(I+1).LT.NAMES(I)) THEN

                  TEMP=NAMES(I)
                  NAMES(I)=NAMES(I+1)
                  NAMES(I+1)=TEMP

              END IF

1         CONTINUE

2 CONTINUE

      DO 3 I=1,25
          PRINT*,NAMES(I)
3 CONTINUE

      END


File Lastna

'ABLE          '
'RIDDERHOLM '
'FREDERICKSO'
'VAN DERHOLM'
'VANAGON      '
'BAKER        '
'JANSON       '
'FRIEDMAN     '
'MASON        '
'LEE          '
'MAYFIELD     '
'JEFFERIES    '
'JUNIPERO     '
'AUGUSTINO    '
'SANCHEZ      '
'ONEIDEN      '
'NICHOLSON    '
'DANFORTH     '
'JASON        '
'FUNUGA       '
'MEYERVITCH '
'ANDERSONVIL'
'MINKOWSKI    '
'YEH          '
'MILLER       '


RUN

  ABLE
  ANDERSONVIL
  AUGUSTINO
  BAKER
  DANFORTH
  FREDERICKSO
  FRIEDMAN
  FUNUGA
  JANSON
  JASON
  JEFFERIES
  JUNIPERO
```

Figure 8.2b   (cont.)

```
LEE
MASON
MAYFIELD
MEYERVITCH
MILLER
MINKOWSKI
NICHOLSON
ONEIDEN
RIDDERHOLM
SANCHEZ
VAN DERHOLM
VANAGON
YEH
```

Figure 8.2b   (cont.)

Sample Problem 8.3

Write a FORTRAN subroutine that displays positive numeric values read from a data file on a horizontal bar graph. Scale the values so that the largest output is fifty units long. Each unit should be represented by an asterisk. Print a vertical axis in column 5 by using periods. The graph should be labeled with a horizontal title which is passed by the calling program. Numbers, ranging from zero to the maximum value, representing the horizontal scale, should lie under the title and appear at every fifth column position along the scale. A dashed line should appear below the numbers to depict the horizontal axis. Output should be displayed at the terminal.

Solution: The algorithm and problem solution are shown in Figures 8.3a and 8.3b. Note how the title is passed from the calling program as a subroutine argument. Also note how a character array, BAR, can be used instead of a string to control character input and output. The array's elements, limited to one character, are initialized (assigned fifty asterisks) within the first DO loop. Only the number of asterisks, M, needed to represent the data will ultimately be displayed:

```
PRINT *,'      .',(BAR(J),J=1,M)
```

M is determined from the scaled data. To find the scale factor, the maximum value of the data in array A, AMAX, is first identified. If it is less than 50, no scaling is performed. (The scale remains set at 1.) If it is greater than 50, a scaling factor, SCALE, is established by finding the nearest integer value, IMAX, greater than AMAX that is a multiple of 50,

```
ITEMP=AMAX/50.0
IMAX=(ITEMP+1)*50
```

and then calculating the scale factor directly from

```
SCALE=50.0/IMAX
```

Working with IMAX instead of AMAX produces a horizontal scale that is always a multiple of 5 instead of some arbitrary number.

Each M is found by rounding to the nearest integer the product of an array element times the scale factor:

```
M=NINT(A(I)*SCALE)
```

If M equals 0, a separate statement is used to print out the vertical axis (a period in column 5):

```
PRINT *,'
```

The scale above the horizontal axis is also found using the scale factor. Scale values are printed out at every fifth column position. For this purpose, a special array, IA, is established having ten values (fifty column positions are available: 10 × 5). The value of IA (I) to be printed out at the terminal is

```
IA(I)=(IMAX/10)*I
```

where I goes from 1 to 10. Sample output is shown in the figure.

Algorithm for Sample Problem 8.3. This subroutine receives N positive array values A(I) from the calling program and plots a horizontal bar graph. Values are scaled, where necessary, to a maximum horizontal scale length of 50.

1. Print A.
2. Do, with index I ranging from 1 to 50, the following:
   a. Set BAR(1),BAR(2),...,BAR(50)=*
3. Set AMAX=A(1),SCALE=1.0
4. Do, with index I ranging from 2 to N, the following:
   a. If A(I)>AMAX then
      i. Set AMAX=A(I)
5. If AMAX>50.0 then
   a. Calculate
      ```
      ITEMP=AMAX/50.0
      IMAX=(ITEMP+1)*50
      SCALE=50.0/IMAX
      ```

6. Print title of graph.
7. Do, with index I ranging from 1 to 10, the following:
   a. Calculate
      ```
      IA(I)=(IMAX/10)*I
      ```
8. Print horizontal scale:
   ```
   IA(1),IA(2),...,IA(10)
   ```
9. Print horizontal axis.

10. Do, with index I ranging from 1 to N, the following:
    a. Calculate
       ```
       M=INT(A(I)*SCALE)
       ```
    b. If M≥1 then
       i. Print BAR(1),BAR(2),...,BAR(M), forming horizontal bar for graph.
       Else
       i. Print dot for vertical axis.

Figure 8.3a

```
            SUBROUTINE GRAPH (TITLE,A,N)

*******************************************************************
*                                                                 *
*    THIS PROGRAM RECEIVES POSITIVE ARRAY VALUES FROM THE CALLING *
*    PROGRAM AND PLOTS A HORIZONTAL BAR GRAPH. VALUES ARE SCALED,  *
*    WHERE NECESSARY, TO A MAXIMUM HORIZONTAL SCALE LENGTH OF 50.  *
*                                                                 *
*******************************************************************

            REAL A(N),AMAX,SCALE
            INTEGER IA(10),I,N,ITEMP,IMAX,M
            CHARACTER BAR(50)
            CHARACTER* (*) TITLE

            PRINT *,'THE VALUES OF [A] ARE:'
            PRINT *
            PRINT 200,(A(I),I=1,N)
            PRINT *

            DO 7 I=1,50
                BAR(I)='*'
        7 CONTINUE

            AMAX=A(1)
            SCALE=1.

            DO 1 I=2,N

                IF(A(I).GT.AMAX) THEN
                    AMAX=A(I)
                END IF

        1 CONTINUE

            IF(AMAX.GT.50.) THEN

                ITEMP=AMAX/50.
                IMAX=(ITEMP+1)*50
                SCALE=50./IMAX

            END IF
C
C    BEGIN PROGRAM OUTPUT.
C
            PRINT*,'      ',TITLE
            PRINT*

C
C    CALCULATE HORIZONTAL SCALE AND DRAW AXIS.
C
            DO 4 I=1,10
                IA(I)=(IMAX/10)*I
        4 CONTINUE
            PRINT 3,(IA(I),I=1,10)
        3 FORMAT('      ',2X,10I5)
            PRINT 5
        5 FORMAT('      .',10('____:'))
C
C    SCALE VALUES AND DRAW BARS.
C
            DO 2 I=1,N
                M=NINT(A(I)*SCALE)
                IF(M.GE.1) THEN
                    PRINT*,'      .',(BAR(J),J=1,M)
```

Figure 8.3b

```
          ELSE
               PRINT*,'     .'

          END IF

    2 CONTINUE

  200 FORMAT(E10.3)

          RETURN
          END


  RUN

  THE VALUES OF [A] ARE:

  .990E+03
  .735E+03
  .800E+01
  .311E+03
  .595E+03

      THIS IS A TEST CASE OF THE BAR GRAPH

         100  200  300  400  500  600  700  800  900 1000
  .____:____:____:____:____:____:____:____:____:____:
  .************************************************
  .***********************************
  .
  .***************
  .*****************************
```

Figure 8.3b   (cont.)

Sample Problem 8.4

Shown in Figure 8.4 is the furnace of Sample Problem 6.8. Write a FORTRAN program that will draw a duplicate of this figure to a scale of 1 inch = 2 feet. Dots, instead of lines, are to be used for the outline of the surfaces. In addition, it is desired that all points within the furnace having temperatures greater than or equal to 390 degrees Fahrenheit be highlighted on the drawing using the letter H.

Solution: The solution to the problem is shown in Figure 8.5. A two-dimensional graphic symbol array C is used for the drawing. Since terminal printers typically have ten columns of output per inch, the graphic array will need forty-one columns. (Eight feet equal 4 inches in this scale, and this corresponds to forty horizontal spaces.) In the vertical direction, printers usually form six rows of output per inch. With the drawing requiring 4 inches of vertical space for an 8-foot length, a total of twenty-five array rows will be needed. Array C is, therefore, given the dimension C(25, 41).

To start the solution, the array A is entered from an auxiliary file, and the graphic symbol array C has blanks assigned to all its elements. When this is completed, the furnace outline is generated. First, the top and bottom boundaries of the furnace are formed; that is, dots are placed in the appropriate elements of the graphic array. Next, the left- and right-hand sidewalls are formed. The procedure is then repeated to create the furnace-chamber outline. When the figure is finished, the A array is searched for elements having temperatures greater than or equal to 390 degrees Fahrenheit. The array A contains the temperature values for the bottom half of the furnace. The temperature values at the top part of the furnace are assumed to be the same as for

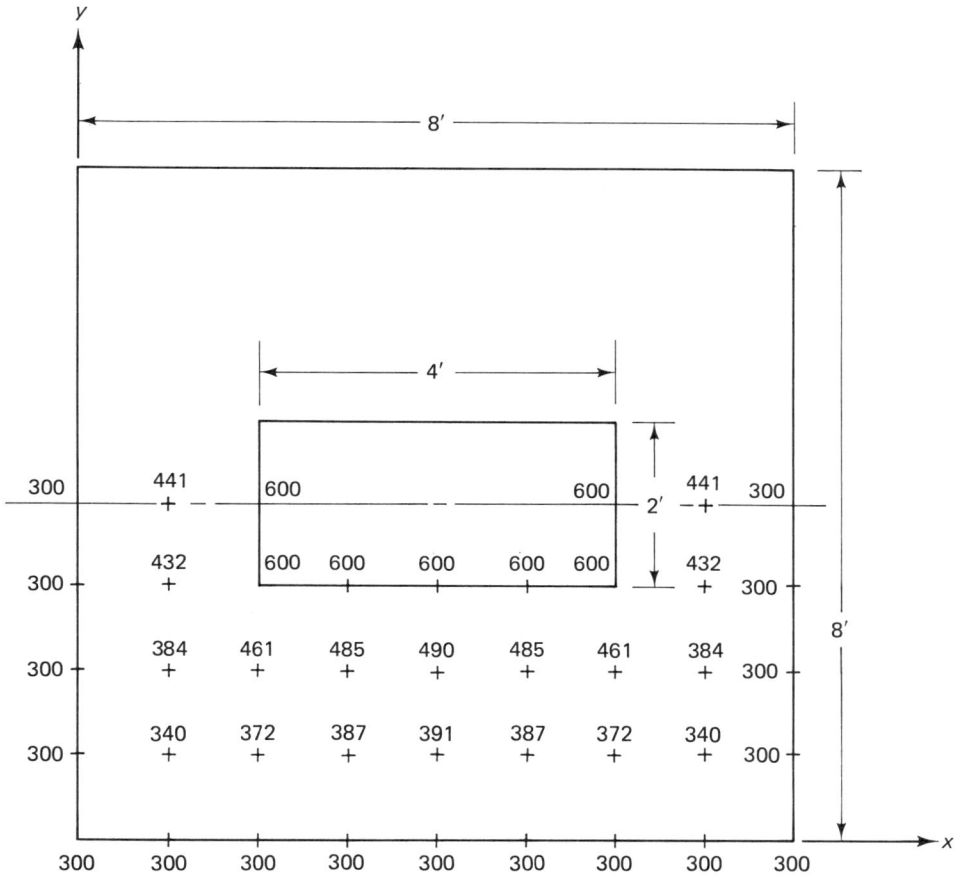

Figure 8.4

```
         PROGRAM DRAW

    ****************************************************************
    *                                                              *
    *    THIS PROGRAM DRAWS THE FURNACE OF FIGURE 6.11, SAMPLE     *
    *    PROBLEM 6.8. THE PICTURE IS SCALED SO THAT ONE INCH       *
    *    REPRESENTS TWO FEET. THE LETTER H IS PLACED AT POINTS ON THE *
    *    DRAWING THAT LOCATE FURNACE TEMPERATURES EQUAL TO OR GREATER *
    *    THAN 390 DEGREES.                                         *
    *                                                              *
    ****************************************************************
         REAL A(5,9)
         CHARACTER*1 C(25,41)
         INTEGER I,J
```

Figure 8.5

```
       OPEN(UNIT=7,FILE='FURNACE')
       REWIND 7

       READ(7,*)((A(I,J),J=1,9),I=1,5)
C
C      THE ARRAY C THAT WILL CONTAIN THE GRAPHIC SYMBOLS FOR THE
C      DRAWING IS FIRST BLANKED OUT.
C
       DO 5 I=1,25

          DO 6 J=1,41
             C(I,J)=' '

     6      CONTINUE

     5 CONTINUE
C
C      THE TOP AND BOTTOM BOUNDARY OF THE FURNACE IS NOW DRAWN.
C
       DO 8 J=1,41
          C(1,J)='.'
          C(25,J)='.'

     8 CONTINUE
C
C      THE LEFT AND RIGHT-HAND SIDE WALLS ARE NOW DRAWN. I STARTS
C      AT 2 AND ENDS AT 24 SO THAT POINTS FROM THE UPPER AND LOWER
C      BOUNDARIES ARE NOT REPEATED.
C
       DO 10 I=2,24
          C(I,1)='.'
          C(I,41)='.'

    10 CONTINUE
C
C      THE TOP AND BOTTOM OF THE FURNACE CHAMBER ARE NOW DRAWN.
C
       DO 11 J=11,31
          C(10,J)='.'
          C(16,J)='.'

    11 CONTINUE
C
C      THE SIDE WALLS OF THE CHAMBER ARE NEXT DRAWN.
C
       DO 12 I=11,15
          C(I,11)='.'
          C(I,31)='.'

    12 CONTINUE
C
C      THE ARRAY A CONTAINS THE TEMPERATURE VALUES FOR THE BOTTOM
C      HALF OF THE FURNACE. SINCE THE TEMPERATURE VALUES OF THE
C      TOP PART OF THE FURNACE ARE ASSUMED TO BE THE SAME AS FOR
C      THE LOWER PART, THE ARRAY A CAN ALSO BE USED TO GENERATE
C      INFORMATION FOR THE UPPER POINTS. A IS SEARCHED BELOW TO
C      DETERMINE THE LOCATION OF ELEMENTS WITH TEMPERATURES
C      GREATER THAN OR EQUAL TO 390 DEGREES. THE LETTER H IS
C      PLACED IN THE GRAPHIC SYMBOL ARRAY C AT THE CORRESPONDING
C      ARRAY POSITION. WHEN THESE ELEMENTS OF THE GRAPHIC ARRAY
C      ARE FOUND, THE HALF PART OF THE A ARRAY THAT WAS OMITTED
C      MUST NOW BE CONSIDERED. IN ADDITION, SINCE SCALING IS USED,
C      THE FOLLOWING CORRESPONDENCE OF ARRAY POSITIONS ARE NOTED:
```

Figure 8.5   (cont.)

```
C   ROW I OF ARRAY A CORRESPONDS TO ROWS 3*I-2 AND 28-3*I OF
C   ARRAY C, COLUMN J OF ARRAY A CORRESPONDS TO COLUMN 5*J-4
C   OF ARRAY C.
C
        DO 4 I=1,5

            DO 3 J=1,9

                IF(A(I,J).GE.390.) THEN
                    C(3*I-2,5*J-4)='H'
                    C(28-3*I,5*J-4)='H'
                END IF

3           CONTINUE

4 CONTINUE

        PRINT 24, ((C(I,J),J=1,41),I=1,25)
24 FORMAT (19X,41A1)

        END
```

Figure 8.5   (cont.)

the bottom part. The array A can therefore be used to generate information for these upper points as well. When an element in the A array is found to have a value greater than or equal to 390, an H is placed in the C array at the corresponding position. For row I of array A, the corresponding rows (one row for the upper half, and one for the lower) of C are $3*I-2$ and $28-3*I$. For column J of array A, the corresponding column position of array C is $5*J-4$. When C has been updated to include this new information, the results are printed out at the terminal, row-wise, and the program ends. Output for this program is shown in Figure 8.6. What advantage does the representation of numeric output with symbols have?

8.2.4 Intrinsic Functions

The FORTRAN library includes several useful intrinsic functions that process character data. The INDEX **function** is an intrinsic function that gives the integer position of the first appearance of a specified string (string 2) found within another string (string 1). Its syntax is

$$\text{INDEX}(\text{string}_1, \text{string}_2)$$

If string 2 is not found, INDEX returns a value of 0. Either string 1 or string 2 can be a character constant, a character variable, or a substring. Some examples of how INDEX can be used follow:

```
N=INDEX(ZERO,'TENSOR')

N=INDEX(ZERO(:8),'ORDER')+5

N=INDEX('CAN YOU FIND','YOU')
```

RUN

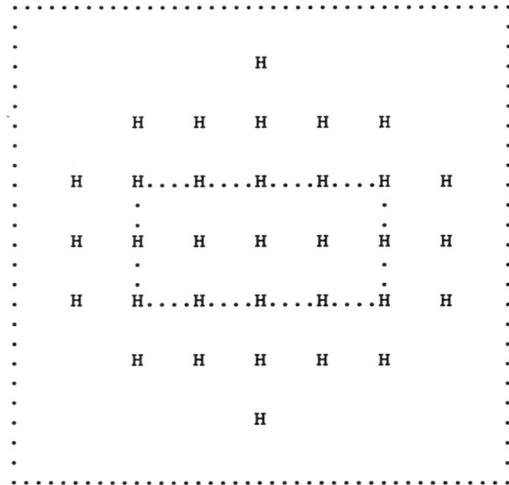

Figure 8.6

If

WE I GHT = ' NEWTONS / METER '

then

N=INDEX(WEIGHT, 'POUNDS') $\Longrightarrow$ N=0
N=INDEX(WEIGHT, '/') $\Longrightarrow$ N=8
N=INDEX(WEIGHT(9:), 'E') $\Longrightarrow$ N=2
WEIGHT(:INDEX(WEIGHT, '/')−1) $\Longrightarrow$ 'NEWTONS'
K//WEIGHT(INDEX(WEIGHT, '/')+1:) $\Longrightarrow$ 'KMETER'

Sample Problem 8.5

When translating roman numerals to arabic numerals, care must be taken for several special cases. Normally, M equals 1000, D equals 500, C equals 100, L equals 50, X equals 10, V equals 5, and I equals 1. However, in certain instances when C, X, or I appears immediately before a letter carrying greater value, it is subtracted. For example:

VI = 6, XI = 11, LX = 60, CX = 110, DC = 600, MC = 1100

but

IV = 4, IX = 9, XL = 40, XC = 90, CD = 400, CM = 900

Write a FORTRAN subroutine that will search a character variable having eighty positions representing a roman numeral for the two instances in which an I might appear immediately to the left of a larger value, that is, IV or IX. Return to the calling program the amount that must be subtracted from the total sum when converting to arabic values.

Solution: The algorithm and problem solution are shown in Figures 8.7a and 8.7b. Remember that in any roman numeral, an I can only appear before a V or an X, not both, Similarly, the appearance of IV or IX can only occur once within any roman numeral string. These considerations simplify the programming effort considerably.

Algorithm for Sample Problem 8.5: First Program. When a roman numeral is converted to arabic numbers, a correction must be made to the converted sum whenever an I immediately precedes a V or an X. This subroutine calculates this correction when converting any roman numeral ROMAN having eighty positions or less.

1. Print ROMAN.

2. Set CORREC=0.0.

3. Calculate
N=INDEX(ROMAN,'IV')

4. If N>0 then
a. Set CORREC=2.0

5. Calculate
N=INDEX(ROMAN,'IX')

6. If N>0 then
a. Set CORREC=2.0

7. Print CORREC.

8. Return to the calling program.

Figure 8.7a

```
        SUBROUTINE EXCEPT (ROMAN,CORREC)

*******************************************************************
*                                                                 *
*    WHEN A ROMAN NUMERAL IS CONVERTED TO HINDU-ARABIC NUMBERS, A  *
*    CORRECTION MUST BE MADE TO THE CONVERTED SUM WHENEVER AN I    *
*    IMMEDIATELY PRECEDES A V OR A X. SUBROUTINE EXCEPT            *
*    CALCULATES THIS CORRECTION WHEN CONVERTING ANY ROMAN NUMERAL  *
*    HAVING 80 POSITIONS OR LESS.                                 *
*                                                                 *
*******************************************************************

        CHARACTER*80 ROMAN,IA*1
        REAL CORREC
        INTEGER I

        PRINT *,'THE ROMAN NUMERAL IS ',ROMAN
        CORREC=0.
        I=1
```

Figure 8.7b

```
C
C    WHILE BLOCK STRUCURE BEGINS
C
     2 IF(I.LT.80) THEN

           IF(ROMAN(I:I).EQ.'I') THEN
              IA=ROMAN(I+1:I+1)

              IF((IA.EQ.'V').OR.(IA.EQ.'X')) THEN
                 CORREC=2.
                 PRINT *,'THE CORRECTION IS ',CORREC
                 RETURN

              END IF

           END IF

           I=I+1
           GO TO 2

        END IF


        RETURN
        END


     RUN

     THE ROMAN NUMERAL IS MDCCLXIV
     THE CORRECTION IS 2.
```

Figure 8.7b   (cont.)

Each time an I appears before a V or an X, a value of 2 must be subtracted from the total sum that was established by adding the translation of each roman-numeral character independently. The variable CORREC carries this information back to the calling program. Alternative algorithm and solution are shown in Figures 8.8a and 8.8b.

Another function that FORTRAN provides is the LEN **function.** The LEN function is used to provide the length of a given character string. It has the form

LEN(string)

where the output of the function is an integer value representing the string's length. For example,

```
PRINT *,LEN(BAR(:8))
N=LEN(BAR//TEST)
```

The function is often used in association with assumed-length specifiers:

```
SUBROUTINE ONE(ABLE,X)
CHARACTER* (*)ABLE
DO 7 I=1,LEN(ABLE)
   .
   .
   .
RETURN
END
```

■ **Algorithm for Sample Problem 8.5: Second Program.** When a roman numeral is converted to arabic numbers, a correction must be made to the converted sum whenever an I immediately precedes a V or an X. This subroutine calculates this correction when converting any roman numeral ROMAN having eighty positions or less.

1. Print ROMAN.

2. Set CORREC=0.0,I=1.

3. While I<80 then
   a. If ROMAN(I:I)='I' then
      i. Set IA=ROMAN(I+1:I+1)
      ii. If IA='V' or IA='X' then
         (a) Set CORREC=2.0
         (b) Print CORREC
         (c) Return to the calling program
   b. Calculate
      I = I + 1

4. Return to the calling program.

Figure 8.8a

```
         SUBROUTINE EXCEPT (ROMAN,CORREC)

*****************************************************************
*                                                               *
*   WHEN A ROMAN NUMERAL IS CONVERTED TO HINDU-ARABIC NUMBERS, A *
*   CORRECTION MUST BE MADE TO THE CONVERTED SUM WHENEVER AN I   *
*   IMMEDIATELY PRECEDES A V OR A X. SUBROUTINE EXCEPT           *
*   CALCULATES THIS CORRECTION WHEN CONVERTING ANY ROMAN NUMERAL *
*   HAVING 80 POSITIONS OR LESS.                                *
*                                                               *
*****************************************************************

         CHARACTER*80 ROMAN
         REAL CORREC
         INTEGER N

         PRINT *,'THE ROMAN NUMERAL IS ',ROMAN
         CORREC=0.
         N=INDEX(ROMAN,'IV')

         IF(N.GT.0) THEN
             CORREC=2.
         END IF

         N=INDEX(ROMAN,'IX')

         IF(N.GT.0) THEN
             CORREC=2.
         END IF

         PRINT *,'THE CORRECTION IS ',CORREC
```

Figure 8.8b

```
          RETURN
          END


     RUN

          THE ROMAN NUMERAL IS MDCCCCXXXIX
          THE CORRECTION IS 2.
```

Figure 8.8b   (cont.)

The position of a character in the collating sequence of the host computer is given by the I CHAR **function.** ICHAR has the form

ICHAR(character)

and has an integer value as output. The inverse function, the CHAR **function,** determines which character occupies a given specific position in the computer's collating sequence. The syntax is

CHAR(integer position)

A character is returned as output.

The **lexical functions,** LGE, LGT, LLE, and LLT, are additional character functions provided in FORTRAN. They are used to guarantee that when comparisons are made between character strings, the ASCII collating sequence is used, no matter which computer is employed. Using for instance, LGT in a program instead of the relational operator .GT. will allow the program to *carry to any other machine* without regard to whether the host computer is ASCII or EBCDIC based. Note how the lexical functions are used in the following statements.

```
     IF (LGT(ABLE(:8),'TEST')) THEN
```

is used instead of

```
     IF (ABLE(:8).GT.'TEST') THEN
```

Or

```
     IF (LLE(BAKER,CUST(3:7))) THEN
```

replaces

```
     IF (BAKER.LE.CUST(3:7)) THEN
```

8.2.5 Character Data Input and Output

List-directed output of character data is accomplished by the statement

PRINT *,'character string'

The character string written between the apostrophes is printed out exactly as it appears. Blanks are not shifted, added, or removed. If an apostrophe is to be part of the output message, a double apostrophe is used.

List-directed input uses the READ statement

READ *,variable list

or

READ(*i*, *)variable list

depending on whether input is from the terminal or from an external file. When the variable list contains a character variable name, the data must be entered, either at the terminal or from a file, enclosed in apostrophes. Just as for output, if an apostrophe is to appear in the string for list-directed input, a double apostrophe must be entered.

If the entered string is larger than the length specified for the character variable, it will be truncated at the right. If the entered string is smaller than the specified length of the character variable, the string will be left-justified, and blanks will be added to the right of the data to fill the string. If several character variables appear in the READ variable list, the character constants must be separated by commas or blanks when entered.

For Formatted input/output operations, the **alphanumeric format specifier** or **descriptor** is used. It has the form

A*w*

or

A

where *w* stands for the field width. If *w* is not specified, it is taken to be the same as the declared length of the variable. For example, the statements

```
      CHARACTER*10 ABLE
      READ(8,7) ABLE
    7 FORMAT(4X,A)
      PRINT 7,ABLE
```

produce the same input or output as

```
      CHARACTER*10 ABLE
      READ(8,7) ABLE
    7 FORMAT(4X,A10)
      PRINT 7,ABLE
```

It should be emphasized that with formatted input, apostrophes are *not* used.

With formatted input, if the field width, *w*, is *less than* the length of the character variable, blanks are added on the right to fill up the string (that is, the character variable is left-justified). For formatted output, only *w* characters appear, and the right-hand side of the string is truncated. If, on the other hand, *w* is specified to be *larger than* the length of the character variable, on input the left-hand side of the data word will be lost and only the rightmost number of field positions equal to the character variable's length will be kept. On output the word is right-justified when the field width w is larger than the character variable's length (leading blanks are placed in the field).

For example, suppose that a data file contains four blanks and the string

　　　＿ ＿ ＿ ＿ ABCD'E'FGHIJ

If the character variable TEST is specified to have eight positions, the statements

```
    CHARACTER*8 TEST
    READ(7,1) TEST
  1 FORMAT(4X,A)
```

will assign TEST the string ABCD'E'F. The statements

```
    CHARACTER*8 TEST
    READ(7,1)TEST
  1 FORMAT(4X,A4)
```

will assign TEST the string ABCD＿ ＿ ＿ ＿. On the other hand, the statements

```
    CHARACTER*8 TEST
    READ(7,1)TEST
  1 FORMAT(4X,A12)
```

will have TEST take on the value 'E'FGHIJ, the rightmost part, the last eight column positions, of the data field. On output, the statements

```
    CHARACTER*10 ABLE
    ABLE='ABCD''E''FGH'
    PRINT 1,ABLE
  1 FORMAT(1X,A)
```

will have

　　　＿ ABCD'E'FGH

appear on the monitor if the first column position is not used for carriage control. The statements

```
    CHARACTER*10 ABLE
    ABLE='ABCD''E''FGH
    PRINT 1,ABLE
  1 FORMAT(1X,A4)
```

will cause to be displayed

　　　＿ ABCD

If the statements

```
    CHARACTER*10 ABLE
    ABLE='ABCD''E''FGH'
    PRINT 1,ABLE
  1 FORMAT(1X,A12)
```

are used, the output would appear as

   _ _ _ABCD'E'FGH

It should be apparent that using the format specifier A without a field width given, when possible, is better programming practice.

8.3 LOGICAL DATA

8.3.1 Logical Constants and Variables

FORTRAN includes two **logical constants.** These constants denote either true or false conditions. To represent logical constants one uses the words .TRUE. or .FALSE. The preceding and following periods must be included with the word, as they inform the compiler that the given symbolic string is a logical constant rather than an ordinary variable name.

A **logical variable name** is a specific address location in memory where logical data can be stored. A LOGICAL **type statement** is used to declare a logical variable. It has the syntax

   LOGICAL logical variable list

The statement is nonexecutable and must precede the first executable statement in the program. The logical variable names must be acceptable FORTRAN names. They are separated on the statement's variable list by commas. For example,

```
   LOGICAL LA,LB,LC,DA(5),SEGEL
```

8.3.2 Logical Operations

Logical values are assigned to a logical variable name by using an assignment statement having the form

   logical variable name=logical expression

Logical expressions have already been discussed in Section 4.2. *A simple logical expression* can be a logical constant, a logical variable, or the result of a relational comparison. In Section 4.2, we learned that the relational operators are

```
.EQ.
.GT.
.GE.
.LT.
.LE.
.NE.
```

To form *compound logical expressions*, the FORTRAN logical operators are employed. They are

```
.AND.
.OR.
.EQV.
.NEQV.
.NOT.
```

The following are some typical logical assignment statements:

```
REAL  X,Y
LOGICAL  A,  B,  C
     .
     .
     .
A=.TRUE.
B=.NOT.A
A=X.GE.Y
C=B.AND..NOT.A
A=((3.2*X).LT.Y).OR.B
```

Remember that each expression on the right-hand side of the equal sign must be, or result in, a logical constant.

To determine the value of complicated logical expressions, all expressions contained within parentheses are evaluated first. Arithmetic expressions are then calculated, after which the truth values of relational expressions are determined. Logical operators are considered last, and in the following order:

1. .NOT.

2. .AND.

3. .OR.

4. .EQV. and .NEQV.

Logical values may also be initialized with PARAMETER or DATA statements. For example, the statements

```
LOGICAL  LA,LB
PARAMETER(LA=.TRUE.,LB=.FALSE.)
```

assign program values to the logical constants LA and LB. Note that the specification statement must precede the PARAMETER statement. The two statements

```
LOGICAL LA,LB
DATA LA,LB/.TRUE.,.FALSE./
```

assign initial logical values to the logical variables LA and LB when the program is first compiled.

8.3.3 Logical Variables as Flags

Logical constants and variables are usually used in engineering programs as a flag or signal that a certain event in the program has occurred. The logical variable acts as a switch that is either open or closed. Take, for example, the following source code:

```
PROGRAM FLAG
LOGICAL LA
LA=.FALSE.
DO 3 I=1,10
    IF (.NOT.LA) THEN
        READ *,DX,X
        PRINT *,'DX = ',DX,'XO = ',X
        LA=.TRUE.
    END IF
    X=X+DX
    PRINT *,'X = ',X
3 CONTINUE
END
```

The program increments by DX an initial value of X, which was entered at the terminal, ten times. The value of DX and XO (the initial value of X) are read and displayed only once, even though they are part of the DO loop, since the switch LA becomes .TRUE. during the first passage through the loop. The ten values for X are then formed and displayed.

8.3.4 Logical Data Input and Output

List-directed output of logical data is accomplished via the statement

```
PRINT *, logical variable
```

Output is displayed as a simple "T" or "F." No periods or other characters are displayed. If several logical variables appear in the PRINT statement, the output values are usually placed so that blanks appear between each item. (Actual spacing is dependent on which computer system is used.)

For list-directed input, the first letter appearing after a blank space, comma, or a period must be a T or an F. Only the two letters, T or F, need be entered, since periods and trailing letters are unessential and are ignored. For example, the statements

```
LOGICAL LA,LB,LC,LD
READ *,LA,LB,LC,LD
PRINT *,LA,LB,LC,LD
```

with the entries

```
.T., F FALSE,.FEL
```

will display

```
T F F F
```

For formatted input/output operations, the **logical format specifier** is used. It has the form

$$Lw$$

where w is the field width. On output, only a "T" or an "F" will be written. If the field width, w, has a value greater than 1, the character "T" or "F" will be placed to the right in the field (that is, it will be right-justified). On input, the first nonblank character entered must be a period, or a "T" or an "F." If a period is used, a "T" or an "F" must immediately follow. Periods are optional and are ignored. Trailing letters appearing in the field are also disregarded. For instance, the statements

```
  LOGICAL LA,LB,LC,LD
  READ(5,1) LA,LB,LC,LD
1 FORMAT(L3,L2,L6,L4)
  WRITE(6,1) LA,LB,LC,LD
```

with the data field

```
.T. F FALSE.FEL
```

will produce the terminal display

```
__T_F_____F___F
```

8.4 DOUBLE-PRECISION DATA

8.4.1 Introduction

Section 2.1 briefly discussed how decimal numbers must first be converted to the binary base-2 system so that the digital computer can process them (see also Appendix A). To represent this numeric data, each computer provides a certain word size, which might be one, two, three, four, eight, or sixteen bytes long. (Eight bits equals one byte.) For a real or decimal number, several bits of the word must be dedicated to represent the exponent (seven bits is usual), one bit must be saved to code the sign, and what remains of the word is used to contain the binary fraction, for example, twenty-four bits for a thirty-two-bit word.

If we assume that the computer has a thirty-two-bit architecture (four-byte word), the twenty-four bits remaining correspond to approximately seven decimal digits. (We have to say approximately because the result is dependent on the actual binary number stored.) Although the number can be very large or very small because of the presence of the exponent, the number's *precision* is limited to seven digits. And although seven-digit significance is more than acceptable for typical engineering work, it should be mentioned that the conversion of a decimal number to binary form usually leads to a truncation error, which may at times be vexatious. For instance, the decimal number 0.1 is represented by the unending binary sequence

$$.000110011001100110011001100110011\ldots$$

Truncated after twenty-four bits, the number, with exponent, becomes

$$.110011001100110011001100 \times 2^{-3}$$

which corresponds to the decimal number

$$.0999999\ldots$$

and exactness has been lost (see Appendix A for another example using a sixteen-bit word). This lack of exactness becomes very noticeable after thousands of arithmetic operations have been performed. "Roundoff error" is the term used to describe the error introduced by the cumulative truncation of results.

One possible remedy to this problem, when long calculation chains must be handled, is to use double the standard word size for key variables. These variables, with twice the storage capacity, are called **double-precision variables.** (Real variables are represented in **single precision.**) For a computer with a thirty-two-bit architecture, a double-precision variable will have sixty-four bits available for storage and representation; that is, two consecutive storage words are combined to form a double-sized word. The second word contributes all its thirty-two bits toward representation of the **mantissa** or **fractional part** of the number. The mantissa then becomes fifty-six bits long. Note that the data's magnitude limits have not increased, since the double-sized word still only has a seven-bit exponent, nor has its binary representation become exact. Accuracy has increased though, and roundoff errors will be reduced in arithmetic operations.

A sixty-four-bit word with a fifty-six-bit fractional part is capable of storing and representing approximately fourteen decimal digits. Although impressive sounding, the reader is reminded that this type of decimal significance is uncalled for in normal engineering work because we cannot measure, build, or design to these tolerances. Employing double-precision calculations would, in general, be misleading. It would also waste expensive computer time, since double-precision operations take longer to perform. The general rule is to not use double precision unless one has to. Sometimes it is necessary, though, and the following material should, therefore, be studied carefully.

8.4.2 Double-Precision Variables and Constants

A **double-precision constant** is represented in exponential form. A "D" replaces the "E" in the exponent. For example,

```
3.2D-6
.196432148732D4
3.21468921712D+2
```

are all double-precision constants. Remember that the internal representation of 3.2E-6 and 3.2D-6 are different. If 3.2E-6, a single-precision constant, is assigned to a double-precision variable, only twenty-four bits of significant (mantissa) information are present, and the second part of the storage word (the extra thirty-two bits) will be filled automatically with zeros. This would not be the result if 3.2D-6 were assigned to a double-precision variable, as the mantissa of this number has full fifty-six-bit representation to begin with; that is, its last thirty-two bits are nonzero. To take complete advantage of double-precision processing, mixed-mode operations should be avoided: only integer values, double-precision constants, and double-precision variables or functions should appear in arithmetic expressions. (Integer values are expressed exactly internally, and thus do not reduce calculation accuracy.)

A double-precision variable is declared in a nonexecutable DOUBLE PRECISION **type statement.** The statement has the form

DOUBLE PRECISION double-precision variable list

Any legal FORTRAN name can be used as a variable name. Each variable name on the list is separated from the others by a comma. Since it is a specification statement, it must appear before the first executable statement of the program. Some examples of DOUBLE PRECISION statements are

```
DOUBLE PRECISION LOW,ABLE

DOUBLE PRECISION ARBET(10,10)
```

To specify that a function has a double-precision value, one can declare it in a FUNCTION statement:

```
DOUBLE PRECISION FUNCTION DBL(X,Y,N)
```

8.4.3 Double-Precision Operations

The double-precision operations available in FORTRAN are identical to those provided in single precision. PARAMETER and DATA statements may be used to initialize variables and constants:

```
DOUBLE PRECISION PI,DBY,DBZ
```

```
PARAMETER(PI=.3141593D1)

DATA DBY,DBZ/1.32146D0,.83214D1/
```

The result of a mixed-mode operation involving double-precision- and real- or integer-type data is a double-precision data value. The real or integer operands are first converted to the double-precision type by the computer, which adds a string of thirty-two 0s to the right of the number's binary internal representation, forming a sixty-four-bit word, before the operation is carried out. (The sole exception occurs when raising a double-precision value to an integer power: the integer need not be corrected by the computer.) If a double-precision value is assigned to a real variable, it will be truncated after the first storage word. Some typical double-precision operations follow.

```
PROGRAM DOUBLE
DOUBLE PRECISION DBX,DBY,DBA
DBA=3.0D2
DBX=(DBA)**2+14.5D0
DBY=(DBX+7.1D0)-DBA
```

8.4.4 Double-Precision Intrinsic Functions

Double-precision values for the common intrinsic functions can be obtained by using double-precision arguments for the applicable generic function (see Appendix B). Usually, for program clarity, the double-precision function name is used instead of the generic name. Most of the basic library functions provided have a generic, real, integer, complex, and double-precision version. A double-precision function is identified by a "D" appearing before the generic name, and only double-precision arguments can be used for these functions. For example,

DINT	DMIN1	DTAN	DTANH
DNINT	DSQRT	DASIN	
DABS	DEXP	DACOS	
DMOD	DLOG	DATAN	
DSIGN	DLOG10	DATAN2	
DDIM	DSIN	DSINH	
DMAX1	DCOS	DCOSH	

Some examples of statement using double-precision library functions follow.

```
DBX=3.2D0*DSIGN(DBY)

DBY=DABS(DBC)+DBX

DBZ=DCOS(DBC)
```

There are three intrinsic library functions that are specifically used for double-precision processing: DBLE, DPROD, and REAL (or SNGL). DBLE converts any argument to a double-precision value. For complex arguments, only the real part is used. DPROD converts the product of two real arguments to a double-precision number. The generic function REAL converts any argument to a real type.

8.4.5 Double-Precision Data Input and Output

List-directed output of double-precision data is accomplished through the statement

PRINT *,double-precision variable

The data is printed in the same way it would appear for real, (single-precision) variables except that more significant digits are displayed (a "D" replaces the "E" in the exponent).

List-directed input uses the READ statement:

READ *,variable list

or

READ(i,*) variable list

Data values can be entered in any format, separated by a blank or a comma.

For formatted input/output operations, the F, E, or **double-precision specifier** can be used. The double-precision specifier has the form

D$w.d$

where w is the field width and d is the number of decimal positions. For formatted output, the field width should make provision for a sign, the decimal point, the decimal portion, the sign and magnitude of the exponent, and the symbol "D" for the exponent. A leading zero may appear before the decimal point if results are negative. The form of output obtained using a D descriptor is almost identical to that produced from E-format code (see Section 3.6.7.) For example, if the double-precision variables DBA = 1.36248742814D0, DBB = −.7143216D3 and DBC = 3.2D−2 are written with the following format,

```
    PRINT 2,DBA,DBB,DBC
  2 FORMAT(1X,2D19.12,D21.7)
```

the terminal output will appear as

```
___.136248742814D+01_−.714321600000D+03_____.3200000D−01
```

Since E and F format can also be used to output double-precision values, the FORMAT statement

```
  2 FORMAT(1X,2E19.12,E21.7)
```

produces almost identical output for the previous data:

```
___.136248742814E+01_-.714321600000E+03_____.3200000E-01
```

Using

```
2 FORMAT(1X,2F19.12,F21.7)
```

produces the results

```
_____1.362487428140__-714.311600000000_____.0320000
```

For formatted input of double-precision data, D, E, or F field descriptors can be used. For the F field descriptor, the following statements,

```
DOUBLE PRECISION DBR,DBS,DBT
READ(8,11) DBR,DBS,DBT
11 FORMAT(F15.10,2X,F10.9,2X,F10.0)
```

referencing the data line

```
-461.7423481248__6413264567__2.64001423
```

will result in the assignments

```
DBR=-461.7423481248
DBS=6.413264567
DBT=2.64001423
```

Using E formatting, the new FORMAT statement,

```
11 FORMAT(E17.13,2X,E12.9,2X,E12.0)
```

and the data line

```
-.4617423481248E3__6413264567E0__.264001243E1
```

will achieve identical results. One can also change "E" to "D" in the data line,

```
-.4617423481248D3__6413264567D0__.264001243D1
```

and use the FORMAT statement

```
11 FORMAT(D17.13,2X,D12.9,2X,D12.0)
```

to enter the same data.

Sample Problem 8.6

Surprisingly, even for the fastest digital computer, using the simple, well-known method of applying determinants to solve simultaneous equations can be shown to take solution times of *years* for larger systems of equations because of the great number of multiplications required. Therefore, other numerical algorithms are needed for the computer to effect solutions. One such method is called the Gaussian elimination method. In this method, equations are combined consecutively to eliminate unknowns.

The final elimination step leaves the last equation written in terms of one unknown. This unknown can be solved for directly, and the others can then be found by back substitution. For this problem, write a FORTRAN subroutine, using double precision, that will solve a set of two simultaneous equations using the Gaussian elimination method.

Analysis: To demonstrate the Gaussian elimination method, we can start with the two simultaneous equations

$$a(1, 1)x(1) + a(1, 2)x(2) = b(1)$$

$$a(2, 1)x(1) + a(2, 2)x(2) = b(2)$$

where $x(1)$ and $x(2)$ are the unknowns, $a(1, 1)$, $a(1, 2)$, $a(2, 1)$, and $a(2, 2)$ are the equation's coefficients, and $b(1)$ and $b(2)$ are the equation's constants. Through Gaussian elimination, the equations can be rewritten in the form

$$a(1, 1)x(1) + a(1, 2)x(2) = b(1)$$

$$0 \times x(1) + \left\{ a(2, 2) - \left[\frac{a(2, 1)}{a(1, 1)} \right] a(1, 2) \right\} x(2) = \left\{ b(2) - \left[\frac{a(2, 1)}{a(1, 1)} \right] b(1) \right\}$$

This yields the immediate result for $x(2)$

$$x(2) = \frac{b(2) - \left[\dfrac{a(2, 1)}{a(1, 1)} \right] b(1)}{a(2, 2) - \left[\dfrac{a(2, 1)}{a(1, 1)} \right] a(1, 2)}$$

and by back substitution, the result for $x(1)$,

$$x(1) = \frac{b(1) - a(1, 2)x(2)}{a(1, 1)}$$

Unfortunately, reduction methods require repetitive calculations for the equations' coefficients, and roundoff error is thus introduced. These roundoff errors can be particularly troublesome when trying to solve simultaneous equations whose coefficients are almost identical. A small change in the coefficients or constants of one equation can produce entirely different results. Such a system is said to be ill-conditioned. Take, for example, the set of equations

$$8.0x(1) + 4.0x(2) = 24.0$$

$$7.99999x(1) + 4.0x(2) = 23.99999$$

which has as a solution $x(1) = 1.0$ and $x(2) = 4.0$. However, the equations

$$8.0x(1) + 4.0x(2) = 24.0$$

$$8.00001x(1) + 4.0x(2) = 23.99999$$

have the result $x(1) = -2.0$ and $x(2) = 10.0$. As a minimal improvement, it is suggested that when working with ill-conditioned systems double precision be used.

Solution: The algorithm and problem solution are shown in Figures 8.9a and 8.9b. The subroutine is written in general terms so that it can readily be expanded (see CP16, this chapter). It should also be noted that the program will not work if A(K, K) is, or

■ **Algorithm for Sample Problem 8.6.** This subroutine finds the solution to
a set of two simultaneous equations by the Gaussian elimination method.
Double precision is used to improve accuracy and reduce roundoff errors.
The coefficient matrix is A, the given constants lie in vector B, and X is the
solution vector.

1. Print A,B.

2. Set K=1,N=2.

3. Do, with index I ranging from K+1 to N, the following:
  a. Do, with index J ranging from K to N, the following:
     i. If A(K,K)=0.0 then
      (a) Print 'Zero value on diagonal.'
      (b) Stop
     ii. Calculate
      A(I,J)=A(I,J)-A(I,K)*A(K,J)/A(K,K)
  b. Calculate
    B(I)=B(I)-A(I,K)*B(K)/A(K,K)

4. Calculate
  X(N)=B(N)/A(N,N)

5. Do, with index I ranging from N-1 to 1 in decrements of 1, the following:
  a. Set X(I)=0.0
  b. Do, with index J ranging from I+1 to N, the following:
     i. Calculate
      X(I)=X(I)+A(I,J)*X(J)
  c. Calculate
    X(I)=(B(I)-X(I))/A(I,I)

6. Print X.

7. Return to the calling program.

Figure 8.9a

```
      SUBROUTINE GAUSS (A,B,X)

*******************************************************************
*                                                                 *
*    THIS SUBROUTINE FINDS THE SOLUTION TO A SET OF TWO           *
*    SIMULTANEOUS EQUATIONS BY THE GAUSSIAN ELIMINATION METHOD.   *
*    DOUBLE PRECISION IS USED TO IMPROVE ACCURACY AND REDUCE      *
*    ROUNDOFF ERRORS.                                             *
*                                                                 *
*******************************************************************

      DOUBLE PRECISION A(2,2),B(2),X(2)
      INTEGER K,N,I

      PRINT 5
      PRINT 6,((A(I,J),J=1,2),I=1,2)
```

Figure 8.9b

```
      PRINT 7
      PRINT 6,(B(I),I=1,2)

      K=1
      N=2

      DO 1 I=K+1,N

          DO 2 J=K+1,N

              IF(A(K,K).EQ.0.0) THEN
                  PRINT*,'ZERO VALUE ON THE DIAGONAL.'
                  STOP

              END IF

              A(I,J)=A(I,J)-A(I,K)*A(K,J)/A(K,K)

    2         CONTINUE
                  B(I)=B(I)-A(I,K)*B(K)/A(K,K)

    1 CONTINUE

      X(N)=B(N)/A(N,N)

      DO 3 I=N-1,1,-1
          X(I)=0.

          DO 4 J=I+1,N
              X(I)=X(I)+A(I,J)*X(J)

    4         CONTINUE

          X(I)=(B(I)-X(I))/A(I,I)

    3 CONTINUE

      PRINT 8
      PRINT 6,(X(I),I=1,2)

    5 FORMAT(' THE VALUE OF [A] IS'/)
    6 FORMAT(2(5X,D14.7))
    7 FORMAT(/' THE VALUE OF [B] IS'/ )
    8 FORMAT(/' THE VALUES FOR X ARE'/)

      RETURN
      END


RUN

 THE VALUE OF [A] IS

        .4001000D+01        .2310000D+01
        .1040000D+01        .3000200D+01

 THE VALUE OF [B] IS

        .1250000D+02        .2780000D+01

 THE VALUES FOR X ARE

        .3237101D+01       -.1955152D+00
```

Figure 8.9b   (cont.)

becomes (in a larger problem), zero. Larger programs usually employ some sort of pivot method to rearrange the equations so that $A(K, K)$ is assigned the largest possible coefficient value. This not only prevents zero from occurring there, but is also helps to reduce program roundoff error (see CP17, this chapter).

The reader should run the subroutine given here with *and without* double precision for the foregoing ill-conditioned problem. The results will prove surprising.

*8.5 COMPLEX DATA

8.5.1 Introduction

The branch of mathematics dealing with complex numbers and variables is extremely important to engineering work. For instance, the response of many physical systems is dependent not only on the magnitude of the driving mechanical or electrical input that is present but also on how frequently the input repeats itself, that is, the driving frequency. The behavior or characteristics of the system are said to be frequency dependent. Formally, one speaks of the "frequency response" or "characteristic response" of the system to a harmonic input (a sine or cosine wave). One can write this mathematically as

$$\text{Characteristic response } (\omega) = \frac{\text{output}}{\text{input}}$$

where ω is the frequency of the harmonic input. Complex analysis makes it possible to derive a simple expression for the characteristic response as a function of frequency directly from the system (differential) equation. In addition, if the input can be represented as a sine or cosine wave (or a series of waves using the methods of Fourier analysis), the output of the system can be found using complex algebra rather than finding the formal time solution to the differential equation (see Sample Problem 8.8). Because of its importance, complex algebra will be reviewed in the next section.

8.5.2 Basic Complex Algebra

A complex number z has the form

$$z = u + iv \tag{8.1}$$

where u and v are real numbers. The variable u is called the **real part,** and v is called the **imaginary part** of the number. The imaginary unit, i, is a number such that $i^2 = -1$. The complex number $u + iv$ can be geometrically represented by a complex vector located in the complex plane that originates from the origin and extends to the point (u, v). The vector has the rectangular components u and v, where u represents the abscissa and v the ordinate of the point referred to by a pair of orthogonal x- and y-axes. The x-axis is called the real axis, and the y-axis is called the imaginary axis (see Figure 8.10). If polar coordinates are used,

$$z = r(\cos \theta + i \sin \theta) \qquad (8.2)$$

where

$$r = |z| = (u^2 + v^2)^{1/2} \qquad (8.3)$$

the magnitude of z, and

$$\theta = \tan^{-1} \frac{v}{u} \qquad (8.4)$$

θ is the angle measured from the positive x-axis to the vector z. Note that

$$u = r \cos \theta \qquad (8.5)$$

and

$$v = r \sin \theta \qquad (8.6)$$

It is also possible to express z in exponential form as

$$z = re^{i\theta} \qquad (8.7)$$

because of the relationship

$$e^{i\theta} = \cos \theta + i \sin \theta \qquad (8.8)$$

The sum or difference of the two complex numbers

$$z = u + iv = r_1 e^{i\theta_1}$$

and

$$w = s + it = r_2 e^{i\theta_2}$$

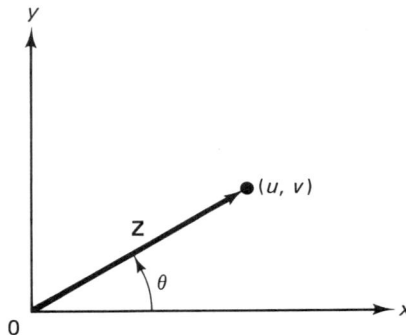

Figure 8.10

is given by

$$z \pm w = (u \pm s) + i(v \pm t) \tag{8.9}$$

The product of the two is defined as

$$z \cdot w = (us - vt) + i(ut + vs) \tag{8.10}$$

or

$$z \cdot w = r_1 r_2 e^{i(\theta_1 + \theta_2)} \tag{8.11}$$

The quotient of z and w is

$$\frac{z}{w} = \frac{(us + vt) + i(vs - ut)}{s^2 + t^2} \tag{8.12}$$

or

$$\frac{z}{w} = \frac{r_1}{r_2} e^{i(\theta_1 - \theta_2)} \tag{8.13}$$

The complex exponential e^z is

$$e^z = e^u(\cos v + i \sin v) \tag{8.14}$$

and z to any power is given by

$$z^a = e^{a \ln z} \tag{8.15}$$

The series representation for functions of complex variables is similar to that for real variables. For example,

$$\sin z = z - \frac{z^3}{3!} + \frac{z^5}{5!} - \cdots$$

and

$$\cos z = 1 - \frac{z^2}{2!} + \frac{z^4}{4!} - \cdots$$

Finally, the complex conjugate $\bar{z}$ of a number z is defined by

$$\bar{z} = u - iv$$

which is a reflection of the complex vector **z** about the real axis.

8.5.3 Complex Variables and Constants

In FORTRAN, **complex constants** are written as

$$(u, v)$$

where u and v are floating-point (single-precision) numbers representing the real and imaginary parts of the complex number, respectively. For example, the complex numbers

$$z_1 = 2.0 + i3.0$$

$$z_2 = -6.23 + i7.413$$

$$z_3 = -3.2 - i3.2$$

would be represented in FORTRAN by

```
Z1=(2.0,3.0)
Z2=(-6.23,7.413)
Z3=(-3.2,-3.2)
```

Complex data requires two consecutive words in memory for representation; therefore, double-precision numbers cannot be used because they would require four storage words.

Complex variables are declared using the COMPLEX **type statement.** The COMPLEX specification statement is nonexecutable and must appear before the first executable statement in the program. The COMPLEX statement has the form

COMPLEX complex variable list

The names on the variable list are separated by commas and must be legal FORTRAN names. For example,

```
COMPLEX A,B
```

or, for a complex array,

```
COMPLEX Z(10)
```

A complex-valued function may be specified through the complex FUNCTION statement,

```
COMPLEX FUNCTION ZETA(R,Y)
COMPLEX R,Y
     .
     .
     .
END
```

or by

```
FUNCTION ZETA(R,Y)
COMPLEX ZETA,R,Y
     .
     .
     .
END
```

8.5.4 Complex Operations

All the FORTRAN arithmetic operators are available for complex operations; however, complex variables can only be raised to an integer power (exponentiation). The result of an expression involving solely complex constants, complex variables, or complex-valued functions is itself complex. The result of a mixed-mode operation involving complex data and real or integer data is also a complex value. In mixed-mode operations, real or integer data is first converted to a complex value having an imaginary part equal to 0 before the operation is performed. For example, if x is real and equal to 2.3 and z is equal to $1.1 + i2.3$,

```
X    ⇒ (2.3,0.0)
```

and

```
Z+X ⇒ (3.4,2.3)
```

When a complex value is assigned to a floating-point or integer variable, the imaginary part of the complex value is discarded, and only the real part is kept. For example, the statements

```
REAL X
COMPLEX Z
INTEGER N
Z=(3.2,4.3)
X=Z
N=Z
```

will result in $X = 3.2$ and $N = 3$.

The only relational operators allowed for complex data are .EQ. and .NE.

8.5.5 Complex Intrinsic Functions

If a complex argument is used with a FORTRAN supplied generic function, the result will be complex (see Appendix B). Usually, for program clarity, the complex function's name is used directly. The basic library functions having a complex version (a "C" precedes the generic name) are

CABS	CLOG
CSQRT	CSIN
CEXP	CCOS

The arguments to these functions must be complex. There are, however, four intrinsic library functions that are specifically provided for complex data: AIMAG, CMPLX, CONJG, and REAL. The first, AIMAG(Z) returns the imaginary part of Z as a floating-point number. The complement to AIMAG(Z) is REAL(Z), which returns the real part of Z as a floating-point value. CMPLX is an important function, since it converts one or two integer, real, or double-precision arguments to a complex value. For instance, if S and T are real variables and Z is declared as complex, the following statements would assign the value (2.1,4.6) to Z.

```
S=2.1
T=4.6
Z=CMPLX(S,T)
```

The reader should verify that the following will *not* work:

```
S=2.1
T=4.6
Z=(S,T)
```

If only one argument is given, it is assumed that the imaginary part of the function is zero. For example,

```
Z=CMPLX(S)
```

implies

```
Z=CMPLX(S,0.0)
```

The last function given, CONJG, converts a complex number to its complex conjugate form.

8.5.6 Complex Data Input and Output

PARAMETER and DATA statements may be used to initialize complex data as follows:

```
COMPLEX ZA,ZB
PARAMETER(ZA=(3.0,2.0))
DATA ZB/(2.013,-6.247)/
```

To input complex data using list-directed format, the following statements are used:

READ *,complex variable list

or

READ(i, *) complex variable list

The data must be entered in pairs, enclosed within parentheses. The first value within the parentheses will become the real part of the complex value; the second value will become the imaginary part. A comma separates the real and imaginary parts. Each complex constant's paired values are separated from any others that may be present by a blank, a comma, or a carriage return. Data values can be entered in any format except double precision.

The PRINT statement can be used for list-directed output of complex data. It has the form

PRINT *,complex variable list

The statement causes complex values to be displayed as two single-precision numbers, separated by commas, and enclosed within parentheses.

Formatted input of complex values requires an F or E descriptor, that is, a total of two descriptors for each complex variable entered. The first field of data described will contain the value for the real part of the complex number; the second field of data described will hold the complex number's imaginary part. The parentheses are *not* used for formatted input of complex values, nor do commas separate entries. For instance, the READ statement for the complex variables ZA and ZB,

```
READ(8,11) ZA,ZB
```

having the associated format statement

```
11 FORMAT(F5.3,1X,F4.0,1X,E8.2,E9.0)
```

and the data line

```
3.200 7.0    .34E1 -0.26E-1
```

will assign ZA $= (3.2, 7.0)$ and ZB $= (3.4, -.026)$.

Formatted output of complex values also requires a pair of E or F descriptors for each complex variable. In addition, it is recommended that descriptive characters such "+I" or "(,)" be included with the output of each complex variable displayed. For the previous data, the statements

```
   PRINT 2,ZA
 2 FORMAT(1X,E8.2,' + I',E8.2)
   PRINT 3,ZB
 3 FORMAT(1X,'(',E8.2,',',E8.2,')')
```

will produce the output

```
 .32E+01 + I .70E+01
 (  .34E+01,-.26E-01)
```

Sample Problem 8.7

Evaluate the products of the complex functions shown for values of ω ranging from 0 to 5 in increments of 0.1.

$$z = (3.0 + i2.0\omega)(1.0 + i\omega)^3$$

Solution: The algorithm and problem solution are given in Figures 8.11a and 8.11b.

■ **Algorithm for Sample Problem 8.7.**

1. Do, with index W ranging from 0.0 to 5.0 in increments of 0.1, the following:
   a. Calculate
   ```
   Z1=CMPLX(3.0,2.0*W)
   Z2=CMPLX(1.0,W)
   Z=Z1*Z2**3
   ```
   b. Print Z.

Figure 8.11a

```
PROGRAM CPROD

COMPLEX Z1,Z2,Z
REAL W

DO 1 W=0.0,5.0001,0.1

    Z1=CMPLX(3.,2.*W)
    Z2=CMPLX(1.,W)
    Z=Z1*Z2**3

    PRINT 2,Z

1 CONTINUE
2 FORMAT(' THE VALUE OF Z IS (',E10.3,',',E10.3,')')

    END

RUN

THE VALUE OF Z IS (  .300E+01,  .000E+00)
THE VALUE OF Z IS (  .285E+01,  .109E+01)
THE VALUE OF Z IS (  .240E+01,  .213E+01)
    .
    .
    .
```

Figure 8.11b

Sample Problem 8.8

The ordinary differential equation representing the system dynamics of a spring-mass system can be written as

$$M\frac{d^2x(t)}{dt^2} + Kx(t) = F \sin \omega t = f(t)$$

where M is the mass, K is the spring constant, $f(t)$ is the input function, t is the time, and $x(t)$ is the displacement measured as a function of time from the unstretched position of the spring (see Figure 8.12). The ratio of the system's displacement (output) to the

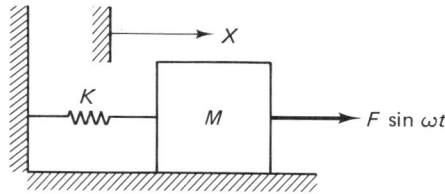

Figure 8.12

applied harmonic force (input) is called the characteristic response of the system. Since it is dependent on the frequency of the force, it is also known as the frequency response of the system. Prove that the characteristic response for this problem is given by

$$\text{Characteristic reponse}(\omega) = \frac{\dfrac{1}{K}}{1 - \left(\dfrac{\omega}{\omega_n}\right)^2}$$

where ω_n is defined as the natural frequency of the system and is equal to $(K/M)^{1/2}$. Write a FORTRAN program that will find the value of the characteristic response for $\omega = A\omega_n$. Use A values from 0 to 4 in increments of 0.1. Exclude the case in which A equals 1. Take $\omega_n = 10.0$ and $K = 1.5 \times 10^5$ newtons/meter. (A system's behavior as ω approaches ω_n or becomes much larger or smaller than ω_n is of particular importance. Note that when $A = 1$, a condition of resonance occurs and the denominator goes to 0.) Although not strictly necessary, since damping is neglected (see CP24 this chapter), treat the characteristic response as a complex variable.

Analysis: If the forcing function is sinusoidal, it can be described by the equation

$$f(t) = F \sin \omega t$$

Here, F is a constant, and ω is the applied frequency. Using complex notation, $f(t)$ can also be represented as Im $(Fe^{i\omega t})$, the imaginary part of $Fe^{i\omega t}$, since $e^{i\omega t} = \cos \omega t + i \sin \omega t$. A particular trial solution, x_p, to the equation can then be taken as

$$x_p = Be^{i\omega t}$$

where only the imaginary part of the final solution is to be retained. Then

$$M\frac{d^2x_p}{dt^2} + Kx_p = f(t) = Fe^{i\omega t}$$

becomes

$$(-M\omega^2 + K)Be^{i\omega t} = Fe^{i\omega t}$$

The characteristic response, the ratio of displacement to applied force, is therefore given by

$$\text{Characteristic response } (\omega) = \frac{x_p}{Fe^{i\omega t}} = \frac{\dfrac{1}{K}}{1 - \dfrac{M\omega^2}{K}}$$

$$= \frac{\dfrac{1}{K}}{1 - \left(\dfrac{\omega}{\omega_n}\right)^2}$$

where ω_n equals $(K/M)^{1/2}$ and is the natural frequency.

Solution: The algorithm and problem solution are shown in Figures 8.13a and 8.13b. The program prevents a condition of resonance from occurring by passing over the calculation when A equals 1. It also provides for the eventuality of roundoff error when A

Algorithm for Sample Problem 8.8. This program finds the characteristic response, CHARES, of a spring-mass system to a sinusoidal input. CHARES is treated as a complex variable. The program avoids the resonant condition, where $A = 1$. Note that the system's spring constant is K.

1. Initialize K.
2. Print heading.
3. Do, with index A ranging from 0.0 to 4.0 in increments of 0.1, the following:
   a. If A<0.9999 or A>1.0001 then
      i. Calculate
      CHARES=CMPLX((1.0/K)/(1.0-A**2))
      ii. Print A,CHARES.

Figure 8.13a

```
          PROGRAM RESPON

     ****************************************************************
     *                                                              *
     *    THIS PROGRAM FINDS THE CHARACTERISTIC RESPONSE, CHARES, OF *
     *    A SPRING-MASS SYSTEM TO A SINUSOIDAL INPUT. CHARES IS      *
     *    TREATED AS A COMPLEX VARIABLE. THE PROGRAM AVOIDS THE      *
     *    RESONANT CONDITION WHERE A=1. NOTE THAT THE SYSTEM'S SPRING *
     *    CONSTANT IS K.                                             *
     *                                                              *
     ****************************************************************

          REAL K,A
          COMPLEX CHARES
```

Figure 8.13b

```
        DATA K/1.5E5/

        PRINT 1

        DO 4 A=0.0,4.0001,0.1

            IF((A.LT.0.9999).OR.(A.GT.1.0001)) THEN
                CHARES=CMPLX ((1./K)/(1.-A**2))

                PRINT 2, A,CHARES

            END IF

    4 CONTINUE

    1 FORMAT(T12,'FREQ.RATIO',27X,'CHAR.RESPONSE',/)
    2 FORMAT(T13,E9.2,21X,E10.3,'  +I',E10.3)

        END

  RUN

        FREQ.RATIO                          CHAR.RESPONSE

            .00E+00                    .667E-05   +I   .000E+00
            .10E+00                    .673E-05   +I   .000E+00
            .20E+00                    .694E-05   +I   .000E+00
            .30E+00                    .733E-05   +I   .000E+00
              .
              .
              .
```

Figure 8.13b   (cont.)

---

is compared; for example, A becomes 0.9999999 or 1.000001 in the DO loop counter, instead of assuming the exact value 1.000000.

REVIEW

This chapter discusses the four remaining data types of FORTRAN: character, logical, double precision, and complex.

Character Data

FORTRAN character data has nonnumeric symbolic form and consists of the following set of characters:

ABCDEFGH I JKLMNOPQRSTUVWXYZ

0123456789

+ – ( ) * / = . ' $: b̲ (a blank character)

Character data values, or constants, are called strings; they are a sequence of symbols taken from the character set. A character constant must be enclosed within apostrophes.

Character variables are declared with a CHARACTER type statement. The allowable forms of the CHARACTER statement are

CHARACTER*n character variable list

CHARACTER character variable list

CHARACTER*n character variable list, character
             variable*m, character variable list

CHARACTER character variable$_1$*n_1, character variable$_2$*n_2, . . .

The character variable list is a list of acceptable FORTRAN variable names separated by commas, and n, m, n_1, and n_2 are the length of the character strings that can be assigned to each variable.

Character arrays are specified by

CHARACTER*n array(m)

or

CHARACTER array(m)*n

where m is the array size and n is the string size.

Character functions are declared with a FUNCTION statement:

CHARACTER*n FUNCTION name(argument list)

When character variables appear as the formal arguments of a subprogram, they must be declared not only in the calling program but in the subprogram as well. In a subprogram, the assumed length specifier (*) may be used in the CHARACTER statement:

CHARACTER* (*) variable list

A substring is a subset of adjacent characters that can be formed from a string. Any substring is defined by the following:

string name($I:J$)

where I and J are integer constants, integer variable names, or integer variable expressions. The position of the first character of the substring is given by I, and J is the position of the last character. Character strings may be linked together using the concatenation operator symbolized by //.

It is possible to compare strings using relational operators. In FORTRAN, the collating sequence for characters is from A to Z for letters, and from 0 to 9 for digits. Blanks precede any character in ASCII or EBCDIC code. Characters are compared with one another from left to right using the established collating sequence.

FORTRAN provides several intrinsic functions for character data. INDEX has the format

INDEX(string$_1$, string$_2$)

INDEX provides the integer position of the first appearance of a specified string (string 2) within another string (string 1). The function LEN has as its output the integer value representing the length of the character string. It is called by

LEN(string)

The integer position of a character in the collating sequence of the host computer is found by ICHAR. ICHAR has the form

ICHAR (character)

The inverse function is CHAR. Its output is the character in the host computer's collating sequence that has the given integer position.

CHAR(integer)

The lexical functions LGE, LGT, LLE, and LLT can be used instead of the relational operators. They specify that comparisons should be made using the ASCII collating sequence, regardless of whether that is the standard character set of the host computer.

List-directed output of character data is accomplished with the PRINT statement:

PRINT *,'character string'

The character string written between the apostrophes is written exactly as it appears. When using list-directed input, the data must also be enclosed within apostrophes.

Formatted input/output operations use the alphanumeric format specifier A. It has the form

A*w* or A

where *w* stands for the field width. If *w* does not appear, it is taken to be equal to the declared length of the character variable. Apostrophes are not used to enclose formatted input.

Logical Data

Logical constants signify true or false conditions. To write a logical constant one uses the words .TRUE. or .FALSE. The preceding and ending periods are a necessary part of the string.

A logical variable name is declared by a LOGICAL type statement. It has the syntax

LOGICAL logical variable list

Logical values are assigned to logical variable names using an assignment statement.

logical variable name=logical expression

A logical expression can be a logical constant, a logical variable, or the result of a relational comparison (which is a logical constant). The relational operators are

.EQ. .GT. .GE. .LT. .LE. .NE.

Logical operators are available to form compound logical expressions. The logical operators are

.AND. .OR. .EQV. .NEQV. .NOT.

List-directed output of logical data is displayed as a simple "T" or "F." No periods or other characters are displayed. For list-directed input, the first letter appearing after a blank space, comma, or a period must be a "T" or an "F."

For formatted input/output operations, the logical format specifier is used. It has the form

Lw

where w is the field width. On output, only a "T" or an "F" will be written. On input, the first nonblank character entered must be a period or a "T" or an "F."

Double-Precision Data

Double-precision data is numeric data that is assigned to two consecutive storage places in memory. Real or floating-point data (single-precision data) is stored in one storage word. Double-precision constants in FORTRAN are given in exponential form with a "D" replacing the "E" in the exponent. A double-precision variable is declared with a DOUBLE PRECISION specification statement. It has the form

DOUBLE PRECISION double-precision variable list

Any legal FORTRAN name can be used as a variable name. Functions are declared in a FUNCTION statement.

DOUBLE PRECISION FUNCTION DBL(X,Y,N)

All the FORTRAN arithmetic operators, as well as all the FORTRAN generic intrinsic functions (see Appendix B), are available for double-precision work. Specific double-precision function names are also provided. These intrinsic functions are identified by the character "D" that precedes the generic name. There are also special library functions set aside solely for double-precision work: DBLE, DPROD, and REAL (or SNGL). DBLE converts any argument to a double-precision value. For complex arguments, only the real part is used. DPROD converts the product of two real arguments to a double-precision number. The generic function REAL converts any argument, including a double-precision argument, to a real type. The specific form SNGL converts a double-precision argument to a real type.

List-directed output of double-precision data is printed out in the same way it would appear for real variables except more significant digits are displayed and a "D" replaces the "E" in the exponent. For list-directed input, data values can be entered in any format.

Formatted input/output operations employ the double-precision specifier, "D." The syntax is

$Dw.d$

where w is the field width and d is the number of decimal positions retained. E and F format can also be used to output double-precision values. For formatted input of double-precision data, D, E, or F field descriptors are employed. If a D format descriptor is used, the constants entered should have a "D" replace the "E" in the exponent.

Complex Data

In FORTRAN complex constants are expressed as (u, v) where u, the real part, and v, the imaginary part are single-precision numbers separated by a comma. The parentheses are required.

Complex-variables are declared by a COMPLEX type statement. Its syntax is

COMPLEX complex variable list

Any legal FORTRAN name may be used as a complex variable name. A complex-valued function is specified by a FUNCTION statement.

```
COMPLEX FUNCTION ZETA(R,Y)
```

In FORTRAN, one can add, subtract, multiply, and divide complex numbers. Exponentiation is also available, but the exponent must be an integer.

Most of the generic functions available in the FORTRAN library can be used with complex arguments. These functions also have a specific complex form that bears the generic name but precedes it with the letter C. There are four library functions that are designed solely to process complex data: AIMAG, CMPLX, CONJG and REAL. AIMAG returns the value of the imaginary part of the complex argument as a single-precision number. REAL is its complement function, returning the real part of the complex argument. CMPLX forms a complex variable from two real arguments. It has two possible forms:

```
CMPLX(X,Y)
```

or

```
CMPLX(X)
```

In the latter case, the imaginary part of the complex variable formed is set to zero.

List-directed output of complex data is displayed as two single-precision numbers, separated by a comma, enclosed within parentheses. List-directed input must also be enclosed within parentheses. The paired values are entered separated by a comma.

Formatted input/output operations of complex data require two F or E descriptors for each variable. The first descriptor is for the real part; the second descriptor is for the imaginary part. When the data is entered for formatted input, parentheses should not be used, nor should commas separate the two parts of the value.

KEY TERMS

AIMAG **function**
alphanumeric format specifier or descriptor
assumed-length specifier
CHAR **function**
character array
character constant
character data
CHARACTER **type statement**
character variable name
character variable expression
CMPLX **function**
collating sequence
complex constant
COMPLEX **type statement**
complex variable
compound logical expression
concatenation
concatenation operator
CONJG **function**
DBLE **function**
double-precision constant
double-precision specifier or descriptor

DOUBLE PRECISION **type statement**
double-precision variable
DPROD **function**
fractional part of a number
ICHAR **function**
imaginary part
INDEX **function**
LEN **function**
lexical functions: LGE, LGT, LLE, **and** LLT
logical constant
logical format specifier or descriptor
LOGICAL **type statement**
logical variable name
mantissa of a number
REAL **function**
real part
simple logical expression
single precision
SNGL **function**
string
substring

REVIEW QUESTIONS

1. How can an apostrophe be represented in a character string?
2. What is the string length of a character variable set to if its length is not specifically listed when it is declared in a CHARACTER type statement?

3. How is the length of a constant string determined?

4. What happens when a character variable is assigned a string length that is less than the variable's declared length?

5. What happens when a character variable is assigned a string length that is more than the variable's declared length?

6. What is the minimum allowable size of a substring?

7. Describe what values are assumed when limiting position indices are omitted in a substring reference.

8. What are the limiting values for the position indices of a substring?

9. May one assign one substring to another substring of the same character variable if their positions overlap?

10. Describe how character strings are compared.

11. Describe how comparisons are made between numbers and letters or special symbols, or letters and special symbols.

12. What is the purpose of the lexical functions?

13. Describe what happens when using list-directed input if the character string entered is longer or shorter than the declared length of the character variable it is assigned to.

14. Are apostrophes used to enclose formatted-input data?

15. For formatted input of character variables, describe what happens when the entered string length is longer or shorter than the declared length of the variable.

16. Repeat question 15 for formatted output.

17. Describe the order in which a complex logical expression is evaluated.

18. Describe how list-directed input of logical data is entered.

19. Describe how list-directed output of logical data appears.

20. Describe what happens for formatted output of logical variables if the field length used, w, is larger than 1.

21. Describe how formatted input of logical data works.

22. When is double-precision numerical work justified?

23. If DA is a double-precision variable, why is the assignment statement DA=0.1D+00 more accurate than the FORTRAN statement DA=0.1?

24. Describe what happens in a mixed-mode operation involving a complex value.

25. What happens when a complex value is assigned to a real or integer variable?

26. If the given READ statements refer to a single data record line, beginning in the first column, which contains

```
              |
     12345678901
     WIND ENERGY
```

indicate what output will appear for the statement blocks shown.

(a)	CHARACTER*4 A	(b)	CHARACTER*4 A	(c)	CHARACTER*11 A
	READ(7,1) A		READ(7,1) A		READ(7,1) A
	1 FORMAT(A)		1 FORMAT(A2)		1 FORMAT(A15)
	PRINT *,A		PRINT *,A		PRINT *,A

27. For the following problem, indicate what output will appear for the given statement blocks.

(a)
```
        CHARACTER*6  A
        A='WIND''S ENERGY'
        WRITE(7,1)  A
      1 FORMAT(1X,A)
```
(b)
```
        CHARACTER*6  A
        A='WIND''S ENERGY'
        WRITE(7,1)  A
      1 FORMAT(1X,A4)
```
(c)
```
        CHARACTER*6  A
        A='WIND''S ENERGY'
        WRITE(7,1)  A
      1 FORMAT(1X,A10)
```

28. For the following statements, indicate what values will be assigned to the character variable shown.

(a)
```
CHARACTER*4  B
B='FIFTY'
PRINT *,B
```
(b)
```
CHARACTER*4  B
B='TWO'
PRINT *,B
```
(c)
```
CHARACTER*5  B
B='TWO'//''''//'S'
PRINT *,B
```

29. Indicate whether the following comparisons are true or false.

(a)
```
A='ABC'
B='ABD'
A.GT.B
```
(b)
```
A=' ABC'
B='ABC'
A.LT.B
```
(c)
```
A='ABCD'
A.LE.'AB E'
```

30. If in the following problem A = 'WAVE' and B = 'POWER', find what output will appear for the character string variable C.

(a)
```
C=A//' '//B
PRINT *,C
```
(b)
```
C=A//' LENGTH'
PRINT *,C
```
(c)
```
C=A(:1)//B(2:3)
PRINT *,C
```

31. If the character string variable RECORD1 contains the constant 'DYNAMICS', find what output will result from the following statements.

(a) `PRINT *,RECORD1(1:)`   (b) `PRINT *,RECORD1(:5)`
(c) `PRINT *,RECORD1(4:4)`

32. Assuming that the character string variable FAR contains the character constant 'PROFESSIONAL ENGINEER', find what output will result from the following commands.

(a) `PRINT *,LEN(FAR)`   (b) `PRINT *,LEN(FAR(INDEX(FAR,' '):))`
(c) `PRINT *,INDEX(FAR,'E')`

33. If the logical variable LA is .TRUE. and logical LB is .FALSE., and the real variables C and D equal 2.1 and 3.2, respectively, determine whether the following compound logical expressions are true or false.

(a) `(.NOT.LA)`   (b) `(.NOT.LB.AND.C.LT.D)`   (c) `(C.EQ.D.OR.LA.OR.LB)`

34. If DA is a double-precision variable equal to 161.2746954 and LA is a logical variable that is .TRUE., what output will result from the following assignment statements?

(a)
```
        WRITE(6,1)  DA
      1 FORMAT(1X,D14.7)
```
(b)
```
        WRITE(6,1)  LA
      1 FORMAT(1X,L4)
```

*35. If the variables A, B, and C are complex, with A = $(1.0, 1.0)$ and B = $(2.0, 2.0)$, find what output will result from the following series of statements.

(a)
```
    C=B-A
    WRITE(6,1) C
  1 FORMAT(1X,2F3.1)
```

(b)
```
    C=B-A
    WRITE(6,1) C
  1 FORMAT(1X,F3.1)
```

(c)
```
    C=A+B
    WRITE(6,1) C
  1 FORMAT(1X,F3.1,' +I ',F3.1)
```

*36. If the complex variable A has a value $3 + i4$, and the complex variable B equals $2 - i6$, find

(a) $A + B$ (b) $B - A$

(c) $A \times B$ (d) A/B

(e) $|A|$ (f) e^A

(g) Write A in exponential form.

COMPUTER PROJECTS

CP1. Modify the program given in Sample Problem 8.2 to accommodate leading blanks in the character data field.

CP2. Write a FORTRAN subroutine that receives any fifty-character string and reverses the string's order.

CP3. Write a FORTRAN program that reads first-, middle-, and last-name entries from a fifty-character data field (excluding leading and trailing apostrophes) and changes the order of the data so that it is printed out at the terminal as follows:

last name, _ first name, _ middle initial

where _ represents a blank character. Assume that the first, middle, and last names are separated in the data field by a blank character. Also assume that there are no leading blanks appearing in the field before the first name.

CP4. Write a FORTRAN subroutine that will delete a character in an eighty-column text line. The character's position in the line (an integer value) is to be given as an argument of the subroutine. After the character is deleted in the line, the remaining text material should be shifted to the left.

CP5. Write a FORTRAN program that will translate an eighty-column line of alphanumeric text to International Morse Code. Print each character's Morse representation on a separate line, starting in column 5; that is, each text line will have eighty rows of output that begin in column 5.

International Morse Code (partial list)

A = .–	G = ––.	M = ––	S = ...	Y = –.––	5 = ...
B = –...	H = ...	N = –.	T = –	Z = ––..	6 = –....
C = –.–.	I = ..	O = –––	U = ..–	1 = .––––	7 = ––...
D = –..	J = .–––	P = .––.	V = ...–	2 = ..–––	8 = –––..
E = .	K = –.–	Q = ––.–	W = .––	3 = ...––	9 = ––––.
F = ..–.	L = .–..	R = .–.	X = –..–	4 = ...–	0 = –––––

CP6. Repeat CP5 but this time write a FORTRAN program that will read eighty lines of Morse code input from a file and convert the message to eighty columns of text output.

CP7. Write a FORTRAN program that will convert roman numerals contained within a fifteen-column data field (the number is left-justified in the field) to an arabic number. The base roman numerals are I = 1, V = 5, X = 10, L = 50, C = 100, D = 500, and M = 1000. Remember that when a smaller numeral appears directly before a larger numeral, the smaller numeral is subtracted; that is, IV = 4, IX = 9, XL = 40, XC = 90, CD = 400, and CM = 900. (See Sample Problem 8.5.)

CP8. Redo CP7 but this time have the FORTRAN program convert an arabic number contained within a three-column data field to a roman numeral.

CP9. Modify the solution to Sample Problem 8.3 so that a vertical scale for the vertical axis is included as well. Numbers on this scale should appear at every fifth row of output and represent the array element's index; that is, 5, 10, 15, ... should appear on the scale, properly placed.

CP10. It is desired to write a subroutine that will plot the function $y = f(x)$. The n values of y and x are to be read in through array arguments. The y-axis, horizontal, should be 121 columns long. The x-axis, vertical, should be placed in column 61. Both axes should be represented by periods. Scaling will be necessary for the y values: the horizontal scale factor should be determined by

$$\text{S.F.} = \frac{|y_{max}|}{60.0}$$

Test your subroutine using the function $y = 60.0 \cos(0.2\pi x)$, which is evaluated in the calling program. Have x vary from 0.0 to 10.0 in increments of 0.1. Use an "X" to plot each data point.

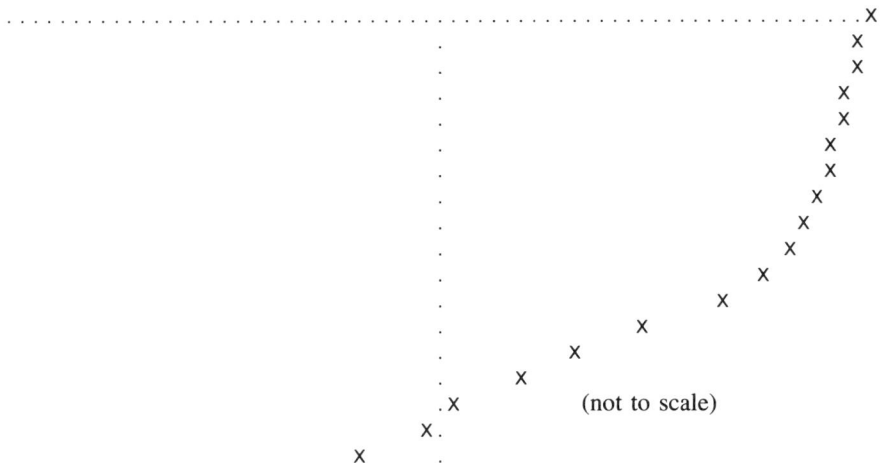

CP11. Repeat CP10 but this time plot two functions on the same graph. Test your subroutine with the functions $y_1 = 60.0 \cos(0.2\pi x)$ and $y_2 = 60.0 \sin(0.2\pi x)$, evaluated in the calling program, for x values ranging from 0.0 to 10.0 in increments of 0.1. Use the symbol "X" for curve y_1, and the symbol "Y" for curve y_2.

CP12. If in CP10 we now want to plot a horizontal bar graph instead, what modifications to the subroutine must be made? Test your results on the function $y = 60.0 \cos(0.2\pi x)$ for x values varying from 0.0 to 10.0 in increments of 0.1. Use an "X" character to fill in the graph.

```
. . . . . . . . . . . . . . . . . . . . . . . . . . . . . XXXXXXXXXXXXXXXXXXXXXXXXXXXXXXXX
                                  XXXXXXXXXXXXXXXXXXXXXXXXXXXXXXX
                                  XXXXXXXXXXXXXXXXXXXXXXXXXXXXX
                                  XXXXXXXXXXXXXXXXXXXXXXXXXXXXX
                                  XXXXXXXXXXXXXXXXXXXXXXXXXXXXX
                                  XXXXXXXXXXXXXXXXXXXXXXXXXXX
                                  XXXXXXXXXXXXXXXXXXXXXXXXXXX
                                  XXXXXXXXXXXXXXXXXXXXXXXXX
                                  XXXXXXXXXXXXXXXXXXXXXXXX
                                  XXXXXXXXXXXXXXXXXXXXXXX
                                  XXXXXXXXXXXXXXXXXXXX
                                  XXXXXXXXXXXXXXXXX
                                  XXXXXXXXXXXXXX
                                  XXXXXXXXXX
                                  XXXXXXX
                                  XX                    (not to scale)
                                 XX
                               XXXXXXX
```

CP13. Modify the solution to Sample Problem 8.3 so that the horizontal axis is 121 columns long and represents the y-axis. The first column of output should be the x-axis, which should be represented by a period. The bar graph symbols "X" should replace the period where appropriate (when data exists). Scaling will be necessary for the y values. The horizontal scale factor is determined by

$$\text{S.F.} = \frac{\text{maximum of data}}{120.0}$$

The number of "X" symbols for the maximum data value is therefore 121, including the x-axis. Test your results on the function $y = 60.0 + 60.0 \cos(0.2\pi x)$, evaluated in the calling program, for x values varying from 0.0 to 10.0 in increments of 0.1.

```
XXXXXXXXXXXXXXXXXXXXXXXXXXXXXXXXXXXXXXXXXXXXXXXXXXXXXXXXXXXXXXXXXXXX
XXXXXXXXXXXXXXXXXXXXXXXXXXXXXXXXXXXXXXXXXXXXXXXXXXXXXXXXXXXXXXXXXXX
XXXXXXXXXXXXXXXXXXXXXXXXXXXXXXXXXXXXXXXXXXXXXXXXXXXXXXXXXXXXXXXXXX
XXXXXXXXXXXXXXXXXXXXXXXXXXXXXXXXXXXXXXXXXXXXXXXXXXXXXXXXXXXXXXXX
XXXXXXXXXXXXXXXXXXXXXXXXXXXXXXXXXXXXXXXXXXXXXXXXXXXXXXXXXXXXXX
XXXXXXXXXXXXXXXXXXXXXXXXXXXXXXXXXXXXXXXXXXXXXXXXXXXXXXXXXXXX
XXXXXXXXXXXXXXXXXXXXXXXXXXXXXXXXXXXXXXXXXXXXXXXXXXXXXXXXXX
XXXXXXXXXXXXXXXXXXXXXXXXXXXXXXXXXXXXXXXXXXXXXXXXXXXXXXXX
XXXXXXXXXXXXXXXXXXXXXXXXXXXXXXXXXXXXXXXXXXXXXXXXXXXXXX
XXXXXXXXXXXXXXXXXXXXXXXXXXXXXXXXXXXXXXXXXXXXXXXXXXX
XXXXXXXXXXXXXXXXXXXXXXXXXXXXXXXXXXXXXXXXXXXXXXXX
XXXXXXXXXXXXXXXXXXXXXXXXXXXXXXXXXXXXXXXXXXXXX
XXXXXXXXXXXXXXXXXXXXXXXXXXXXXXXXXXXXXXXXX
XXXXXXXXXXXXXXXXXXXXXXXXXXXXXXXXXXXXX
XXXXXXXXXXXXXXXXXXXXXXXXXXXXXXXXXXX        (not to scale)
```

CP14. Modify Sample Problem 4.10 so that the Newton-Raphson method is performed with double precision.

CP15. The voltage equations around the two circuit loops shown can be written as

$$60.0i_1 - 40.0i_2 = 100.0$$

$$-40.0i_1 + 160.0i_2 = 0.0$$

Use the subroutine of Sample Problem 8.6 to solve for the unknown currents i_1 and i_2.

CP16. Modify Sample Problem 8.6 so that the subroutine can treat n equations using the Gaussian elimination method; that is, K now varies from 1 to N $-$ 1. Use adjustable dimensions for A, B, and X.

CP17. Modify the solution to CP16 so that if for any K, element $A(K, K)$ equals 0, a row interchange is performed with the first succeeding row (I greater than K) whose element $A(I, K)$ is nonzero. (In practice, the largest element $A(I, K)$ is found first and, through a row interchange, is placed on the main diagonal, so that it becomes $A(K, K)$. Dividing by the largest element possible — that is, $A(K, K)$ — will reduce roundoff error. The placing of the largest element on the main diagonal through a row interchange is called partial pivoting.)

*CP18. Modify the solution to Sample Problem 4.6 so that it can accommodate complex coefficients.

*CP19. To find the root of a complex number, one can first write the number in exponential form

$$z = u + iv = re^{i\theta}$$

where

$$e^{i\theta} = \cos\theta + i\sin\theta \qquad r = (u^2 + v^2)^{1/2}$$

and

$$\theta = \tan^{-1}\frac{v}{u}$$

The root then is

$$(z)^{1/n} = r^{1/n}(e^{i(\theta+2\pi K)/n}) \qquad K = 0, 1, 2, 3, \ldots, n - 1$$

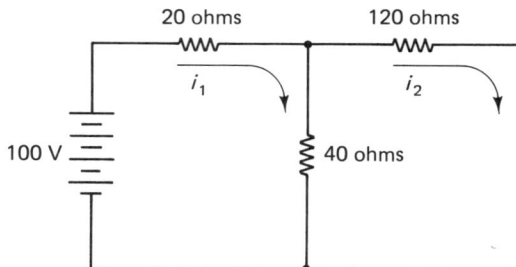

CP 8.15

Write a FORTRAN program that will find the root of any complex number. Test your program by finding the roots of $(-1)^{1/6}$. Label your answers appropriately.

*CP20. The series expansion for sin z where z is complex is given by

$$\sin z = z - \frac{z^3}{3!} + \frac{z^5}{5!} - \frac{z^7}{7!} + \frac{z^9}{9!} - \cdots \qquad (|z| < \infty)$$

Write a FORTRAN program to evaluate sin z, keeping the first five terms of the series. Check your results for the case in which z equals $1 + i2$. Use CSIN(Z) as well as the identity

$$\sin z = \sin u \cosh v + i \cos u \sinh v \qquad z = u + iv$$

to validate your answer.

*CP21. For the electrical circuit shown, use the computer to calculate what the current flow will be every 0.0003 seconds for a total time of 0.018 seconds.

IN OHMS

$z_1 = 3.0$, resistance

$z_2 = i4.0$, inductive impedance

$z_3 = 20.0$, resistance

$z_4 = i37.7$, inductive impedance

$z_5 = 10.0$, resistance

$z_6 = -i53.1$, capacitive impedance

and $E = 230.0 \sin(120.0\pi t) = 230.0 \sin(376.99t)$ volts. Note that series impedances add as follows:

$$z_C = z_A + z_B$$

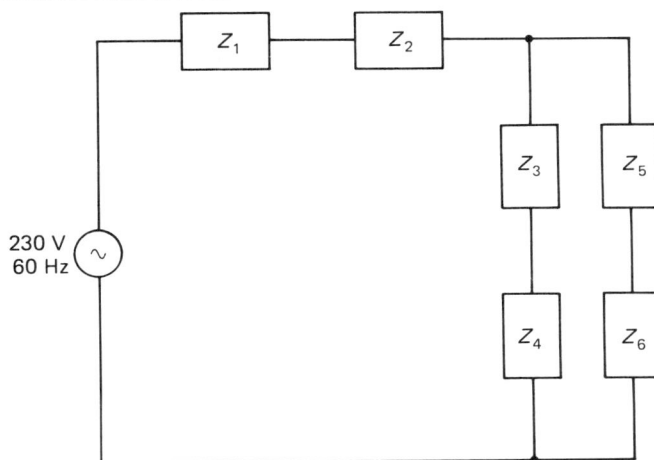

CP 8.21

and parallel impedances add as

$$\frac{1}{z_C} = \frac{1}{z_A} + \frac{1}{z_B}$$

The solution is then given by

$$\text{current} = Im\left(\frac{230.0e^{i(376.99t)}}{z_{\text{total}}}\right)$$

where *Im* stands for the imaginary part. Use the A I MAG intrinsic function.

*CP22. A closed-loop control system regulates a system's response or behavior by monitoring and then comparing the actual output of the system with the system's input, which describes the desired results. If the output response does not match the input, corrective measures are taken by the system until the error becomes zero. By using this feedback comparison, differences between output and input can be automatically corrected for. In the accompanying figure, a typical feedback control system is represented in block diagram form. The system's ratio of $\theta_{\text{out}}/\theta_{\text{in}}$, which is frequency dependent, can be given by

$$\frac{\theta_{\text{out}}}{\theta_{\text{in}}} = \frac{G}{1 + GH}$$

where all quantities are complex. If the output is not to become unstable, *GH* cannot equal $(-1.0, 0.0)$ at any applied frequency.

$$GH = \frac{10(4 + i\omega)}{i\omega(4 - \omega^2 + i\omega)(6 + i\omega)}$$

determine its magnitude (see CABS) and phase angle (the arctangent of the imaginary part divided by the real part) for the circular frequencies ranging from 0.5 to 50 radians/second in increments of 0.5. Express your answer for the magnitude in decibels (db).

$$\text{number of decibels} = 20 \log_{10}|GH|$$

Have the computer print out a table showing frequency, db, and phase angle. Is the

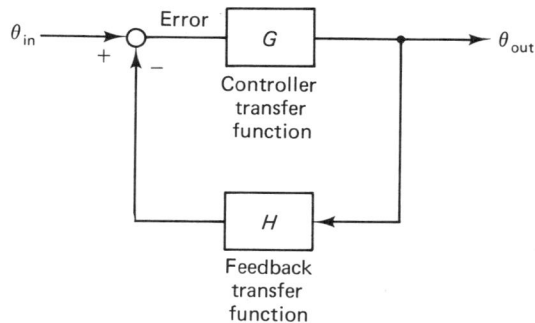

CP 8.22

system stable? (That is, does a condition for *GH* occur in which its magnitude is 0 db and the phase angle is 180 degrees simultaneously?)

*CP23. For a viscously damped mechanical system with harmonic excitation, the equation of motion is given by

$$M\frac{d^2x}{dt^2} + C\frac{dx}{dt} + Kx = F_0 \sin \omega t$$

Using complex algebra, the solution for *x* becomes

$$x = Im\left(\frac{\dfrac{F_0}{M}e^{i\omega t}}{\left(-\omega^2 + \dfrac{K}{M}\right) + i\dfrac{C}{M}\omega}\right)$$

where *Im* stands for imaginary part (see A I MAG). Write a computer program that will solve for *x* given *K/M* equal to 1.51E5, $\omega = 188.5$, $C/M = 4.92\text{E}1$, and $F_0/M = 220.3$. The output should be printed out every 0.0008 seconds to a maximum time of 0.04 seconds.

*CP24. In CP23 the magnification factor, M.F., the ratio of the amplitude of the *x* motion to its static deflection (that is, the deflection if the load was not varying) is given by,

$$\text{M.F.} = \left|\frac{\dfrac{K}{M}}{\left(-\omega^2 + \dfrac{K}{M}\right) + i\dfrac{C}{M}\omega}\right|$$

The phase angle between the applied load and the *x* motion is

$$\phi = \tan^{-1}\frac{\dfrac{C}{M}\omega}{\left(-\omega^2 + \dfrac{K}{M}\right)}$$

For frequencies ranging from 0 to 400 radians/second in increments of 5 radians/second, have the computer calculate the magnification ratio and phase angle. (Use the CABS and ATAN2 intrinsic functions). Output should appear as a table showing ω, M.F., and ϕ. At what frequency is the amplitude the greatest?

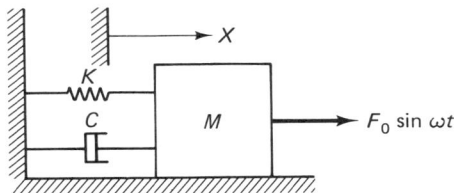

CP 8.23

CHAPTER 9

File Management

Computers being used to automate production. At the GMAD-Orion and Wentsville plants, nearly 4700 welds are applied to each automobile by computer-controlled robots. *(Photo courtesy of General Motors Corporation.)*

9.1 INTRODUCTION

In Chapter 1 we discussed the six basic components of a digital computer. We learned that auxiliary external memory is "permanent" in contrast with internal memory, which is lost when the user logs off. We saw that programs can be stored in external memory and later entered (into internal memory) when they are needed for processing.

Permanent data storage in the form of external data files was introduced in Section 2.8. They are absolutely essential for handling large quantities of data. Besides the key advantages offered by having a data set entered simply, reliably, and as often as needed without retyping, external data files provide a means of making the same data available to any other calling program on the system. The files are even accessible to programs on other machines if the data has been entered for formatted input operations. (List-directed input/output operations are system dependent and will not carry to another machine.)

The different formats available for entering and reading data to and from an external file were given in Chapter 3 when input and output was discussed. It is the purpose of this chapter to review this material and to cover in detail the various means of external file management that FORTRAN provides. We will be mainly interested in how files are created, accessed, and updated. To better understand these operations, we first need to describe the physical nature of auxiliary memory.

9.2 AUXILIARY EXTERNAL MEMORY

To create a file in auxiliary external memory, data is first entered from a source into the machine's internal memory. The source could be input from the terminal, another auxiliary file, or the program's results. From internal memory, data is then character coded and sent as electronic output to some auxiliary memory device, where it is stored. To access a file in external memory, the process is reversed. In that case, auxiliary storage provides the input data that is loaded and sent electronically to the machine's internal storage for subsequent processing.

Usually, the auxiliary external files are stored on magnetic disks or magnetic tape. Data is read or written using read/write heads that encode or interpret magnetic patterns in the disk's or tape's magnetic surface coating. Data is entered one record at a time. Each record may contain several fields. When data is written at a specific location on the disk or tape, the previous information stored there is lost. Reading information from external memory does not destroy the file's contents, however.

On magnetic disks, information is physically organized and indexed on tracks, or concentrically spaced circles, laid out on the disk. The disk rotates at high speed in a disk drive, and its information is accessed with a read/write head that is attached to a moving access arm. The arm moves in and out over the magnetic surface.

There are two standard means available for the arm to access a record. If the access arm begins at a file's first assigned track and then reads or writes record entries in consecutive order, going from the first track to the next, we speak of **sequential access,** and the file is called a **sequential file.** To reach any record, all previous records must be passed through first. The sample programs thus far discussed in this book have only accessed sequential files. If, on the other hand, the file's records are numbered (the first record has number 1, and so on) the disk drive's access arm can move directly to individual records of a file and access them without having to pass through the other records first. This means of access is called direct access, and the file is called a **direct-access** or **random-access file.** Records are organized in a direct-access file so that they all have the same length.

A tape drive is used to enter or read information on magnetic tapes. A tape drive basically consists of rotating supply and take-up reels to move the tape along, and a stationary head that reads or writes magnetic patterns. Tape drives can only access file records sequentially, since their read/write head is fixed. Records are stored on the tape one after the other, and to access an individual file record, the head must start at the beginning of the file and process all preceding data.

9.3 OPENING AND CLOSING A FILE

To connect an external file from auxiliary memory to a FORTRAN program unit, an OPEN statement is used. It is an executable statement that is commonly placed in the beginning of the program. The OPEN statement was introduced, in abbreviated form, in Section 2.8. We will now introduce its full syntax.

OPEN(UNIT=*IU*,FILE=*name*,ACCESS=*ACC*,STATUS=*STA*,
        FORM=*FM*,IOSTAT=*JERR*,ERR=*LABEL*,RECL=*LREC*,
        BLANK=*BLNK*)

UNIT   The integer expression *IU* specifies the integer unit code number of the device that is to be connected to the referenced file. The physical unit that is actually associated with each number is computer dependent. The string 'UNIT=' is optional, but if it is omitted, the integer unit number must be given as the first item to appear in the OPEN statement list. The order of other parameters specified in the OPEN list is unimportant. It might be mentioned that two standard input/output devices are automatically connected to the program when it is executed. The sequential input and output data files sent or entered using these standard devices do *not* have to be opened formally via an OPEN statement. One device is associated with input and is typically the terminal keyboard; the other device is dedicated for output and is commonly the terminal monitor and a printer. As described earlier in Chapter 3, an asterisk may be used instead of the system-defined integer unit number to reference them.

FILE   The character expression *name* specifies the name of the file to be opened. Every file must have a unique name. If several files are to be opened at once, each file must have a different associated unit number and a separate OPEN statement. Any legal FORTRAN name may be used for a file.

ACCESS   This parameter specifies the method of access for the file. The character expression *ACC* must have the value 'SEQUENTIAL' or 'DIRECT'. If this parameter is omitted, the computer will assume that sequential access is required.

STATUS   For this optional parameter, the character expression *STA* must have one of the following values:

> 'OLD' — the file already exists.

> 'NEW' — the file does not exist currently.

> 'SCRATCH' — the file is not to be saved upon program termination. A file that will be scratched or discarded may not be specified by a FILE parameter.

> 'UNKNOWN' — the file status is unknown. If the STATUS parameter is omitted, the file is assumed to have UNKNOWN status and none of the other classifications apply.

FORM   The character expression *FM* must have the value 'FORMATTED' or 'UNFORMATTED'. If the file stored in auxiliary external memory is in ASCII or EBCDIC code — that is, has external character form — it is spoken of as a **formatted file.** All the input/output operations we have thus far performed in the book have dealt with formatted files. Note that this file classification also includes those files that are read or written using list-directed input or output. Data from formatted files must first be converted to machine internal form before it can be processed. Similarly, internal code must first be converted to external form when it is sent to external storage.

To save processing time, it is possible to save information in the machine's more compact internal code. A file created in such a manner is called an **unformatted file.** When referenced by a READ or a WRITE statement, format identifiers are omitted (see Sections 3.4 and 3.7). Unformatted files, which are coded in machine language, cannot be displayed or entered at the terminal, since they do not have character form. They also cannot carry to another machine, since each machine has a different internal code to represent data. For this reason, we will continue to treat only formatted files. The default value for sequential files is FORM='FORMATTED'. If the parameter is missing for direct-access files, the default value is FORM='UNFORMATTED'.

IOSTAT   The integer variable name given to *JERR* will be assigned an integer value. If no error occurs when opening a file, *JERR* is assigned the value 0. If the system fails to open a file (for example, the file cannot be found), it will be assigned a positive value that corresponds to a numbered error message listed in the computer-system reference manual. IOSTAT is an optional parameter. If it is omitted, an execution error occurring while opening a file will cause immediate program termination.

ERR   ERR is an optional parameter. If it is included, any error that might occur while opening a file will cause program execution to be transferred to the statement numbered *LABEL*. If it is omitted, any execution error that might occur while opening a file will cause immediate program termination. *LABEL* is an integer constant provided by the programmer.

RECL   This parameter specifies the record length for direct-access or sequential files. *LREC* should be given as an integer variable or as a positive integer constant. It is a required parameter for direct-access files, but is usually omitted when accessing sequential files. (The system defines the number of characters allowed per record for a sequential file by default.)

BLANK   The character expression *BLNK* must have the value 'ZERO' and 'NULL'. If the word 'NULL' appears, blanks appearing in formatted input fields are usually ignored; a numeric field of all blanks is treated as a zero. If the word 'ZERO' appears, any blanks, except leading blanks, are treated as zeros. If the parameter is omitted, a value of 'NULL' is assumed.

Some examples of OPEN statements follows:

```
      CHARACTER*6,  NAME,STAT
      OPEN(9,FILE='TESTI')
      OPEN(UNIT=7,FILE='TEST2',STATUS='OLD')
      OPEN(UNIT=3,FILE='TEST3',ACCESS='SEQUENTIAL',
     +   FORM='FORMATTED',STATUS='NEW')
      OPEN(UNIT=8,FILE='TEST4',STATUS='NEW',
     +   IOSTAT=NERR)
      IF(NERR.NE.0) PRINT *,NERR
      OPEN(UNIT=10,STATUS='SCRATCH')
      OPEN(UNIT=9,FILE='TEST5',STATUS='NEW',
     +   ERR=112)
  112 PRINT *,'THE FILE TEST5 CANNOT BE OPENED'
      .
      .
      .
      READ *,IU,NAME,STAT
      OPEN(UNIT=IU,FILE=NAME,STATUS=STAT)
```

When a file is to be disconnected before the program terminates, a CLOSE statement is used. All files that are not closed by a formal CLOSE statement are closed automatically when the program terminates normally. Closing a file during program execution releases that unit device and file for other purposes. The CLOSE statement has the syntax

CLOSE(UNIT=*IU*,STATUS=*STA*,IOSTAT=*JERR*,ERR=*LABEL*)

UNIT   The integer expression *IU* specifies the integer unit code number of the device that is to be disconnected. The string 'UNIT=' is optional.

IOSTAT   The integer variable name given to *JERR* will be assigned an integer value. If no error occurs when closing a file, *JERR* is assigned the value 0. If the system fails to close a file, it will be assigned a positive value that corresponds to a numbered error message listed in the computer-system reference manual. IOSTAT is an optional parameter. If it is omitted, an execution error occurring while closing a file will cause immediate program termination.

ERR   ERR is an optional parameter. If it is included, any error that might occur while closing a file will cause program execution to be transferred to the statement numbered *LABEL*. If it is omitted, an execution error that might occur while closing a file will cause immediate program termination. *LABEL* is an integer constant provided by the programmer.

STATUS   The character expression *STA* must have the value 'KEEP' or 'DE-LETE'. When the word 'KEEP' is used, the file is kept after the CLOSE statement is executed. If the word 'DELETE' is used, the file is discarded when the CLOSE statement is performed. The word 'KEEP' cannot be used for a file that has already been defined as a 'SCRATCH' file (see OPEN statement). The default value is STATUS='KEEP' except for a 'SCRATCH' file, for which the default value is STATUS='DELETE'.

Some examples of the use of CLOSE statements follow:

```
CLOSE(UNIT=7)

CLOSE(7)

CLOSE(UNIT=7,STATUS='DELETE')
```

9.4 THE INQUIRE STATEMENT

The INQUIRE **statement** may be used to determine the properties of a file or the current status of a specified unit number. The syntax of the INQUIRE statement is

$$\text{INQUIRE} \left(\begin{array}{c} \text{UNIT}=IU \\ \text{or} \\ \text{FILE}=name \end{array} \quad ,\text{inquiry list} \right)$$

Either a unit number, *IU*, or a file name, *name,* is given in the statement, but not both. The inquiry list contains optional inquiry specifiers that relate to unit or file status or properties. When the INQUIRE statement is executed the variable names associated with the inquiry specifiers are assigned the values shown in Table 9.1. IOSTAT=*JERR* and ERR=*LABEL* can also be included in the inquiry list, although they do not give information about unit or file status. They function in the same manner as described in previous sections (see OPEN and CLOSE statements). Some examples of how INQUIRE statements are used follow:

```
CHARACTER*11 FORMT
LOGICAL OP,OEX
    .
    .
    .
INQUIRE(FILE='TEST',OPENED=OP)
    .
    .
    .
IF (.NOT.OP) THEN
    OPEN (UNIT=7,FILE='TEST')
END IF
    .
    .
    .
```

TABLE 9.1   INQUIRY SPECIFIERS

Inquiry specifier	Values assigned
ACCESS = character variable	'SEQUENTIAL' or 'DIRECT'
BLANK = character variable	'NULL' or 'ZERO'
DIRECT = character variable	'YES', 'NO', or 'UNKNOWN'. For a file inquiry only.
ERR = integer variable or constant	Statement label of error routine.
EXIST = logical variable	.TRUE. or .FALSE.
FORM = character variable	'FORMATTED' or 'UNFORMATTED'
FORMATTED = character variable	'YES', 'NO', or 'UNKNOWN'. For a file inquiry only.
IOSTAT = integer variable	Error code.
NAME = character variable	Name of the file at a specified unit. (Not applicable to a 'SCRATCH' file.) For a unit inquiry only.
NAMED = logical variable	.TRUE. if the file at a specified unit has a name; .FALSE. otherwise. (Not applicable to a 'SCRATCH' file.) For a unit inquiry only.
NEXTREC = integer variable	Next record number in a direct-access file. Undefined if the file is not available for direct access.
NUMBER = integer variable	File's unit number. (Not applicable to a 'SCRATCH' file.) For a file inquiry only.
OPENED = logical variable	.TRUE. or .FALSE.
RECL = integer variable	Integer value of the record length for a direct-access file. Undefined if the file is not available for direct access.
SEQUENTIAL = character variable	'YES', 'NO', or 'UNKNOWN'. For a file inquiry only.
UNFORMATTED = character variable	'YES', 'NO', or 'UNKNOWN'. For a file inquiry only.

```
INQUIRE(UNIT=8,EXIST=OEX)
   .
   .
   .
IF  (.NOT.OEX)  THEN
   OPEN(UNIT=8,FILE='TEST',STATUS='OLD')
END IF
   .
   .
   .
INQUIRE(FILE='TEST2',FORM=FORMT)
PRINT *,'FILE TEST2 IS ',FORMT
```

9.5 FILE-POSITIONING STATEMENTS

When working with sequential files, it is sometimes necessary to move the read/ write head relative to the current record being processed. There are three FORTRAN commands that effect sequential file positioning: REWIND, BACKSPACE, and ENDFILE. The REWIND statement positions the file such that the read/write head lies at the file's first record. On some computer systems, a REWIND statement is required before an input file can be read. The REWIND statement has the syntax

REWIND(UNIT=*IU*,IOSTAT=*JERR*,ERR=*LABEL*)

The parameters IOSTAT and ERR are optional and relate only to the rewind operation (see OPEN statement). The character string 'UNIT=' can also be omitted. For example,

```
REWIND(UNIT=7)
REWIND(7)
    .
    .
    .
REWIND(UNIT=3,IOSTAT=MISS)
IF (MISS.NE.0) THEN
      PRINT *,FAILURE TO REWIND, ERROR CODE IS', MISS
END IF
```

When auxiliary file units are disconnected by a CLOSE statement or because of program termination, their files are not automatically rewound. (The object file is always rewound, since it has primary status. See Section 2.8.) Therefore, it is good practice to place a unit REWIND statement immediately after the file is opened, that is, after the pertinent OPEN statement. If this is not done, an "END-OF-FILE EN-COUNTERED" termination error is likely to occur. The REWIND statement is also useful if it is necessary to read over the same file, within a program unit, from the very beginning.

The BACKSPACE command rewinds the file to the last record read or written; that is, the computer goes back one record in the referenced file. The syntax of the BACKSPACE command is

BACKSPACE(UNIT=*IU*,IOSTAT=*JERR*,ERR=*LABEL*)

The parameters IOSTAT and ERR are optional and relate only to the file being positioned backward one record. The character string 'UNIT=' is also optional. For example,

```
BACKSPACE(UNIT=3)

BACKSPACE(3)
```

Study the following group of statements:

```
  READ(7,2) K
2 FORMAT(15X,I1)
  J=3*K
  BACKSPACE(7)
  READ(7,3) S(J)
3 FORMAT(E14.7)
```

associated with the data record

```
__.1234567E+02  3
```

For the first READ statement, only the integer K is read. It has the value 3. The index J is then calculated and the file is backed up one record. The *same* record is then accessed once more, although this time only the value in the first field (that is, the value for S(9)) is read because of the different FORMAT statement.

The last control statement, the ENDFILE statement, can be used to place an end-of-file special record at the end of a sequential file. When data file units are disconnected by a CLOSE statement or because of normal program termination, an end-of-file marker is automatically placed by the system. If a READ statement tries to read beyond the last data record in a file where no end-of-file record is present, a termination error will occur. The syntax of the ENDFILE statement is

ENDFILE(UNIT=*IU*,IOSTAT=*JERR*,ERR=*LABEL*)

Again, IOSTAT and ERR are optional, referring now to the ENDFILE operation. The character string 'UNIT=' can also be omitted.

```
ENDFILE(UNIT=7)
```

```
ENDFILE(7)
```

9.6 FILE INPUT/OUTPUT STATEMENTS

In Chapter 3 the READ and WRITE statements were introduced, but in abbreviated form. The complete syntax of the READ statement is

READ(UNIT=*IU*,FMT=*NFORM*,IOSTAT=*JERR*,ERR=*LABEL*,
END=*LEND*,REC=*NREC*) variable list

The WRITE statement appears as

WRITE(UNIT=*IU*,FMT=*NFORM*,IOSTAT=*JERR*,ERR=*LABEL*,
REC=*NREC*) variable list

In both statements, the words 'UNIT=' and 'FMT=' are optional. If the string 'FMT=' is omitted, the word 'UNIT=' must also be omitted. In that case, *IU*, the integer unit code, must be the first value enclosed within the parentheses, and *NFORM*, the format identifier, must immediately follow *IU*.

If an asterisk (∗) is used to replace the integer unit code number, *IU*, the main input/output units will be used: the terminal for input, the monitor or a printer for output. To specify list-directed format, an asterisk is substituted for the format identifier, *NFORM*. (List-directed format cannot be used with a direct-access file, nor may a record be written in a file that is longer than the record length specified in the file's OPEN statement.) If the file is not formatted, the FMT parameter is omitted.

IOSTAT and ERR parameters are also optional. They both refer to input/output errors. If *JERR* is less than zero, an end-of-file record has been reached; if it equals zero, the operation was successful. If *JERR* is greater than zero, the error condition corresponding to that code number can be referred to in the system manual. As before, *LABEL* is a statement number referring to an executable statement to which control should be transferred if an input/output operation error occurs. A typical input/output error is having a character appear in a field assigned to a REAL or an INTEGER variable.

The END parameter is used to specify where control should pass if an end-of-file record is reached during an input operation. *LEND* is the statement number of the next executable statement. The parameter is optional, but should be included if the file's length is unknown; an execution error will result if a READ statement missing an END parameter encounters an end-of-file marker. (See also ENDFILE.)

The REC parameter is used to specify the record number, *NREC*, to be read or entered in a direct-access file. *NREC* must be a positive nonzero integer constant, variable, or integer expression. The REC parameter must be included when direct-access files are processed. When a record is written, a positive integer record number is assigned to it using the REC parameter. When it is read, the same record number is referred to by using the READ statement's REC clause. This means that when working with direct-access files, an END parameter cannot appear in the READ statement. File positioning — that is, REWIND, BACKSPACE, and ENDFILE — also has no effect.

Some typical input/output statements follow.

```
READ(9,7,ERR=50) X,Y

READ(3,6,END=100) I,A

READ(3,*,END=100) I,A

READ(UNIT=2,FMT=17,END=5) U,W

WRITE(6,4,REC=2*NPART+1) R,S,I

WRITE(10,3,REC=3,IOSTAT=MISS) R,T
```

An unformatted file might be accessed as follows:

```
WRITE(UNIT=7,ERR=50) I,J,S
```

Sample Problem 9.1

Updating an external file is not an easy matter. Updating information, as well as the information on the file to be renewed, must be brought into the computer as input data for the modifying program. The modification and processing are then performed in internal memory, after which the results are sent as output to auxiliary storage. Sequential files are entered record by record, and updated in that manner, since the entire file is typically too large to bring in all at once and store internally beforehand. The results of the changes are usually sent to a *new* updated file instead of being stored in the old master file.

Since their records are not numbered, updating a sequential file also requires that at least one identification field, or key field, be included in all the master file's records. If the new updating information is also similarly keyed, its information can be readily placed without further difficulty. Files are usually sorted so that their records have key fields placed in an ascending or descending order.

In this problem, develop an updating program for an inventory file. Assume that the master file and the updating material have key fields arranged in ascending numeric order. The updating material should be entered via an auxiliary file, which for this example will be called the transaction file. If a key field in the transaction file is not present in the master file, an error message should be printed. The result of the updating should be stored in a new master file. All auxiliary files should have sequential access. Verify the program using the following information:

MASTER FILE — BRAKE PARTS INVENTORY

Part auto	Year	Price	Stock
0050 AUDI	1974–1978	6.50	10
0060 AUDI	1968–1973	17.33	8
1502 BMW	1975–	20.43	6
1602 BMW	1969–1975	24.30	7
2100 DATSUN	1972–1975	19.66	11
2124 FIAT	1968–1975	12.84	5
2616 MAZDA	1971–	9.13	6
3190 MERCED	1963–1965	43.80	7
3200 MERCED	1968–1976	41.16	8
4104 PEUGOT	1973–	7.59	4
4105 RENAUL	1973–	11.30	5
5099 SAAB	1975–	17.65	4
6070 TOYOTA	1979–	20.75	8
6402 VOLKSWAG	1973–1979	23.99	7
7043 VOLVO	1980–	22.45	6

TRANSACTION FILE — BRAKE PARTS

Part no.	From stock	Added to stock
0060	2	7
1602	5	1
1603	3	4
3200	6	6
4104	8	4

The first field shown in both files is the key field. In the transaction file, the next field over accounts for the amount withdrawn from inventory; the last field is the amount added to the inventory. For the master file, the last two fields represent the price and the amount present in inventory, respectively.

Solution: The algorithm and problem solution are shown in Figures 9.1a and 9.1b. A new master file, NMAST, is created while updating the old master file. Updating information is presented by the transaction file. Notice that the program has two FORTRAN WHILE loops that allow the files to run out if one is larger than the other. If the transaction file is smaller than the master file, and thus empties first, the new master file copies the old master file without further changes until the old master file runs out. If the transaction file is larger than the master file, error messages are printed out until it too is empty.

Algorithm for Sample Problem 9.1. This program updates an inventory file that has sequential access. It is assumed that the master and transaction files' key fields are found in ascending order.

1. Open files MASTER,TRAN,NMAST.
2. Read NPART,ISUB,IADD from file TRAN.
3. Read MPART,AUTO,PRICE,INV from file MASTER.
4. While neither the end of file TRAN nor the end of file MASTER has been reached, do the following:
   a. If NPART<MPART then
      i. Print 'NPART is not on the list.'
      ii. READ a new NPART,ISUB,IADD from file TRAN
      Else if NPART>MPART then
      i. Write MPART,AUTO,PRICE,INV in file NMAST
      ii. Print MPART,AUTO,PRICE,INV
      iii. Read a new MPART,AUTO,PRICE,INV from file MASTER.
      Else
      i. Calculate
         INV=INV−ISUB+IADD
      ii. Write MPART,AUTO,PRICE,INV in file NMAST.
      iii. Print MPART,AUTO,PRICE,INV
      iv. Read a new NPART,ISUB,IADD from file TRAN
      v. Read a new MPART,AUTO,PRICE,INV from file MASTER
5. While the end of file MASTER has not been reached, do the following:
   a. Read a new MPART,AUTO,PRICE,INV from file MASTER
   b. Write MPART,AUTO,PRICE,INV in file NMAST
   c. Print MPART,AUTO,PRICE,INV
6. While the end of file TRAN has not been reached, do the following:
   a. Print 'NPART is not on the list.'
   b. Read a new NPART,ISUB,IADD from file TRAN

Figure 9.1a

```
      PROGRAM UPDATE

******************************************************************
*                                                                *
*   THIS PROGRAM UPDATES AN INVENTORY FILE THAT HAS SEQUENTIAL    *
*   ACCESS. IT IS ASSUMED THAT THE MASTER AND TRANSACTION FILE'S  *
*   KEY FIELDS ARE FOUND IN ASCENDING ORDER.                      *
*                                                                *
******************************************************************

      CHARACTER*16 AUTO
      REAL PRICE
      INTEGER NPART,ISUB,IADD,MPART,INV

      OPEN(UNIT=3, FILE='MASTER', ACCESS='SEQUENTIAL',
    + FORM='FORMATTED', STATUS='OLD')
      OPEN(UNIT=4, FILE='TRAN', ACCESS='SEQUENTIAL',
    + FORM='FORMATTED', STATUS='OLD')
      REWIND (3)
      REWIND (4)
      OPEN(UNIT=7, FILE='NMAST', ACCESS='SEQUENTIAL',
    + FORM='FORMATTED',STATUS='NEW')

      READ(4,7) NPART,ISUB,IADD
      READ(3,8) MPART,AUTO,PRICE,INV
C
C   IF NPART IS LESS THAN MPART, AN ERROR MESSAGE IS PRINTED
C   OUT, AND A NEW VALUE OF NPART IS READ. IF THIS FILE BECOMES
C   EMPTY, CONTROL TRANSFERS TO THE LOOP BELOW WHICH EMPTIES
C   THE REMAINING FILE, THE MASTER FILE. IF NPART IS GREATER
C   THAN MPART, MPART IS ENTERED INTO THE UPDATED MASTER LIST,
C   AND A NEW VALUE OF MPART IS READ. IF THE FILE FOR MPART
C   BECOMES EMPTY, CONTROL TRANSFERS TO THE LOOP BELOW WHICH
C   EMPTIES THE TRANSACTION FILE INSTEAD. IF NPART EQUALS
C   MPART, THE INVENTORY VALUE (INV) IS CHANGED, THE UPDATED
C   INFORMATION IS ENTERED INTO THE NEW MASTER FILE, AND NEW
C   VALUES OF NPART ANMD MPART ARE READ UNTIL ONE FILE BECOMES
C   EMPTY. WHEN THAT HAPPENS, A CONTROL TRANSFER IS MADE TO
C   EMPTY THE REMAINING FILE, AS DISCUSSED.
C
  100 IF(NPART.LT.MPART) THEN

         PRINT*,'PART ',NPART,' IS NOT ON THE LIST.'
         READ(4,7,END=130) NPART,ISUB,IADD

         GO TO 100

      ELSE IF(NPART.GT.MPART) THEN

         WRITE(7,8) MPART,AUTO,PRICE,INV
         PRINT 8,MPART,AUTO,PRICE,INV
         READ(3,8,END=140) MPART,AUTO,PRICE,INV

         GO TO 100

7043 VOLVO  1980-      22.45  6

File Tran

0060   2   7
1602   5   1
1603   3   4
3200   6   6
4104   8   4
```

Figure 9.1b

```
RUN

    50 AUDI    1974-1978  6.50 10
    60 AUDI    1968-1973 17.33 13
  1502 BMW     1975-      20.43  6
  1602 BMW     1967-1975 24.30  3
   PART 1603 IS NOT ON THE LIST.
  2100 DATSON 1972-1975 19.66 11
  2124 FIAT    1968-1975 12.84  5
  2616 MAZDA   1971-      9.13  6
  3190 MERCED 1963-1965 43.80  7
  3200 MERCED 1968-1976 41.16  8
  4104 PEUGOT 1971-      7.59  0
  4105 RENAUL 1973-     11.30  5
  5099 SAAB    1975-     17.65  4
  6070 TOYOTA 1979-     20.75  8
  6402 VOLKSW 1973-1979 23.99  7
  7043 VOLVO   1980-     22.45  6

File Nmast

    50 AUDI    1974-1978  6.50 10
    60 AUDI    1968-1973 17.33 13
  1502 BMW     1975-      20.43  6
  1602 BMW     1967-1975 24.30  3
  2100 DATSON 1972-1975 19.66 11
  2124 FIAT    1968-1975 12.84  5
  2616 MAZDA   1971-      9.13  6
  3190 MERCED 1963-1965 43.80  7
  3200 MERCED 1968-1976 41.16  8
  4104 PEUGOT 1971-      7.59  0
  4105 RENAUL 1973-     11.30  5
  5099 SAAB    1975-     17.65  4
  6070 TOYOTA 1979-     20.75  8
  6402 VOLKSW 1973-1979 23.99  7
  7043 VOLVO   1980-     22.45  6

        ELSE

            INV=INV-ISUB+IADD

            WRITE(7,8) MPART,AUTO,PRICE,INV
            PRINT 8,MPART,AUTO,PRICE,INV
            READ(4,7,END=130) NPART,ISUB,IADD
            READ(3,8,END=140) MPART,AUTO,PRICE,INV

            GO TO 100

        END IF


C
C THE FOLLOWING WHILE LOOP IS TRAVERSED WHEN THE TRANSACTION
C FILE IS EMPTY. THE LOOP EMPTIES THE MASTER FILE.
C
   130 READ(3,8,END=150) MPART,AUTO,PRICE,INV
        WRITE(7,8) MPART,AUTO,PRICE,INV
        PRINT 8,MPART,AUTO,PRICE,INV

        GO TO 130
C
C THE FOLLOWING WHILE LOOP IS TRAVERSED WHEN THE MASTER FILE
C IS EMPTY. THE LOOP EMPTIES THE TRANSACTION FILE.
C
   140 PRINT*,'PART ',NPART,' IS NOT ON THE LIST.'
        READ(4,7,END=150) NPART,ISUB,IADD
```

Figure 9.1b   (cont.)

```
      GO TO 140

  7 FORMAT(I4,2I3)
  8 FORMAT(I4,1X,A,F6.2,I3)

150 STOP
    END


File Master

0050 AUDI   1974-1978  6.50 10
0060 AUDI   1968-1973 17.33  8
1502 BMW    1975-      20.43  6
1602 BMW    1967-1975 24.30  7
2100 DATSON 1972-1975 19.66 11
2124 FIAT   1968-1975 12.84  5
2616 MAZDA  1971-      9.13   6
3190 MERCED 1963-1965 43.80  7
3200 MERCED 1968-1976 41.16  8
4104 PEUGOT 1971-      7.59   4
4105 RENAUL 1973-     11.30  5
5099 SAAB   1975-     17.65  4
6070 TOYOTA 1979-     20.75  8
6402 VOLKSW 1973-1979 23.99  7
```

Figure 9.1b   (cont.)

Sample Problem 9.2

Write a program that will merge two sequential files. Assume that their key fields are numeric and are placed in ascending order and that values in file A (ALPHA) do not duplicate those in file B (BETA). The merged result should be stored in a new file named MERGED. The sample data in the following tables should be used to test the program. The key field contains the part number.

FILE A	
Part name	Part no.
Battery cable	12
Tires	30
Gas tank	42
Exhaust pipe	98
Radiator	99
Brake fluid	112
Fan belt	150

FILE B	
Part name	Part no.
Muffler	46
Distributor	48
Spark plugs	59
Points	160
Gas filter	215
Bearings	223

Solution: The algorithm and problem solution are shown in Figures 9.2a and 9.2b.

Sample Problem 9.3

We introduced internal sorting in Chapter 6. An **external sort** is required when dealing with large quantities of data. Write a program that will perform an external sort on a sequential access file. Assume that the key field is numeric (the part number again) and is to be placed in ascending order. Results should be stored in a new file named

■   **Algorithm for Sample Problem 9.2.** This program merges two sequential files. The key fields are assumed to be found in ascending order.

1. Open files ALPHA,BETA,MERGED.
2. Read PARTA,KEYA from file ALPHA.
3. Read PARTB,KEYB from file BETA.
4. While neither the end of file ALPHA nor the end of file BETA has been reached, do the following:
   a. If KEYA<KEYB then
       i. Write PARTA,KEYA in file MERGED
      ii. Print PARTA,KEYA
     iii. Read a new PARTA,KEYA from file ALPHA
      Else
       i. Write PARTB,KEYB in file MERGED
      ii. Print PARTB,KEYB
     iii. Read a new PARTB,KEYB from file BETA
5. While the end of file BETA has not been reached, do the following:
   a. Write PARTB,KEYB in file MERGED
   b. Print PARTB,KEYB
   c. Read a new PARTB,KEYB from file BETA
6. While the end of file ALPHA has not been reached, do the following:
   a. Write PARTA,KEYA in file MERGED
   b. Print PARTA,KEYA
   c. Read a new PARTA,KEYA from file ALPHA

Figure 9.2a

```
          PROGRAM MERGE

      *****************************************************************
      *                                                               *
      *    THIS PROGRAM MERGES TWO SEQUENTIAL FILES. THE KEY FIELDS ARE *
      *    ASSUMED TO BE FOUND IN ASCENDING ORDER.                    *
      *                                                               *
      *****************************************************************

          CHARACTER*14 PARTA, PARTB
          INTEGER KEYA,KEYB

          OPEN(UNIT=3, FILE='ALPHA', STATUS='OLD')
          OPEN(UNIT=4, FILE='BETA', STATUS='OLD')
          OPEN(UNIT=7, FILE='MERGED',STATUS='NEW')
          REWIND (3)
          REWIND (4)

          READ(3,10) PARTA,KEYA
          READ(4,10) PARTB,KEYB
      C
```

Figure 9.2b

```
C   IF KEYA IS LESS THAN KEYB, PARTA IS ENTERED INTO THE NEW
C   MASTER LIST, AND NEW VALUES OF KEYA AND PARTA ARE READ FROM
C   FILE ALPHA. IF THIS FILE BECOMES EMPTY, CONTROL TRANSFERS
C   TO THE LOOP BELOW WHICH EMPTIES THE REMAINING FILE, FILE
C   BETA. IF KEYA IS GREATER THAN KEYB, KEYB IS ENTERED INTO
C   THE NEW MASTER LIST, AND NEW VALUES OF KEYB AND PARTB ARE
C   READ FROM FILE BETA. IF THIS FILE BECOMES EMPTY, CONTROL
C   TRANSFERS TO THE LOOP BELOW WHICH EMPTIES THE REMAINING
C   FILE, FILE ALPHA. KEYA AND KEYB WERE ASSUMED TO BE UNEQUAL
C   IN THIS PROBLEM.
C
    30 IF(KEYA.LT.KEYB) THEN
            WRITE(7,10) PARTA,KEYA
            PRINT 10,PARTA,KEYA
            READ(3,10,END=20) PARTA,KEYA

            GO TO 30

       ELSE
            WRITE(7,10) PARTB,KEYB
            PRINT 10, PARTB,KEYB
            READ(4,10,END=40) PARTB,KEYB

            GO TO 30

       END IF
C
C   THIS WHILE LOOP IS DESIGNED TO EMPTY FILE BETA.
C
    20 WRITE(7,10) PARTB,KEYB
       PRINT 10,PARTB,KEYB
       READ(4,10,END=50) PARTB,KEYB

       GO TO 20

C
C   THIS WHILE LOOP IS DESIGNED TO EMPTY FILE ALPHA.
C
    40 WRITE(7,10) PARTA,KEYA
       PRINT 10, PARTA,KEYA
       READ(3,10,END=50) PARTA,KEYA

       GO TO 40

    50 STOP
    10 FORMAT(A,2X,I3)
       END
```

```
File Alpha

BATTERY CABLE     12
TIRES             30
GAS TANK          42
EXHAUST PIPE      98
RADIATOR          99
BRAKE FLUID      112
FAN BELT         150

File Beta

MUFFLER           46
DISTRIBUTOR       48
SPARK PLUGS       59
POINTS           160
GAS FILTER       215
BEARINGS         223
```

Figure 9.2b   (cont.)

```
RUN

BATTERY CABLE    12
TIRES            30
GAS TANK         42
MUFFLER          46
DISTRIBUTOR      48
SPARK PLUGS      59
EXHAUST PIPE     98
RADIATOR         99
BRAKE FLUID     112
FAN BELT        150
POINTS          160
GAS FILTER      215
BEARINGS        223

File Merged

BATTERY CABLE    12
TIRES            30
GAS TANK         42
MUFFLER          46
DISTRIBUTOR      48
SPARK PLUGS      59
EXHAUST PIPE     98
RADIATOR         99
BRAKE FLUID     112
FAN BELT        150
POINTS          160
GAS FILTER      215
BEARINGS        223
```

Figure 9.2b   (cont.)

XSORT. Test the program on the following data:

FILE A1

Part name	Part no.
Fan belt	150
Bearings	223
Gas filter	215
Muffler	46
Brake fluid	112
Radiator	99
Distributor	48
Tires	30
Gas tank	42
Points	160
Spark plugs	59
Battery cable	12
Exhaust pipe	98

Solution: One possible solution to the problem that also illustrates most of what we have covered thus far in this chapter is shown in Figures 9.3a and 9.3b. In the program, the original file A1 remains unchanged. It is first copied into a scratch file at UNIT=4 while searching for the lowest key value, KEYLO, and its associated part name, PARTLO. The file at unit number 4 does not receive all the records from A1, however. Instead, the one record line containing the specific values of KEYLO and PARTLO winds up being sent to a new file at UNIT=7 called XSORT. The scratch file at unit number 4 receives an end-of-file mark and is rewound. This file now takes on the role of the main file for the rest of the program.

Algorithm for Sample Problem 9.3. This program performs an external sort on a sequential file. Its key numeric field is to be placed in ascending order.

1. Open files A1,XSORT.
2. Open scratch files: FILE4,FILE8.
3. Read PARTLO,KEYLO from file A1.
4. While the end of file A1 has not been reached, do the following:
   a. Read a new PART,KEY from file A1
      i. If KEY<KEYLO then
         (a) Set CTEMP=PARTLO,ITEMP=KEYLO
         (b) Set PARTLO=PART,KEYLO=KEY
         (c) Set PART=CTEMP,KEY=ITEMP
         (d) Write PART,KEY in FILE4
      ii. Write PART,KEY in FILE4
5. Write PARTLO,KEYLO in file XSORT.
6. Print PARTLO,KEYLO.
7. Place end-of-file mark in FILE4 and rewind it.
8. While the end of FILE4 has not been reached, (which will signal the end of the program) do the following:
   a. Read a new PARTLO,KEYLO from FILE4
   b. While the end of FILE4 has not been reached, do the following:
      i. Read a new PART,KEY from FILE4
      ii. If KEY<KEYLO then
         (a) Set CTEMP=PARTLO,ITEMP=KEYLO
         (b) Set PARTLO=PART,KEYLO=KEY
         (c) Set PART=CTEMP,KEY=ITEMP
         (d) Write PART,KEY in FILE8
      iii. Write PART,KEY in FILE8
   c. Write PARTLO,KEYLO in file XSORT
   d. Print PARTLO,KEYLO
   e. Place end-of-file mark in FILE8 and rewind it
   f. Rewind FILE4
   g. While the end of FILE8 has not been reached, do the following:
      i. Read PART,KEY from FILE8
      ii. Write PART,KEY in FILE4
   h. Place end-of-file mark in FILE4 and rewind it
   i. Rewind FILE8

Figure 9.3a

```
         PROGRAM XSORT

*****************************************************************
*                                                               *
*   THIS PROGRAM PERFORMS AN EXTERNAL SORT ON A SEQUENTIAL FILE. *
*   ITS KEY NUMERIC FIELD IS TO BE PLACED IN ASCENDING ORDER.   *
*                                                               *
*****************************************************************

         CHARACTER*14 PARTLO,PART,CTEMP
         INTEGER KEYLO,KEY,ITEMP

         OPEN(UNIT=3, FILE='A1',STATUS='OLD')
         OPEN(UNIT=4, STATUS='SCRATCH')
         OPEN(UNIT=8, STATUS='SCRATCH')
         OPEN(UNIT=7, FILE='XSORT',STATUS='NEW')
         REWIND 3

         READ(3,10) PARTLO,KEYLO
C
C        THE FOLLOWING LOOP CREATES A SCRATCH FILE AT UNIT 4. THE
C        SCRATCH FILE AT UNIT 4 CONTAINS THE SAME INFORMATION AS THE
C        FILE TO BE SORTED AT UNIT 3 ('A1') EXCEPT THAT THE LOWEST
C        KEY VALUE AND ITS PART NAME ARE MISSING. THAT INFORMATION
C        GOES TO UNIT 7 WHERE THE SORTED FILE ('XSORT') WILL RESIDE
C        WHEN THE PROGRAM ENDS.
C
      40 READ(3,10,END=20) PART,KEY

         IF(KEY.LT.KEYLO) THEN
             CTEMP=PARTLO
             ITEMP=KEYLO
             PARTLO=PART
             KEYLO=KEY
             PART=CTEMP
             KEY=ITEMP

             WRITE(4,10) PART,KEY

             GO TO 40

         END IF

         WRITE(4,10) PART,KEY

         GO TO 40

      20 WRITE(7,10) PARTLO,KEYLO
         PRINT 10, PARTLO,KEYLO
C
C        THE SCRATCH FILE AT UNIT 4 RECEIVES AN EOF MARK AND IS
C        REWOUND.
C
         ENDFILE (4)
         REWIND (4)
C
C        A SECOND SCRATCH FILE AT UNIT 8 IS CREATED. NOW THE
C        RECENTLY CREATED (SCRATCH) FILE AT UNIT 4 IS USED AS A
C        SOURCE TO WORK WITH. THE FILE AT UNIT 4 IS SEARCHED FOR ITS
C        LOWEST REMAINING KEY VALUE AND ITS PART NAME. THIS
C        INFORMATION IS AGAIN SENT TO FILE XSORT AT UNIT 7. THE NEW
C        SCRATCH FILE AT UNIT 8 WINDS UP CONTAINING THE SAME
C        INFORMATION AS THE SCRATCH FILE AT UNIT 4 EXCEPT THAT IT IS
C        MISSING THIS LOWEST KEY VALUE AND PART NAME.
C
```

Figure 9.3b

```
      70 READ(4,10,END=80) PARTLO,KEYLO

      50      READ(4,10,END=60) PART,KEY
              IF(KEY.LT.KEYLO) THEN
                  CTEMP=PARTLO
                  ITEMP=KEYLO
                  PARTLO=PART
                  KEYLO=KEY
                  PART=CTEMP
                  KEY=ITEMP

                  WRITE(8,10) PART,KEY

                  GO TO 50

              END IF

              WRITE(8,10) PART,KEY

              GO TO 50

      60      WRITE(7,10) PARTLO,KEYLO
              PRINT 10, PARTLO,KEYLO
C
C      THE SCRATCH FILE AT UNIT 8 RECEIVES AN EOF MARK AND IS
C      REWOUND.
C
              ENDFILE (8)
              REWIND (8)
C
C      THE SCRATCH FILE AT UNIT 4 IS REWOUND, AND IN THE FOLLOWING
C      LOOP, THE OLD CONTENTS OF THE SCRATCH FILE FOUND AT UNIT 4
C      IS LOST AND REPLACED WITH THE CONTENTS OF THE NEW SCRATCH
C      FILE AT UNIT 8. THE NEW SCRATCH FILE AT UNIT 4 WINDS UP
C      CONTAINING THE SAME INFORMATION AS ITS PREDECESSOR EXCEPT
C      THAT IT IS MISSING ITS LOWEST KEY VALUE AND PART NAME, AND
C      HAS BECOME SMALLER BY ONE RECORD: THE RECORD CONTAINING THE
C      LOWEST KEY VALUE AND PART NAME.
C
              REWIND (4)

     120      READ(8,10,END=110) PART,KEY
              WRITE(4,10) PART,KEY

              GO TO 120
     110      ENDFILE (4)
C
C      THE FILES AT UNITS 4 AND 8 ARE REWOUND FOR THE NEXT CYCLE.
C      CONTROL PASSES TO STATEMENT NUMBER 70. THIS OUTER LOOP IS
C      REPEATED UNTIL THE FILE AT UNIT 4 IS EMPTY. AT THAT TIME,
C      CONTROL IS TRANSFERRED TO STATEMENT NUMBER 80 AND THE
C      PROGRAM STOPS. FINAL RESULTS ARE FOUND IN FILE XSORT.
C
              REWIND (4)
              REWIND (8)

          GO TO 70

      80 STOP
      10 FORMAT(A,2X,I3)
         END


File A1

FAN BELT         150
BEARINGS         223
```

Figure 9.3b   (cont.)

```
GAS FILTER        215
MUFFLER            46
BRAKE FLUID       112
RADIATOR           99
DISTRIBUTOR        48
TIRES              30
GAS TANK           42
POINTS            160
SPARK PLUGS        59
BATTERY CABLE      12
EXHAUST PIPE       98


RUN

BATTERY CABLE      12
TIRES              30
GAS TANK           42
MUFFLER            46
DISTRIBUTOR        48
SPARK PLUGS        59
EXHAUST PIPE       98
RADIATOR           99
BRAKE FLUID       112
FAN BELT          150
POINTS            160
GAS FILTER        215
BEARINGS          223


File Xsort

BATTERY CABLE      12
TIRES              30
GAS TANK           42
MUFFLER            46
DISTRIBUTOR        48
SPARK PLUGS        59
EXHAUST PIPE       98
RADIATOR           99
BRAKE FLUID       112
FAN BELT          150
POINTS            160
GAS FILTER        215
BEARINGS          223
```

Figure 9.3b   (cont.)

The next part of the solution requires that the scratch file at unit number 4 be copied into another scratch file at UNIT=8. Once again, all values are transmitted except the one record line that contains the lowest values of KEYLO and PARTLO. This record is sent to file XSORT instead. Note that the file at unit number 8 is smaller, by one record line, than the file at unit number 4. An end-of-file mark is placed in this file as well. Both scratch files are rewound, and the contents of the file at unit number 8 are sent to the file at unit number 4. As a result, the file at unit number 4 becomes smaller, by one record line, than it was in the previous cycle. A new end-of-file record is placed in the file at unit number 4, and both files are once again rewound. The process now repeats, with each pass through the loop finding the file at unit number 4 smaller by one record line (that which contains the current KEYLO and PARTLO values.) Similarly, each pass through the loop causes the file XSORT to have the current values of KEYLO and PARTLO entered (in ascending order). The process repeats until the file at unit number 4 is empty.

9.7 INTERNAL FILES

A form of **internal file** is also available in FORTRAN. An internal file is considered to be a group of internal storage locations in which data is stored in character form. The area or file is given the name of the character variable or substring or character array or element that will be stored there. Data is sent to or retrieved from the internal file by means of READ or WRITE statements, but no input/output unit is referred to, since all transfers involve internal computer memory (the internal file name is given instead of the unit number). Specifying the file name in a READ or a WRITE statement establishes it as an internal file, and the character value assigned to the character variable or array name referenced becomes the file's contents. When forming an internal file from a character array, each array element is considered to hold a file record. The length of each record is limited to the declared length of the character variable or character array. In the following examples, it is assumed that the variable ABLE and the array CARRAY are of the character type. In addition, CARRAY must be an array having at least four elements.

Internal file name	Input/output operations
ABLE	READ(ABLE,4) I
ABLE(7:12)	READ(ABLE(7:12),4) J
CARRAY	WRITE(CARRAY,5) X,Y,W,U
CARRAY(2)	WRITE(CARRAY (2),5)Y

Internal files are very useful for editing purposes. They also allow conversion of data from numeric to character type and vice versa. For example, suppose the character variable PART has the value

```
PART='21364.A'
```

If it is desired to assign the numeric value 21364 to an integer variable J, one may establish an internal file that is a substring of PART to effect the transfer. In this example, PART(1:5) contains the necessary information. Then the statements

```
    READ(PART(1:5),2)  J
2 FORMAT(I5)
```

will assign J the correct numeric value.

If J were given first, and PART was to receive the character constant '21364.A', one could write

```
    WRITE(PART(1:5),2)  J
2 FORMAT(I5)
    PART=PART(1:5)//'.A'
```

Note that the numeric value J is written into the internal file as character data.

Only formatted input/output operations are allowed with internal files. List-directed input/output cannot be used. Similarly, the commands available for auxiliary files — that is, file positioning and file status statements (OPEN, CLOSE, and INQUIRE) — cannot be used for internal files.

9.8 DIRECT-ACCESS FILES

A sequential file must process every preceding record, one by one, before the desired record is reached, but a direct-access file may access a specific file location directly. This is accomplished by numbering all record lines in the file. These numbers are referenced when reading or writing an entry. Records may be read or written in any order. The integer record numbering, however, must start at the value 1 and then increase. *The record numbers do not have to appear in the actual data fields,* but they are recorded in the machine's internal bookkeeping system. In a sequential file, key fields are used for updating data, sorting, and so forth. For direct-access files, the key field is the record number itself. Sometimes a key field must be coded first — for example, if it is alphabetic — so that integer record numbers can be used to represent it.

With a sequential file, the read/write head's current position is very important. After a record is accessed, the head "moves" to the next file record. The FORTRAN commands REWIND, BACKSPACE, and ENDFILE are used to help position the file relative to the head. Because file position is not important for direct-access files, the aforementioned positioning commands will not work for them. There is no end-of-file special record written on a direct-access file either. To be cognizant of a file's being empty, a flag of some sort should be included in the last record's data field. It is also a good policy to have an ERR parameter included in the READ statements of a direct-access file so that termination does not occur if the record number sought does not exist.

To connect a direct-access file, the OPEN statement is used (see Section 9.3). All the statement's parameters have the same meaning as before. Now ACCESS= 'DIRECT' and RECL=*LREC* must be specified. The record length, *LREC*, must be the same for all records in the file. For formatted files, the record length is an integer-value constant or variable specifying the number of characters allowed in each record of the file. For instance, if all READ/WRITE direct-access file operations in a program will use the program FORMAT statement

```
7 FORMAT(2I4,5X,2E14.7,A10)
```

the record length, *LREC*, should be set to at least fifty-one characters. The OPEN statement for the file might appear as

```
OPEN(UNIT=3,FILE='DIRACC',ACCESS='DIRECT',
+ STATUS='NEW',FORM='FORMATTED',RECL=51)
```

Direct-access files are disconnected via the CLOSE statement (see Section 9.3). The statement's parameters and syntax remain unchanged. To close the foregoing file, one can write, for example,

```
CLOSE(UNIT=3)
```

or

```
CLOSE(UNIT=3,STATUS='KEEP')
```

To read a direct-access file, the READ statement described in Section 9.6 is used. A record number must be given as one of the parameters; that is, REC=*NREC* must be specified. The record number can be an integer constant, an integer variable, or an integer expression. For example, referring again to file DIRACC, if we wish to read the second record in the file, we can use the following statements:

```
CHARACTER*10,WORD
OPEN(UNIT=3,FILE='DIRACC',ACCESS='DIRECT',
+ STATUS='OLD',FORM='FORMATTED',RECL=51)
  .
  .
  .
READ(3,7,REC=2)  J,K,X,Y,WORD
7 FORMAT(2I4,5X,2E14.7,A10)
```

Note again that the record length given in the OPEN statement matches the format specifiers for the READ statement. Since the record length of each entry is controlled, you cannot use list-directed input/output operations for a direct-access file. The record number can be represented as an integer variable as well. For example,

```
L=2
READ(3,7,REC=L)  J,K,X,Y,WORD
  .
  .
  .
READ(3,7,REC=4*L)  J,K,X,Y,WORD
```

To write an entry in a direct-access file the WRITE statement described in Section 9.6 is employed. Again, a record number must be given as one of the parameters. The record number can be an integer constant, an integer variable, or an integer expression. For example,

```
OPEN(UNIT=3,FILE='RESEAR',ACCESS='DIRECT',
+ STATUS='NEW',FORM='FORMATTED',RECL=30)
WRITE(4,8,REC=1)  I,J,EXAMP
8 FORMAT(2(5X,I3)E14.7)
```

Note that the format specification agrees with the total allowed record length of 30. Other possible forms for the WRITE statement are

```
WRITE(4,8,REC=K)  I,J,EXAMP
```

```
WRITE(4,8,REC=2*M+8)  I,J,EXAMP
```

Sample Problem 9.4

It is desired to transform the two sequential files MASTER and TRAN of Sample Problem 9.1 to the direct-access format. Write a program that accomplishes this, creating the two direct-access files DMAST and DTRAN. Place a flag value of 9999 in the part number's field to represent the last entry of each file. (Direct-access files have no end-of-file mark.)

Solution: The algorithm and problem solution are given in Figures 9.4a and 9.4b. Pay close attention to the formats given. The FORMAT statements having numbers 7 and 8 were taken directly from Sample Problem 9.1 for the sequential files. These field lengths determine how long the record length should be, in this case 30 and 10.

A means of printing the results at the terminal is included in the program as well. The counters N and M help in determining how long the file is and therefore how great the DO loop's final index should be in order to print out the file.

Sample Problem 9.5

From the data given in Sample Problem 9.2, create two direct-access files, DIRA and DIRB at the terminal. Omit the numeric fields, but include a flag value at the end of each file.

Algorithm for Sample Problem 9.4. This program reads the two sequential files MASTER and TRAN and then converts them to direct-access files. The created files are called DMAST and DTRAN.

1. Set N=0.
2. While the end of file MASTER has not been reached, do the following:
   a. Read MPART,AUTO,PRICE,INV from file MASTER
   b. Calculate
      N=N+1
   c. Write MPART,AUTO,PRICE,INV in record number N of file DMAST
3. Set M=0.
4. While the end of file TRAN has not been reached, do the following:
   a. Read NPART,ISUB,IADD from file TRAN
   b. Calculate
      M=M+1
   c. Write NPART,ISUB,IADD in record number M of file DTRAN
5. Set K=9999.
6. Write K,ISUB,IADD in record number M+1 of file DTRAN
7. Write K,AUTO,PRICE,INV in record number N+1 of file DMAST
8. Do, with index I ranging from 1 to N+1, the following:
   a. Read MPART,AUTO,PRICE,INV from record number I of file DMAST
   b. Print MPART,AUTO,PRICE,INV
9. Do, with index I ranging from 1 to M+1, the following:
   a. Read NPART,ISUB,IADD from record number I of file DTRAN
   b. Print NPART,ISUB,IADD

Figure 9.4a

```
      PROGRAM DCREAT

****************************************************************
*                                                              *
*  THIS PROGRAM READS THE TWO SEQUENTIAL FILES AT UNITS 3 AND  *
*  4, AND THEN CONVERTS THEM TO DIRECT-ACCESS FILES. THE FILES *
*  ARE STORED AT UNITS 7 AND 8.                                *
*                                                              *
****************************************************************

      CHARACTER*16, AUTO
      INTEGER MPART,INVOICE,NPART,ISUB,IADD,K,N,M
      REAL PRICE

      OPEN(UNIT=3, FILE='MASTER', ACCESS='SEQUENTIAL',
     + STATUS='OLD')
      OPEN(UNIT=4, FILE='TRAN', ACCESS='SEQUENTIAL',
     + STATUS='OLD')
      OPEN(UNIT=7, FILE='DMAST', ACCESS='DIRECT',
     + FORM='FORMATTED', RECL=30, STATUS='NEW')
      OPEN(UNIT=8, FILE='DTRAN', ACCESS='DIRECT',
     + FORM='FORMATTED', RECL=10, STATUS='NEW')
      REWIND (3)
      REWIND (4)

      N=0

   11 READ(3,8,END=20) MPART,AUTO,PRICE,INV

      N=N+1

      WRITE(7,8,REC=N) MPART,AUTO,PRICE,INV

      GO TO 11

   20 M=0

   30 READ(4,7,END=40) NPART,ISUB,IADD

      M=M+1

      WRITE(8,7,REC=M) NPART,ISUB,IADD

      GO TO 30
C
C  A FLAG IS SET AT THE END OF BOTH DIRECT-ACCESS FILES.
C
   40 K=9999

      WRITE(8,7,REC=M+1) K,ISUB,IADD
      WRITE(7,8,REC=N+1) K,AUTO,PRICE,INV
C
C  RESULTS ARE NOW CHECKED.
C
      DO 25 I=1,N+1

         READ(7,8,REC=I) MPART,AUTO,PRICE,INV
         PRINT 8,MPART,AUTO,PRICE,INV

   25 CONTINUE

      DO 35 I=1,M+1
         READ(8,7,REC=I) NPART,ISUB,IADD
         PRINT 7,NPART,ISUB,IADD
```

Figure 9.4b

```
   35 CONTINUE

    8 FORMAT(I4,1X,A,F6.2,I3)
C
C  RECORD LENGTH NEEDED IS 30.
C
    7 FORMAT(I4,2I3)
C
C  RECORD LENGTH NEEDED IS 10.
C
       END


RUN

    50 AUDI    1974-1978  6.50 10
    60 AUDI    1968-1973 17.33  8
  1502 BMW     1975-     20.43  6
  1602 BMW     1967-1975 24.30  7
  2100 DATSON  1972-1975 19.66 11
  2124 FIAT    1968-1975 12.84  5
  2616 MAZDA   1971-      9.13  6
  3190 MERCED  1963-1965 43.80  7
  3200 MERCED  1968-1976 41.16  8
  4104 PEUGOT  1971-      7.59  4
  4105 RENAUL  1973-     11.30  5
  5099 SAAB    1975-     17.65  4
  6070 TOYOTA  1979-     20.75  8
  6402 VOLKSW  1973-1979 23.99  7
  7043 VOLVO   1980-     22.45  6
  9999 VOLVO   1980-     22.45  6
    60  2  7
  1602  5  1
  1603  3  4
  3200  6  6
  4104  8  4
  9999  8  4
```

Figure 9.4b   (cont.)

Solution: The algorithm and problem solution are shown in Figures 9.5a and 9.5b. Note how the word END entered by the user becomes the flag value.

Sample Problem 9.6

Use the direct-access file DTRAN created in Sample Problem 9.4 to update the master file DMAST. The result of the updating should be placed back in the old master file.

Solution: The algorithm and problem solution are shown in Figures 9.6a and 9.6b.

■ **Algorithm for Sample Problem 9.5.** This program creates two direct-access files, DIRA and DIRB, from data entered at the terminal.

 1. Set N=0.

 2. Enter PARTNA or 'END'.

 3. While PARTNA≠'END' then
   a. Calculate
      N=N+1

Figure 9.5a

b. Write PARTNA in record N of file DIRA

c. Enter PARTNA or 'END'

4. Write the last value of PARTNA, which equals 'END', in recorded number N+1 of file DIRA.

5. Set K=0.

6. Enter PARTNA or 'END'.

7. While PARTNA≠'END' then
   a. Calculate
      K=K+1
   b. Write PARTNA in record K of file DIRB
   c. Enter PARTNA or 'END'

8. Write the last value of PARTNA, which equals 'END', in record number K+1 of file DIRB.

9. Check results by printing files DIRA and DIRB at the terminal.

Figure 9.5a   (cont.)

```
          PROGRAM DTERM

     ****************************************************************
     *                                                              *
     *   THIS PROGRAM CREATES TWO DIRECT-ACCESS FILES: DIRA AND DIRB, *
     *   FROM DATA ENTERED AT THE TERMINAL.                         *
     *                                                              *
     ****************************************************************
          CHARACTER*14 PARTNA
          INTEGER N,K,I

          OPEN(UNIT=3, FILE='DIRA', ACCESS='DIRECT', STATUS='NEW',
         + FORM='FORMATTED', RECL=14)
          OPEN(UNIT=4, FILE='DIRB', ACCESS='DIRECT', STATUS='NEW',
         + FORM='FORMATTED', RECL=14)

          N=0

          PRINT*,'ENTER THE PART NAME FOR FILE DIRA ENCLOSED WITHIN',
         + ' APOSTROPHES.'
          PRINT*,'IF YOU HAVE NO MORE VALUES, ENTER ''END''.'
          READ*,PARTNA

        5 IF(PARTNA.NE.'END') THEN

             N=N+1

             WRITE(3,7,REC=N) PARTNA
             PRINT*,'ENTER THE PART NAME FOR FILE DIRA ENCLOSED',
         +      ' WITHIN APOSTROPHES.'
             PRINT*,'IF YOU HAVE NO MORE VALUES, ENTER ''END''.'
             READ*,PARTNA

             GO TO 5

          END IF
```

Figure 9.5b

```
C
C   THE FLAG IS THE CHARACTER STRING 'END' WHICH WAS ENTERED BY
C   THE USER.
C
        WRITE(3,7,REC=N+1) PARTNA

        K=0

        PRINT*,'ENTER THE PART NAME FOR FILE DIRB ENCLOSED WITHIN',
      + ' APOSTROPHES.'
        PRINT*,'IF YOU HAVE NO MORE VALUES, ENTER ''END''.'
        READ*,PARTNA

   10 IF(PARTNA.NE.'END') THEN
          K=K+1

          WRITE(4,7,REC=K) PARTNA
          PRINT*,'ENTER THE PART NAME FOR FILE DIRB ENCLOSED',
      +       ' WITHIN APOSTROPHES.'
          PRINT*,'IF YOU HAVE NO MORE VALUES, ENTER ''END''.'
          READ*,PARTNA

          GO TO 10

      END IF
C
C   THE FLAG IS THE CHARACTER STRING 'END' WHICH WAS ENTERED BY
C   THE USER.
C
        WRITE(4,7,REC=K+1) PARTNA

        PRINT *
        PRINT *,'THE CONTENTS OF DIRA ARE:'
        PRINT *
        DO 15 I=1,N+1
            READ(3,7,REC=I)PARTNA
            PRINT 33,I,PARTNA

   15 CONTINUE

        PRINT *
        PRINT *,'THE CONTENTS OF DIRB ARE:'
        PRINT *
        DO 17 I=1,K+1
            READ(4,7,REC=I)PARTNA
            PRINT 33,I,PARTNA

   17 CONTINUE

    7 FORMAT(A)
   33 FORMAT(I3,2X,A)
      END


RUN

    ENTER THE PART NAME FOR FILE DIRA ENCLOSED WITHIN APOSTROPHES.
    IF YOU HAVE NO MORE VALUES, ENTER 'END'.
  ? 'BATTERY CABLE'
    ENTER THE PART NAME FOR FILE DIRA ENCLOSED WITHIN APOSTROPHES.
    IF YOU HAVE NO MORE VALUES, ENTER 'END'.
  ? 'TIRES'
        .
        .
        .
    ENTER THE PART NAME FOR FILE DIRA ENCLOSED WITHIN APOSTROPHES.
    IF YOU HAVE NO MORE VALUES, ENTER 'END'.
```

Figure 9.5b   (cont.)

```
? 'END'
 ENTER THE PART NAME FOR FILE DIRB ENCLOSED WITHIN APOSTROPHES.
 IF YOU HAVE NO MORE VALUES, ENTER 'END'.
? 'MUFFLER'
 ENTER THE PART NAME FOR FILE DIRB ENCLOSED WITHIN APOSTROPHES.
 IF YOU HAVE NO MORE VALUES, ENTER 'END'.
? 'DISTRIBUTOR'
        .
        .
        .
 ENTER THE PART NAME FOR FILE DIRB ENCLOSED WITHIN APOSTROPHES.
 IF YOU HAVE NO MORE VALUES, ENTER 'END'.
? 'END'

 THE CONTENTS OF DIRA ARE:

    1   BATTERY CABLE
    2   TIRES
    3   GAS TANK
    4   EXHAUST PIPE
    5   RADIATOR
    6   BRAKE FLUID
    7   FAN BELT
    8   END

 THE CONTENTS OF DIRB ARE:

    1   MUFFLER
    2   DISTRIBUTOR
    3   SPARK PLUGS
    4   POINTS
    5   GAS
    6   BEARINGS
    7   END
```

Figure 9.5b   (cont.)

Algorithm for Sample Problem 9.6. This program updates the direct-access file DMAST with the information contained in the direct-access file DTRAN. The updating information is entered directly into file DMAST without creating a new master file.

1. Set I=1.
2. Read NPART,ISUB,IADD from record number I fo file DTRAN.
3. If NPART=9999, stop the program.
4. Set J=1.
5. Read MPART,AUTO,PRICE,INV from record number J of file DMAST.
6. If MPART=9999 then
   a. Print message 'NPART was not found.'
   b. Calculate
      I = I + 1
   c. Return to step 2

Figure 9.6a

Else if MPART=NPART then
a. Calculate
   INV=INV-ISUB+IADD
b. Write MPART,AUTO,PRICE,INV in record number J of file DMAST
c. Calculate
   I=I+1
d. Return to step 2
Else
a. Calculate
   J=J+1
b. Return to step 5

Figure 9.6a   (cont.)

```
            PROGRAM UPDATE

****************************************************************
*                                                              *
*   THIS PROGRAM UPDATES THE DIRECT-ACCESS FILE DMAST WITH THE *
*   INFORMATION CONTAINED IN THE DIRECT-ACCESS FILE DTRANS. THE*
*   UPDATING INFORMATION IS ENTERED DIRECTLY INTO FILE DMAST   *
*   WITHOUT CREATING A NEW MASTER FILE.                        *
*                                                              *
****************************************************************
            CHARACTER*16 AUTO
            INTEGER I,J,INV,ISUB,IADD,MPART,NPART
            REAL PRICE

            OPEN(UNIT=3, FILE='DMAST', ACCESS='DIRECT', STATUS='OLD',
           + FORM='FORMATTED', RECL=30)
            OPEN(UNIT=4, FILE='DTRAN', ACCESS='DIRECT', STATUS='OLD',
           + FORM='FORMATTED', RECL=10)

            I=1

         20 READ(4,15,REC=I) NPART,ISUB,IADD

            IF(NPART.EQ.9999) STOP

            J=1

         30 READ(3,17,REC=J) MPART,AUTO,PRICE,INV

            IF(MPART.EQ.9999) THEN
                 PRINT*,'PART ',NPART,' WAS NOT FOUND.'

                 I=I+1

                 GO TO 20

            ELSE IF(MPART.EQ.NPART) THEN

                 INV=INV-ISUB+IADD

                 WRITE(3,17,REC=J) MPART,AUTO,PRICE,INV
```

Figure 9.6b

```
          I=I+1

          GO TO 20

      ELSE

          J=J+1

          GO TO 30

      END IF
   15 FORMAT(I4,2I3)

   17 FORMAT(I4,1X,A,F6.2,I3)
      END
```

Figure 9.6b   (cont.)

REVIEW

Auxiliary files are stored externally on magnetic disks or tape. They are connected to a program via an OPEN statement. The OPEN statement has the syntax

OPEN (UNIT = *IU* [unit number device code],
        FILE = file *name* ['SCRATCH' files are not given a name],
        ACCESS = 'SEQUENTIAL'* or 'DIRECT',
        STATUS = 'OLD', 'NEW', 'SCRATCH', or 'UNKNOWN',
        FORM = 'FORMATTED' or 'UNFORMATTED',[†]
        IOSTAT = *JERR* [integer-value system error code],
        ERR = *LABEL* [statement label for error routine],
        RECL = *LREC* [record length mandatory for direct-access files],
        BLANK = 'ZERO' or 'NULL')

The minimum amount of information required for the OPEN statement is the unit code number and the FILE parameter:

```
OPEN(UNIT=7,FILE='NAME')
```

When a file is to be disconnected before the program terminates, a CLOSE statement is used. Otherwise, the system releases and closes all files automatically when the program terminates normally. The CLOSE statement has the syntax

---

*The underlined words are the default values.

[†]The default value for FORM is 'FORMATTED' for sequential files, and 'UNFORMATTED' for direct-access files.

CLOSE (UNIT = *IU* [unit number device code],
    STATUS = 'KEEP'* or 'DELETE' ['SCRATCH' files cannot
        be saved],
    IOSTAT = *JERR* [integer-value system error code],
    ERR = *LABEL* [statement label for error routine])

All the parameters in the CLOSE statement are optional except the unit number device code.

CLOSE(*IU*)

The INQUIRE statement is used to question file status. The statement has the syntax

INQUIRE(UNIT=*IU*,inquiry list)

or

INQUIRE(FILE=*name*,inquiry list)

The inquiry list can include all or some of the inquiry parameters shown in Table 9.1: ACCESS, BLANK, DIRECT, EXIST, FORM, FORMATTED, NAME, NAMED, NEXTREC, NUMBER, OPENED, RECL, SEQUENTIAL, and UNFORMATTED. In addition, the IOSTAT and ERR parameters can be included in the list. When the INQUIRE statement is executed, the variable name given for each inquiry parameter listed receives a standard value that depends on the current status of the file. The value assigned to the variable can then be displayed at the terminal.

To position sequential files relative to the read/write head, one uses the REWIND, BACKSPACE, or ENDFILE commands. REWIND rewinds the file to the very first data record, whereas BACKSPACE steps the file back one record. The statement ENDFILE places an end-of-file (EOF) special record in the file during program execution. This EOF marker is needed for all sequential files; it is placed there automatically when the program terminates normally and the file is closed. The syntaxes of all three statements are similar:

REWIND(UNIT=*IU*,IOSTAT=*JERR*,ERR=*LABEL*)
BACKSPACE (UNIT=*IU*,IOSTAT=*JERR*,ERR=*LABEL*)
ENDFILE (UNIT=*IU*,IOSTAT=*JERR*,ERR=*LABEL*)

All the parameters in these statements are optional except the unit number device code.

*The underlined words are the default values.

REWIND(*IU*)

BACKSPACE(*IU*)

ENDFILE(*IU*)

Input/output operations are initiated using the READ or WRITE statements. The syntax for the READ statement is

READ (UNIT = *IU* [unit number device code],

       FMT = *NFORM* [statement label for FORMAT statement],

       IOSTAT = *JERR* [integer-value system error code],

       ERR = *LABEL* [statement label for error routine],

       END = *LEND* [statement label to which control should pass if an end-of-file special record has been read; only used with sequential files],

       REC = *NREC* [record number; used only with direct-access files])

The character strings 'UNIT =' and 'FORMAT =' are optional. If they are omitted, the unit number device code, *IU*, must precede the statement number, *NFORM*, of the referenced FORMAT statement. The parameters IOSTAT, ERR, and END are optional. The parameter END is used with sequential files to prevent automatic program termination if an end-of-file marker is encountered. The given statement number, *LEND*, refers to the executable statement that will be processed next in the event that an EOF mark has been reached. The parameter REC is used only with direct-access files and refers to the record number that is to be read.

The WRITE statement has the syntax

WRITE (UNIT = *IU* [unit number device code],

       FMT = *NFORM* [statement label for FORMAT statement],

       IOSTAT = *JERR* [integer-value system error code],

       ERR = *LABEL* [statement label for error routine],

       REC = *NREC* [record number; used only with direct-access files])

As with the READ statement, the parameter REC must be used with direct-access files to specify the record number, NREC, that is to be written.

Internal files are established when a character variable name, substring, character array, or character array element is used in place of the unit device code number in a READ or a WRITE statement. The character value assigned to the name, substring, and so forth, is considered to be the file's contents and can be accessed or written upon, thus allowing internal transfer of data. All input/output operations involving internal files must be formatted.

Direct-access files have each record entry numbered. The record numbers are then referred to by the READ and WRITE statements that call out the file. Records can be read or written in any order. Integer record numbering must start at the value 1 and then increase. Because accessing of records is accomplished directly with numbering, the REWIND, BACKSPACE, and ENDFILE commands cannot be used with this type of file. Instead of an end-of-record mark, a data flag is usually used in one of the fields. When connecting a direct-access file, the OPEN statement is used. The parameter ACCESS is set to 'DIRECT'. In addition, the record length, *LREC*, must be given and must be the same for the entire file. Because each record has the same length, list-directed input/output operations are not allowed.

KEY TERMS

BACKSPACE **statement**

direct-access file

ENDF I LE **statement**

external sort

formatted file

I NQU I RE **statement**

internal file

random-access file

sequential access

sequential file

unformatted file

REVIEW QUESTIONS

1. Are the contents of a data file left empty after a program reads its information?
2. Discuss the difference between direct and sequential-file access.
3. What is a random-access file?
4. What is an integer unit code number?
5. Can direct-access files be stored on magnetic tape? Why?
6. Explain the difference between a formatted and an unformatted file.
7. Describe the three positioning statements for sequential files.
8. What is the difference between the IOSTAT and ERR parameters?
9. What is the purpose of a record number?
10. What is the significance of the record length, *LREC*, of a direct-access file? How is it measured for a formatted file?
11. Can a direct-access file be rewound?
12. What is meant by a key field?
13. List and describe all the possible parameters for the
    - **(a)** OPEN statement
    - **(b)** BACKSPACE statement
    - **(c)** READ statement
    - **(d)** REWIND statement
    - **(e)** WRITE statement
    - **(f)** ENDFILE statement
    - **(g)** INQUIRE statement
    - **(h)** CLOSE statement
14. What happens on your computer when a system LIST command (copy file at the terminal) is used with a direct-access file?

COMPUTER PROJECTS

CP1. Modify the solution to Sample Problem 9.1 so that the program instead updates a file that has a key field, along with the key field of the transaction file, found in descending order.

CP2. Modify the solution to Sample Problem 9.2 so that it instead merges data files whose key fields are in descending order.

CP3. Modify the solution to Sample Problem 9.2 so that it can merge files whose key fields are names (character strings) found left-justified in ascending alphabetic order. Assume that the character fields are sixteen characters wide.

CP4. Modify the solution to Sample Problem 9.3 so that the program instead performs an external sort of the key numeric field in descending order.

CP5. Repeat CP4, this time sorting a key field of names (character strings) into descending alphabetic order. The names are found left-justified in a data field sixteen characters wide.

CP6. Modify the solution to Sample Problem 9.6 so that the key field used is the character string assigned to the variable AUTO. Assume that the word END, appearing in the key field, is the flag signaling the end of that data file.

CHAPTER 10

Special Topics

Shown in the picture are the nozzle partitions of a turbine diaphragm. The purpose of a diaphragm is to redirect steam flow to the next turbine wheel along the shaft while at the same time increasing the steam's velocity. The nozzle's efficiency is a function of many factors. Hundreds of hours of computer analysis were spent on the design of the diaphragm shown, which is for an 1100-megawatt turbine-generator. The turbine is located at a Swedish nuclear power plant. *(Photo courtesy of ABB Stal AB.)*

10.1 INTRODUCTION

To complete this study of FORTRAN 77 there are several more statements and FORMAT descriptors we must discuss. They have been set aside for inclusion in this last chapter because they are statements or descriptors that, though useful, are not commonly encountered.

10.2 SPECIFICATION STATEMENTS

Specification statements are used to declare characteristics of FORTRAN variables. They are nonexecutable statements that must appear before the first executable statement or statement functions of the program. The specification statements we have already discussed are DIMENSION, PARAMETER, COMMON, SAVE, EXTER-

NAL, INTRINSIC, and the six type statements, which are INTEGER, REAL, DOUBLE PRECISION, COMPLEX, LOGICAL, and CHARACTER. In this section we introduce the last two FORTRAN specification statements, namely, the IMPLICIT **statement** and the EQUIVALENCE **statement.**

10.2.1 The IMPLICIT Statement

FORTRAN has a means of implicit or default typing of variable names. Any variable name that begins with the letters A through H or O through Z is typed as a real variable name, and any variable name beginning with I through N will be considered an integer variable. The IMPLICIT statement is used to alter the standard default groupings. It has the syntax

IMPLICIT type(single letter or range of letters separated by a hyphen)

For example,

IMPLICIT REAL(I)

IMPLICIT INTEGER(H,O-P)

IMPLICIT CHARACTER*20 (M,R-Z),REAL(I)

The statement IMPLICIT REAL(I) specifies that *in addition* to the standard default typing provided, all variable names beginning with the letter I will also be treated as real variables. IMPLICIT INTEGER(H,O-P) specifies that variable names beginning with H or with any letter between O and P (as well as $I, J, K, L, M,$ and N) will be assigned integer data only.

The IMPLICIT statement must *precede* all other specification statements except the PARAMETER statement. If explicit typing information appears afterward, either through a specification statement or a type FUNCTION statement, it overrides the IMPLICIT statement. For example the statements

```
PROGRAM TYPE
IMPLICIT INTEGER(H,O-P)
REAL PROSE,IRRAD
```

will classify all variables beginning with the letter P, except the single variable name PROSE, as the integer type. The variable name PROSE, as well as the variable name IRRAD, are classified as real.

An IMPLICIT statement may be used locally in a subroutine or a function subprogram and will affect only that unit.

10.2.2 The EQUIVALENCE Statement

The COMMON statement (see Section 7.6) allows variables named in different program units to share common storage areas. It provides a means of exchanging data between program modules without having to formally use subprogram arguments.

The EQUIVALENCE statement, on the other hand, allows several variable names within the *same* program unit to share a common storage location, and thus the same value. The EQUIVALENCE statement has the syntax

EQUIVALENCE(variable list$_1$),(variable list$_2$), . . .

where each variable list is a list of variable names, array names, array element names, or character substrings separated by commas and enclosed within parentheses. The variable lists in an EQUIVALENCE statement are to share a common storage location. For example,

```
EQUIVALENCE(X,Y),(I1,I2,K)
```

This statement causes the real data for X and Y to have the same value, since they are assigned to the same memory location. The same will be true for the group of integer variables I1, I2, and K.

The EQUIVALENCE list provides a quick means of equating the values of different variables that appear in a program. In the early days of computing, the EQUIVALENCE statement was used to conserve storage area. Variable names that did not appear further in a program unit would have their storage area reassigned to a newly introduced variable via the EQUIVALENCE statement. With the memory size of today's computer, storage capacity is hardly a factor in program design.

If character variable names appear in a list, they may only be made equivalent to other character entries. The lengths of the defined character variables can be different, but this is not advisable, since information would be lost. It is also not recommended to equate data of different types, for example, integer and real, since equivalencing does not effect type conversion and truncation problems will arise. It should be pointed out that elements of the same array cannot be equivalenced. Similarly, one should not equate an array name to a simple variable name, since the variable cannot become an array or vice versa by equivalencing. Further, formal arguments for a subprogram unit may not be equivalenced, nor may separate items that appear in the same COMMON block or in different COMMON blocks within the same program unit.

Regardless of whether an EQUIVALENCE statement references the array name or just an array element, the entire array is sent to common storage. For example, the following statements,

```
REAL A(4),B(2,2),C(4)
EQUIVALENCE(A,B,C)
```

will produce the following associations

$$A(1) = B(1, 1) = C(1)$$
$$A(2) = B(2, 1) = C(2)$$
$$A(3) = B(1, 2) = C(3)$$
$$A(4) = B(2, 2) = C(4)$$

where B is stored in a column-wise manner. The statements

```
REAL  A(4),B(2,2)C(4)
EQUIVALENCE(A(1)B,(1,1),C(1))
```

or

```
REAL  A(4),B(2,2),C(4)
EQUIVALENCE(A(2),B(2,1),C(2))
```

or

```
REAL  A(4),B(2,2)C(4)
EQUIVALENCE(A(3),B(1,2),C(3))
```

will produce identical results even though only one array element is called out. Note what happens when array elements having different subscript values are equivalenced.

```
REAL  A(4),B(2,2),C(4)
EQUIVALENCE(A(2),B(1,1),C(3),Z)
```

All array elements are still involved but the associations have now shifted

$$C(1)$$
$$A(1) = \qquad C(2)$$
$$A(2) = B(1, 1) = C(3) = Z$$
$$A(3) = B(2, 1) = C(4)$$
$$A(4) = B(1, 2)$$
$$B(2, 2)$$

Here, $A(1)$ is set equal to $C(2)$; $A(2)$ is set equal to $B(1, 1)$, $C(3)$, and Z; $A(3)$ is set equal to $B(2, 1)$ and $C(4)$; and $A(4)$ is set equal to $B(1, 2)$. Elements $C(1)$ and $B(2, 2)$ have no associated values.

If character variable names are equivalenced, they share the first character storage position. For example, the statements

```
CHARACTER*4   CA,CB,CC*6
EQUIVALENCE  (CA,CB,CC)
```

will produce the associations

$$CA(1:1) = CB(1:1) = CC(1:1)$$
$$CA(2:2) = CB(2:2) = CC(2:2)$$
$$CA(3:3) = CB(3:3) = CC(3:3)$$
$$CA(4:4) = CB(4:4) = CC(4:4)$$
$$CC(5:5)$$
$$CC(6:6)$$

If substrings are referenced in an EQUIVALENCE statement, character positions of the separate substrings are matched (aligned) and set equal to one another. As with arrays, the entire word is brought into common storage, not just the substring, even though the character variable name is not explicitly called out. For example, the statements

```
CHARACTER*4  CA,CB,CC*6
EQUIVALENCE(CA(2:),CC(3:5))
```

will cause the associations

$$CC(1:1)$$
$$CA(1:1) = CC(2:2)$$
$$CA(2:2) = CC(3:3)$$
$$CA(3:3) = CC(4:4)$$
$$CA(4:4) = CC(5:5)$$
$$CC(6:6)$$

With character arrays, associations begin with the first character position of the first array element, or with the first character position of a specified array element or the array element's referenced substring. The entire character array is equivalenced as before. For example,

```
CHARACTER*3  CA(2),CB(3)*2,CC*4,CD
EQUIVALENCE  (CA,CB),(CB(2),CC),(CB(2)(2:),CD)
```

causes the associations

$$CA(1)(1:1) = CB(1)(1:1)$$
$$CA(1)(2:2) = CB(1)(2:2)$$
$$CA(1)(3:3) = CB(2)(1:1) = CC(1:1)$$
$$CA(2)(1:1) = CB(2)(2:2) = CC(2:2) = CD(1:1)$$
$$CA(2)(2:2) = CB(3)(1:1) = CC(3:3) = CD(2:2)$$
$$CA(2)(3:3) = CB(3)(2:2) = CC(4:4) = CD(3:3)$$

When a variable or array appears in both an EQUIVALENCE and a COMMON statement within the same program unit, care must be taken that the common block is not inadvertently extended backward, that is, storage is added before the first storage word. The following program actions are permissible,

```
PROGRAM MAIN
REAL  A(3,2),B(2,3)
EQUIVALENCE(A,B)
COMMON  A
```

.
.
.
```
SUBROUTINE SUB(X,Y)
REAL  Z(6),X,Y
COMMON  Z
```

and lead to the associations

$$A(1, 1) = B(1, 1) = Z(1)$$
$$A(2, 1) = B(2, 1) = Z(2)$$
$$A(3, 1) = B(1, 2) = Z(3)$$
$$A(1, 2) = B(2, 2) = Z(4)$$
$$A(2, 2) = B(1, 3) = Z(5)$$
$$A(3, 2) = B(2, 3) = Z(6)$$

The combined statements

```
PROGRAM  MAIN
REAL  A,B
COMMON  A,B
     .
     .
     .
SUBROUTINE  SUB(X,Y)
REAL  Z(3),X,Y,R,S,
COMMON  R,S
EQUIVALENCE(S,Z(2))
```

are also permissible, leading to the following associations and a blank common area
that has been extended from two to three elements:

$$A = R = Z(1)$$
$$B = S = Z(2)$$
$$Z(3)$$

The following actions are *not* allowed, however,

```
PROGRAM  MAIN
REAL  A,B
COMMON  A,B
     .
     .
     .
SUBROUTINE  SUB(X,Y)
REAL  Z(3),R,S
```

```
COMMON R,S
EQUIVALENCE(R,Z(2))
```

since a storage location for Z(1) would be required before the first storage word of the blank common block.

10.3 INPUT AND OUTPUT

In Chapter 3 we discussed formatted input/output operations. The I, F, and E format descriptors were introduced at that time to accommodate integer and real numeric data. In addition, we learned how to include literal descriptors enclosed within apostrophes in the FORMAT specifications. These character string messages accompanied the output. In this section we will learn about several other field descriptors that add capabilities that make input/output operations less toilsome. They are the G, Ee, P, BN, BP, S, SP, SS, and H descriptors.

10.3.1 The G Descriptor

The G **descriptor** is used with real or double-precision numeric values. The G stands for "generalized." On output, the data with G format is first converted to a floating-point number, which will then appear with or without its exponent, that is, will have either F or E format, depending on the number's absolute magnitude. The G descriptor has the syntax

$Gw.d$

In this expression, w represents the field width and d specifies the number of decimal places to be kept to the right of the decimal point for E format, or the number of significant digits to appear with F format.

What determines whether E or F format is used is ultimately the number's magnitude. If the data is very large or very small, E format must be employed. If the data is of medium size, F format is used. The governing rules are that if the number's exponent is negative or larger than d (smaller than 0.1 or larger than 10^d), the number will appear with E format. If the number's exponent is greater than or equal to zero but less than or equal to d, the number is written right-justified, in F code, having d significant digits followed by four trailing blanks. The number appears in a field $w - 4$ wide. The four trailing blanks after the number give a total field width of w. The four trailing blank spaces are used, otherwise, to carry exponential information for the alternative E formatted display. It is, therefore, important for output purposes that the field width, w, be greater than or equal to $d + 7$. The seven extra spaces are used for sign, possible leading zero, decimal point, and the four spaces reserved for exponent information. The following table illustrates the type of output produced using a format of G12.5.

Number	Output (G12.5)
	ǀ
	1 2 3 4 5 6 7 8 9 0 1 2
.099999	.99999E-01
.99999	.99999
99.999	99.999
99999	99999.
999990	.99999E+06

On input, the G descriptor performs as an E descriptor.

10.3.2 The E*e* Descriptor

For G or E format, it is also possible by appending the E*e* **descriptor** to specify how many positions, *e*, are to be used for the display of the variable's exponent. The syntax is

> E*w.d*E*e* or G*w.d*E*e*

The number of positions normally reserved for the exponent, excluding the sign position, is two, and the field is usually selected so that *w* is greater than or equal to *d* + 7. When an E*e* descriptor is used, however, the field width should be chosen so that it is greater than or equal to *d* + *e* + 5. When using G*w.d*E*e* format and when the value to be displayed is of medium size (that is, will be represented by F format), the number of blanks appearing at the end of the record will now be *e* + 2 instead of four, and the data will be written in *w* − (*e* + 2) character positions. The E*e* descriptor has no effect on input.

10.3.3 The P Descriptor

The P **descriptor** establishes scaling of real data that is presented with D, E, F, or G format. It is useful for representing percentages or scaled values such as milliamperes on micrometers. The P descriptor has the form

> *n*P

where *n* is a signed or unsigned integer constant equal to the scale factor. When a scale factor of *n* is employed, the data is first multiplied by a factor of 10^n — that is, the decimal point is shifted *n* places to right — before it is sent to the output device. When *n* is negative, the decimal point is shifted toward the left. Upon input, the scale factor of *n* causes the data to be divided by 10^n if *no* exponent is present in the data field. *When an exponent is present, the scale factor is ignored.* The various forms the P descriptor may assume are

nPD$w.d$
nPE$w.d$
nPE$w.d$Ee
nPF$w.d$
nPG$w.d$
nP

Initially, a scale factor of zero is established for all descriptors in a FORMAT statement when they are first processed. When one of the edit descriptors present is preceded by a nonzero scale factor, all that FORMAT statement's subsequent specifiers will be assumed to also have this value. It will require a formal action to reestablish a zero scale factor again. For example, the FORMAT statement

```
10  FORMAT(3P,E12.5,F9.2,G12.5,0P,3E12.5)
```

establishes a scale factor of 3 (or 10^3) for all output descriptors following the 3P specifier. The descriptor 0P returns the scale factor to zero for the last repeated specification of E12.5. The FORMAT statement

```
10  FORMAT(3P,E12.5,0PG12.5,-1PF9.2)
```

establishes a scale factor of 3 for the E12.5 specification and zero scaling for the G specification. The data presented in F code will be multiplied by 10^{-1}.

To illustrate how the P descriptor works, assume that the value assigned to X is 3.141592654. Upon output, and with the following P descriptors, X will appear as shown:

Edit descriptors	Displayed output
-1PF11.4	.3142
F11.4	3.1416
1PF11.4	31.4159
-1PE11.4	.0314E+02
E11.4	.3142E+01
1PE11.4	3.1416E+00
2PE11.4	31.416E-01
-3PD16.9	.000314159D+04
3PD16.9	314.1592654D-02

It can be seen that when using F descriptors, the position of the decimal point in the field is fixed while the number is multiplied by 10^n. The number is consequently shifted to the right or left relative to the decimal point depending on whether

n is positive or negative, and the number's magnitude changes. For E or D descriptors the decimal point is first moved right or left n places (depending on sign), and the exponent is adjusted to compensate for the shift. The magnitude of the number does not change. Leading zeros after the decimal point will be required when n is negative. The number of decimal digits, d, appearing on output is adjusted as follows:

$$\text{when } n \text{ is less than or equal to } 1: d \Rightarrow d$$

$$\text{when } n \text{ is greater than } 1: d \Rightarrow d - n + 1$$

When the data is presented using G format, the scale factor will only have effect when the numbers fall in the range in which the E descriptor is normally used, that is, less than 0.1 or greater than 10^d. Otherwise, the scale factor will be ignored.

To demonstrate how the P descriptor affects input, assume that the following data record is given:

.31416E0

The table below shows the values that will ultimately be assigned to variable X after using the indicated nPF edit descriptors.

Edit descriptor	Assigned x value
F6.0	.31416
−1PF6.0	3.1416
1PF6.0	.031416
1PF8.0	.31416

Notice that in the last case, when the field was extended to include the exponent, the P descriptor was ignored.

10.3.4 The BN and BZ Descriptors

On input of numeric data using I, F, E, D, or G format, blanks (other than leading blanks) are either ignored or treated as zeros, depending on which computer system is used. To insure that a program and its data will carry to another machine, the BN or the BZ **descriptor** is used. Once a BN (blank null) specifier is encountered in a format specification, all succeeding blanks appearing in the numeric fields associated with that FORMAT statement are ignored (the null condition); that is, they are assumed not to be present, and the individual fields are then right-justified. If a field contains all blanks, it is considered to have a value of zero. When a BZ (blank zero) specifier is met, all subsequent blanks that appear in the numeric data fields (except leading blanks) are interpreted as zeros. If a BN descriptor is later encountered within that format specification, a blank-null condition is reestablished. After the FORMAT statement ceases to be in effect, the system's null/zero default returns. For example, with the data record

```
5____.321E4_
```

the statements

```
   READ(5,7)  I,X
 7 FORMAT(BN,I3,E9.3)
```

will assign I=5 and X=.321E4. However, the statement

```
 7 FORMAT(BZ,I3,E9.3)
```

will assign I=500 and X=.321E40. The FORMAT statement

```
 7 FORMAT(BZ,I3,BN,E9.3)
```

will have I=500 and X=.321E4 again.

10.3.5 The S, SP, and SS Descriptors

In most computer systems, upon output the compiler suppresses a leading printed plus sign for positive numeric values. It is possible to have the plus sign appear with the output by using the SP **descriptor.** Once the SP (sign positive) descriptor is encountered, all positive succeeding numeric data fields associated with that given FORMAT statement will bear a positive sign (provided that their field widths are large enough). When the next FORMAT statement is encountered, the output of data will return to the system's default mode. The SS **descriptor** (sign suppression descriptor) suppresses the leading plus sign. It will not eliminate the plus sign that the exponent carries for E-formatted output. The S **descriptor** (system descriptor) returns sign control to the system default mode. For example, if $X = 3.14159$, and $Y = 290.643$, the statement

```
   PRINT 6,X,Y
 6 FORMAT(1X,E13.6,3X,SP,E13.6)
```

will produce the printed line

```
__.314159E+03____+.290643E+03
```

for a system in which the positive sign is normally suppressed.

10.3.6 The H Descriptor

The H **descriptor** was used in earlier versions of FORTRAN to present literal character strings (that is, character strings not associated with character variables) for output. The syntax of the H descriptor, also known as the Hollerith descriptor, is

nH character string

The *n* stands for the number of characters, including blanks, to be written. Since it is a tedious, error prone task to count characters in a message, the H descriptor is rarely used nowadays. It has been effectively replaced by the more efficient practice of enclosing a character string within apostrophes. The two following FORMAT statements are therefore equivalent:

```
6 FORMAT(1X,'THIS IS A TEST',E13.6)

6 FORMAT(IX,14HTHIS IS A TEST,E13.6)
```

10.4 THE PAUSE STATEMENT

A PAUSE **statement** may be used to temporarily stop a program so that results may be checked before proceeding further or terminating the run. The PAUSE statement has the syntax

PAUSE constant

where the constant can be a string of 1 to 5 decimal digits or a character string message having seventy characters or less.

When the PAUSE statement is met in a program, execution of the program halts, the constant's value is usually displayed, and the computer awaits further instructions from the console or local terminal. The code or sequence of entered messages needed to restart or terminate a program is system dependent. Before using the PAUSE statement, check to see that it has not been intentionally disabled for interactive processing on your local time-sharing network.

10.5 FUNCTIONS AND SUBROUTINE STATEMENTS

10.5.1 The ENTRY Statement

The usual entry into a subroutine or function subprogram is at the beginning of the program unit. It may be desirable at times to enter the subprogram at a different point. For this purpose, the ENTRY **statement** has been introduced in FORTRAN. It has the format

ENTRY name(argument list)

where the name is the label given to the new entry point and the argument list is a formal list of variables, arrays, and so forth, passed from the calling program. To call out the subprogram and enter at an alternative entry point, the name or label is used in the calling statement rather than the subprogram's name. The actual arguments of the CALL statement must match in number and in type with the formal arguments of the ENTRY statement, but the arguments used in the ENTRY statement may be different from that of the subprogram's argument list.

With an entry into a subprogram at an alternative point, control passes to the very first executable statement following the referenced ENTRY statement. ENTRY is a nonexecutable statement that may be placed anywhere inside a subprogram unit except within the range of a DO loop or an IF-block structure. For example, the following statements effect multiple entry into a subroutine called TEST:

```
PROGRAM MAIN
COMPLEX ROOTA
READ *,A,B,C
DISC=B**2-4.0*A*C
IF (DISC.LT.0.) THEN
     DISC=-DISC
     CALL TESTA(DISC,A,B,ROOTA)
END IF
CALL TEST(DISC,A,B,ROOT)
  .
  .
  .
END

SUBROUTINE TEST(R,S,T,U)
COMPLEX ZA
U=(-T+SQRT(R))/(2.0*S)
RETURN
ENTRY TESTA(W,X,Y,ZA)
ZA=CMPLX(-Y/(2.0*X),SQRT(W)/(2.0*X))
RETURN
END
```

To demonstrate how a function subprogram is called, the foregoing statements can be modified as follows:

```
PROGRAM MAIN
REAL A,B,C,Y1,Z,ROOTA,ROOT,DISC
COMPLEX Y
READ *,A,B,C
DISC=B**2-4.0*A*C
IF (DISC.LT.0.) THEN
     Y1=ROOTA(DISC,A)
     Y=CMPLX(-B/(2.0*A),Y1)
  .
  .
  .
STOP
END IF
```

```
Z=ROOT(DISC,A,B)
    .
    .
    .
END

REAL  FUNCTION  ROOT(R,S,T)
REAL  R,S,T
ROOT=(-T+SQRT(R))/2.0*S)
RETURN
REAL  ENTRY  ROOTA  (W,X)
W=-W
ROOTA=SQRT(W)/(2.0*X)
RETURN
END
```

Notice how the function subprogram either calculates a complex or a real root depending on where entry is made in the subprogram.

10.5.2 Alternate Returns

There is also a means provided in FORTRAN to return control to the calling program at points other than the statement appearing immediately after the referring CALL statement. The method works only with subroutines. To establish an **alternate return,** the subroutine statement RETURN n is used, where n is an integer value, integer variable, or integer expression. The value that n assumes must point to the main program statement number given in the CALL or ENTRY statement to which control will be transferred upon return. The associated CALL statement must then have the syntax

    CALL subroutine name(argument list,$*n_1$,$*n_2$, . . .)

where n_1, n_2, and so on, are statement numbers in the calling program to which return can be made. If n equals 1, return is to statement number n_1; if n equals 2, return is to statement number n_2; and so on. The following statements illustrate how alternate returns can be effected:

```
PROGRAM MAIN
    .
    .
    .
CALL SUB(A,B,C,*10,*20)
    .
    .
    .
10 PRINT 7,A
    .
    .
```

```
          .
      STOP
 20   PRINT  8,B,C
          .

          .

          .


      END

      SUBROUTINE  SUB(X,Y,Z,*,*)
      IF  (X.GT.3000.0)  THEN
          .

          .

          .

           RETURN  1
      END  IF
          .

          .

          .
      RETURN  2
      END
```

Note the appearance of asterisks as formal arguments in the SUBROUTINE statement. In this subroutine, if X is greater than 3000.0, control is returned to statement
number 10 of the calling program. (See the actual argument list of the CALL statement.) Otherwise, control passes to statement number 20, that is, the second statement number given in the argument list (RETURN 2).

The use of alternate returns to the calling program is not recommended, since
it will lead to disjointed, hard-to-follow programs. In addition, their use is not con-
sistent with structured-programming principles. (See the discussion on structured-
programming principles in Section 4.4.)

REVIEW

This last chapter introduces the less commonly used statements in FORTRAN. The
two new specification statements discussed are the IMPLICIT statement and the
EQUIVALENCE statement. The IMPLICIT statement is nonexecutable and appears
before all specification and executable statements in a program unit. It is used to
alter and extend the FORTRAN language's classification of variable names based on
implicit first-letter typing. It has the syntax

IMPLICIT type(single letter or range of letters separated by a hyphen)

Explicit typing overrides the IMPLICIT statement.

The EQUIVALENCE statement is another specification statement that assigns
common storage to a group of listed variables, arrays, array elements, or character

substrings within the *same* program unit. It has the syntax

EQUIVALENCE(variable list$_1$),(variable list$_2$), . . .

The variables and arrays or array elements on each group list should be the same in type and size to avoid truncation errors. When a variable or array appears within the same program unit in a COMMON statement as well as in an EQUIVALENCE statement, care must be taken to see that the common block is not inadvertently extended backward, that is, that storage is added before the first storage word.

Input/output operations are also discussed in this chapter. The G (generalized) field descriptor is introduced. When it is used with real data for output purposes, the data will appear either in F or E format, depending on the data's absolute magnitude. The syntax of the G descriptor is

G$w.d$

where w represents the field width and d represents the number of decimal digits to appear with E format, or the number of significant digits to appear with F format.

The Ee descriptor provides for user control of the number of exponential positions, e, that appear with the exponent when a G or E format is used for output. It has the syntax

G$w.d$Ee or E$w.d$Ee

The number of positions normally reserved for the exponent is two. If e is specified, the field width, w, should then be selected greater than or equal to $d + e + 5$.

The P descriptor allows for scaling of numeric data. By employing a scale factor, quantities such as percentages, or milli- or micro-quantities, can be readily presented. The various forms of the P descriptor are

nPD$w.d$

nPE$w.d$

NPE$w.d$Ee

nPG$w.d$

nPF$w.d$

nP

where n is a signed or unsigned integer constant equal to the scale factor. When a scale factor of n is employed, the data is multiplied by 10^n before being sent to an output device. Upon input, the data is divided by 10^n if no exponent is present in the data field. If an exponent is present, the scale factor is ignored. The scale factor for each FORMAT statement is initially set to zero. When an edit descriptor is preceded by a nonzero scale factor, all the succeeding edit descriptors in that statement then have that scale factor assigned.

The BN (blank null) and BZ (blank zero) descriptors define how blanks appearing in an input data field should be interpreted by the computer. Once a BN descriptor is encountered in a FORMAT statement, all subsequent blanks appearing in the numeric fields associated with that statement are ignored, and the individual fields are right-justified. A field of all blanks is considered to have a zero value. When a BZ descriptor is encountered, all blanks (except leading blanks) are interpreted as zeros. After the FORMAT statement is processed, control returns to the null/zero system default.

The S, SP, and SS descriptors allow the programmer to choose between including the positive sign or omitting it for display of positive numeric output. The SP (sign positive) descriptor, once encountered in a FORMAT statement, will cause all succeeding positive output fields related to that FORMAT statement to appear with a plus sign. The SP descriptor ceases to be in effect after that FORMAT statement is processed or if an SS (sign suppression) descriptor is met later in the same FORMAT statement. Sign suppression omits the appearance of the plus sign for positive data output but has no effect on the display of the exponent's sign for G or E editing. The S (system) descriptor returns control to the system default.

The H, or Hollerith, descriptor, is used to present literal character messages that are not associated with character variables. It appears in the FORMAT statement as

nH character string

where n stands for the number of characters, including blanks, to appear in the message.

The PAUSE statement, a control statement, allows program execution to be temporarily halted so that intermediate results can be studied and the program either terminated or continued. The code, or series of messages, needed to be entered at the terminal or control console to return control to the computer is system dependent. The form the PAUSE statement takes is

PAUSE constant

where the constant can be a string of one to five decimal digits or a character string having seventy characters.

ENTRY and RETURN statements are also discussed in this chapter. The subprogram ENTRY statement has the syntax

ENTRY name(argument list)

where the name is the label given to the alternative entry point in a subprogram, and the formal argument list is a list of variables, arrays, and so forth, passed from the calling program. To call out a subprogram using a different entry point, the ENTRY name is used in the calling statement rather than the subprogram's name. The actual arguments assigned must match in number and in type with the specified formal argument list of the ENTRY statement.

To effect an alternate return to the calling program from a subroutine, the alternate return statement

RETURN n

may be used. The integer value n acts as a pointer to the main program's statement

number, given in the CALL statement's actual argument list, to which control will
be transferred upon return. The associated CALL statement then appears as

> CALL subroutine name(argument list,$*n_1,*n_2, \ldots$)

where n_1, n_2, and so on, are statement numbers in the calling program to which it is
possible to return. The SUBROUTINE statement then appears as

> SUBROUTINE subroutine name(argument list,*,*, . . .)

where the asterisks formally substitute for the statement numbers.

KEY TERMS

alternate return	H **descriptor**
BN **and** BZ **descriptors**	IMPLICIT **statement**
Ee **descriptor**	P **descriptor**
ENTRY **statement**	PAUSE **statement**
EQUIVALENCE **statement**	RETURN n **statement**
G **descriptor**	S, SP, **and** SS **descriptors**

REVIEW QUESTIONS

1. For the numbers shown, which are assigned to the variable X, determine what output will
   appear for X using a G11.4 format.

   (a) .012345   (b) .12345
   (c) 1.2345   (d) 12.345
   (e) 123.45   (f) .12345E4
   (g) 12345.

2. Repeat Problem 1 with a format of G13.6.

3. For the number .4135618, determine what output will result using the following edit
   descriptors:
   (a) -2PF10.3   (b) -1PF10.3
   (c) 0PF10.3   (d) 1PF10.3
   (e) 2PF10.3

4. Repeat Problem 3 using E editors:
   (a) -2PE10.3   (b) -1PE10.3
   (c) 0PE10.3   (d) 1PE10.3
   (e) 2PE10.3

5. Repeat Problem 3 using D editors:
   (a) -2PD13.6   (b) -1PD13.6
   (c) 0PD13.6   (d) 1PD13.6
   (e) 2PD13.6

6. What scale factor would you use to display results in milliamperes? Would the factor be positive or negative?

7. What values will be stored in memory when the data record

      4.135E-1

is accessed using the following edit descriptors?

(a) -2PF5.0   (b) -1PF5.0

(c)   F4.0   (d) 1PF5.0

(e) 2PF8.0

8. For the data line

      4.135E-1___3__

and the READ statement

      READ(7,3)X,K

determine what values will be entered for X and Y using the following FORMAT statements:

(a) 3 FORMAT(BZ,E9.0,1X,I4)

(b) 3 FORMAT(BZ,E9.0,1X,BN,I4)

(c) 3 FORMAT(E9.0,1X,BZ,I4)

Binary Data Representation

The ladle of steel shown at the top of the photo contains 295 tons of liquid steel. It is poured into a conduit that feeds the continuous slab caster shown at the bottom of the photo. The casters produce 8-inch-thick slabs of steel, which can be cut to any desired length. Six digital computers are used to control the production of liquid steel and the continuous-casting process at the plant. *(Photo courtesy of Wheeling Pittsburgh Steel Corporation.)*

In the discussion of machine language in Chapter 1, it was noted that the alphanumeric data received from the terminal had to be given binary representation first—that is, converted to binary code—before it could be processed by the computer. The data received is either of the numeric or the character type. This section describes how alphanumeric data is represented in any digital device.

NUMERIC DATA

We are very familiar with the decimal representation of numbers. Decimal notation uses the base 10 to form and express quantities. There are only ten digits in this system: 0, 1, 2, 3, 4, 5, 6, 7, 8, and 9. When a number becomes larger than 9 or smaller than 0, it is represented by the same set of decimal digits placed in different relative positions to the left or right of the decimal point. It is the position of a digit relative to the decimal point that determines its numeric size. In this manner, using only ten digits, any number can be represented. The first position to the left of the decimal point tells us how many basic units, or ones (10^0) are present. The second position tells us how may tens (10^1), the third position tells us how many hundreds (10^2), and so forth. Similarly, to the right of the decimal point we have tenths (10^{-1}), hundredths (10^{-2}), and so on. The number 8732.14, or 8732.14_{10}, can be thought of as

$$8 \times 10^3 + 7 \times 10^2 + 3 \times 10^1 + 2 \times 10^0 + 1 \times 10^{-1} + 4 \times 10^{-2}$$

4th pos.   3rd pos.   2nd pos.   1st pos. $\wedge$ 1st pos.   2nd pos.

$|\longleftarrow$ ———— left of decimal point ————$\longrightarrow |$ $\leftarrow$ right of decimal point $\rightarrow|$

The digits, in their relative positions, become coefficients of powers of 10.

One can conjecture that the reason the decimal system is so near and dear to us is because we long ago began to count with our ten fingers (and even toes during stressful school exams). It is not the only number system we can use, however. Binary representation, a base-2 number system, is more fundamental. There are only two digits or states in this system: 0 or 1. Something is either present or not. Numeric size is expressed in the binary system by the binary digit's relative placement with respect to the binary point. The first position to the left tells us how many ones (2^0) are present. The second position to the left tells us how many twos (2^1), the third position tells us how many fours (2^2), and so on. The positions to the right of the binary point represent halves (2^{-1}), quarters (2^{-2}), and so on. Each binary digit in a number becomes a coefficient of a power of 2. The binary number 1011.11, or 1011.11_2, can be thought of as

$$1 \times 2^3 + 0 \times 2^2 + 1 \times 2^1 + 1 \times 2^0 + 1 \times 2^{-1} + 1 \times 2^{-2}$$

4th pos.   3rd pos.   2nd pos.   1st pos. $\wedge$ 1st pos.   2nd pos.

$|\longleftarrow$ ———— left of binary point ————$\longrightarrow |$ $\leftarrow$ right of binary point$\rightarrow|$

$$1011.11_2 = (8 + 2 + 1 + 0.5 + 0.25)_{10} = 11.75_{10}$$

Similarly,

$$110111.1011_2 = (32 + 16 + 4 + 2 + 1 + 0.5 + 0.125 + 0.0625)_{10}$$
$$= 55.6875_{10}$$

To convert a decimal whole number to binary, one can repeatedly divide the number by 2 until a 0 quotient results. The remainders, in successive order, form the binary number written from right to left. For example, to convert 110_{10} to binary:

	REMAINDERS
$110 \div 2$	0
$55 \div 2$	1
$27 \div 2$	1
$13 \div 2$	1
$6 \div 2$	0

$$3 \div 2 \qquad 1$$
$$1 \div 2 \qquad 1$$
$$0$$

which equals 1101110.

To convert a decimal fraction to binary representation, one can multiply the fractional part of the number repeatedly by 2. The carryover, the first position to the left of the decimal point, in successive order, becomes the binary form of the number written from left to right from the binary point. The carryover is not used to form the next number. For example, to convert 0.8125_{10} to binary form:

CARRYOVERS

$$.8125 \times 2 = 1.6250 \qquad 1$$
$$(1).6250 \times 2 = 1.2500 \qquad 1$$
$$(1).2500 \times 2 = 0.5000 \qquad 0 \qquad = .1101_2$$
$$(0).5000 \times 2 = 1.0000 \qquad 1$$
$$(0).0000$$

Then

$$110.8125_{10} = 1101110.1101_2 \quad \text{or} \quad (.11011101101)_2 \times 2^7$$

where .11011101101 is the mantissa, or fractional part, and 7 (111_2) is the exponent. Using the mantissa and exponential form to store numbers in the computer allows for very large as well as very small number representation. Each decimal number that is entered into the computer is stored in an address in memory using a fixed number of bytes (8 bits = 1 byte). The number of bytes used varies from computer to computer. For instance, a computer might be designed to have a two-byte numeric data word. This means that the computer word would be limited to having only sixteen bits, or places, to store information per word. For example, the number 110.8125_{10} might be stored in this hypothetical computer as

$$0 \quad 1 \quad 1 \quad 0 \quad 1 \quad 1 \quad 1 \quad 0 \quad 1 \quad 1 \quad 0 \quad 1 \quad 0 \quad 1 \quad 1 \quad 1$$

Sign bit $|\longleftarrow$ Mantissa = 11 bits$\longrightarrow|$ Exponent = 4 bits,
"0" = plus   first bit is sign
"1" = minus   0111 = +7

Other representation is possible. Note that because of the fixed word size, error is introduced when the decimal number cannot be exactly represented using the given number of bytes of storage. For example, the aforementioned computer storage word with sixteen bits of storage can represent only eleven bits of information:

$$16 - 1 \text{ (sign)} - 4 \text{ (exponent)} = 11 \text{ bits}$$

Since the binary form of the decimal number must be truncated to fit eleven positions, information is lost. During long calculation chains, this loss accumulates, leading to calculation error that is called roundoff error. A partial solution to this problem is to use larger storage words, perhaps a thirty-two-bit memory word containing four bytes, although even this solution has limitations. The simple decimal number 0.1_{10} in binary code is $.000110011001100110011\ldots_2$, which is an unending sequence. Somewhere the sequence must be cut off, and this can introduce error. For the smaller sixteen-bit word discussed before, with eleven bits of significant information, this number might be stored as

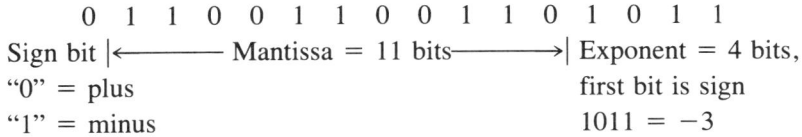

$$0 \quad 1 \quad 1 \quad 0 \quad 0 \quad 1 \quad 1 \quad 0 \quad 0 \quad 1 \quad 1 \quad 0 \quad 1 \quad 0 \quad 1 \quad 1$$

Sign bit $|\longleftarrow$ Mantissa $= 11$ bits$\longrightarrow|$ Exponent $= 4$ bits,

"0" = plus first bit is sign

"1" = minus $1011 = -3$

which in decimal form is $.0999755859\ldots_{10}$

Because it is so difficult to work in binary, two alternative numbering systems are used to express binary data: octal (to the base 8) and hexadecimal (to the base 16). (The computer still uses binary code internally.) The octal system contains the eight digits $0, 1, 2, 3, 4, 5, 6,$ and 7; and the hexadecimal system has the digits $0, 1, 2, 3, 4, 5, 6, 7, 8, 9,$ A (10), B (11), C (12), D (13), E (14), and F (15). The decimal number 589_{10} equals $1001001101_2 = 1115_8 = 24D_{16}$, where

$$1115_8 = 1 \times 8^3 + 1 \times 8^2 + 1 \times 8^1 + 5 \times 8^0$$

and

$$24D_{16} = 2 \times 16^2 + 4 \times 16^1 + 13 \times 16^0$$

Table A.1 lists the decimal, binary, octal and hexadecimal conversions for numbers up to 50_{10}.

TABLE A.1

Decimal	Binary	Octal	Hexadecimal
0	0	0	0
1	1	1	1
2	10	2	2
3	11	3	3
4	100	4	4
5	101	5	5
6	110	6	6
7	111	7	7
8	1000	10	8
9	1001	11	9
10	1010	12	A
11	1011	13	B
12	1100	14	C
13	1101	15	D
14	1110	16	E
15	1111	17	F
16	10000	20	10
17	10001	21	11

TABLE A.1   (continued)

Decimal	Binary	Octal	Hexadecimal
18	10010	22	12
19	10011	23	13
20	10100	24	14
21	10101	25	15
22	10110	26	16
23	10111	27	17
24	11000	30	18
25	11001	31	19
26	11010	32	1A
27	11011	33	1B
28	11100	34	1C
29	11101	35	1D
30	11110	36	1E
31	11111	37	1F
32	100000	40	20
33	100001	41	21
34	100010	42	22
35	100011	43	23
36	100100	44	24
37	100101	45	25
38	100110	46	26
39	100111	47	27
40	101000	50	28
41	101001	51	29
42	101010	52	2A
43	101011	53	2B
44	101100	54	2C
45	101101	55	2D
46	101110	56	2E
47	101111	57	2F
48	110000	60	30
49	110001	61	31
50	110010	62	32

CHARACTER DATA

In addition to numeric data, the computer must also handle character data. Since this information will also be stored in binary form, it must not be confused by the computer with numeric data. A preceding FORMAT message in the program is used to tell the CPU how the data is to be treated.

There are two standard codes that have been developed for character data: ASCII (American Standard Code for Information Interchange) and EBCDIC (Extended Binary Coded Decimal Interchange Code). Table A.2 lists a FORTRAN character set that includes the twenty-six capital letters of the alphabet, ten digits, a blank, and twelve symbols in decimal, binary, and hexadecimal code for ASCII and EBCDIC representation. Note that characters can be represented by one byte.

TABLE A.2 FORTRAN Character Set

Character	Decimal	EBCDIC	Hexadecimal	Decimal	ASCII	Hexadecimal
A	193	1100 0001	C1	65	0100 0001	41
B	194	1100 0010	C2	66	0100 0010	42
C	195	1100 0011	C3	67	0100 0011	43
D	196	1100 0100	C4	68	0100 0100	44
E	197	1100 0101	C5	69	0100 0101	45
F	198	1100 0110	C6	70	0100 0110	46
G	199	1100 0111	C7	71	0100 0111	47
H	200	1100 1000	C8	72	0100 1000	48
I	201	1100 1001	C9	73	0100 1001	49
J	209	1101 0001	D1	74	0100 1010	4A
K	210	1101 0010	D2	75	0100 1011	4B
L	211	1101 0011	D3	76	0100 1100	4C
M	212	1101 0100	D4	77	0100 1101	4D
N	213	1101 0101	D5	78	0100 1110	4E
O	214	1101 0110	D6	79	0100 1111	4F
P	215	1101 0111	D7	80	0101 0000	50
Q	216	1101 1000	D8	81	0101 0001	51
R	217	1101 1001	D9	82	0101 0010	52
S	226	1110 0010	E2	83	0101 0011	53
T	227	1110 0011	E3	84	0101 0100	54
U	228	1110 0100	E4	85	0101 0101	55
V	229	1110 0101	E5	86	0101 0110	56
W	230	1110 0110	E6	87	0101 0111	57
X	231	1110 0111	E7	88	0101 1000	58
Y	232	1110 1000	E8	89	0101 1001	59
Z	233	1110 1001	E9	90	0101 1010	5A
$	91	0101 1011	5B	36	0010 0100	24
*	92	0101 1100	5C	42	0010 1010	2A
'	125	0111 1101	7D	39	0010 0111	27
blank	64	0100 0000	40	32	0010 0000	20
(	77	0100 1101	4D	40	0010 1000	28
)	93	0101 1101	5D	41	0010 1001	29
+	78	0100 1110	4E	43	0010 1011	2B
,	107	0110 1011	6B	44	0010 1100	2C
−	96	0110 0000	60	45	0010 1101	2D
.	75	0100 1011	4B	46	0010 1110	2E
/	97	0110 0001	61	47	0010 1111	2F
=	126	0111 1110	7E	61	0011 1101	3D
:	122	0111 1010	7A	58	0011 1010	3A
0	240	1111 0000	F0	48	0011 0000	30
1	241	1111 0001	F1	49	0011 0001	31
2	242	1111 0010	F2	50	0011 0010	32
3	243	1111 0011	F3	51	0011 0011	33
4	244	1111 0100	F4	52	0011 0100	34
5	245	1111 0101	F5	53	0011 0101	35
6	246	1111 0110	F6	54	0011 0110	36
7	247	1111 0111	F7	55	0011 0111	37
8	248	1111 1000	F8	56	0011 1000	38
9	249	1111 1001	F9	57	0011 1001	39

Intrinsic Functions

Description of function	Generic name**	Specific name	Number of arguments	Type of argument	Type of function
Conversion of argument type to integer	INT	—	1	Integer	Integer
		IFIX		Real	Integer
		IDINT		Double	Integer
		—	(Real part)	Complex	Integer
Conversion of argument type to real	REAL	FLOAT	1	Integer	Real
		—		Real	Real
		SNGL		Double	Real
		—	(Real part)	Complex	Real
Conversion of argument type to double precision	DBLE	—	1	Integer	Double
		—		Real	Double
		—		Double	Double
		—	(Real part)	Complex	Double
Conversion of argument type to complex $CMPLX(a_1) =$ $a_1 + 0i$	CMPLX	—	1 or 2	Integer	Complex
		—		Real	Complex
		—		Double	Complex
$CMPLX(a_1, a_2) =$ $a_1 + a_2 i$	—	—		Complex	Complex
Conversion of character type to integer	—	CHAR	1	Character	Integer
Conversion of integer value to character argument	—	ICHAR	1	Integer	Character
Truncation: $int(a)$	AINT	AINT	1	Real	Real
		DINT		Double	Double
Rounding to nearest integer: $int(a+0.5)$ or $int(a-0.5)$ if a is negative	ANINT	ANINT	1	Real	Real
		DNINT		Double	Double
Rounding to nearest integer	NINT	NINT	1	Real	Integer
		IDNINT		Double	Integer
Absolute value	ABS	IABS	1	Integer	Integer
		—		Real	Real
		DABS		Double	Double
		CABS		Complex	Real
Remaindering: $a_1 - int(a_1/a_2) \times a_2$	MOD	—	2	Integer	Integer
		AMOD		Real	Real
		DMOD		Double	Double

Description of function	Generic name**	Specific name	Number of arguments	Type of argument	Type of function
Transfer of sign: from a_2 to a_1. If $a_2 = 0$, a_1 is positive.	SIGN	ISIGN — DSIGN	2	Integer Real Double	Integer Real Double
Positive difference. Zero if a_1.LT.a_2.	DIM	IDIM — DDIM	2	Integer Real Double	Integer Real Double
Double-precision product	—	DPROD	2	Real	Double
Maximum value	MAX	MAX0 AMAX1 DMAX1 AMAX0 MAX1	≥ 2	Integer Real Double Integer Real	Integer Real Double Real Integer
Minimum value	MIN	MIN0 AMIN1 DMIN1 AMIN0 MIN1	≥ 2	Integer Real Double Integer Real	Integer Real Double Real Integer
Length of character item	—	LEN	1	Character	Integer
Index of substring	—	INDEX	2	Character	Integer
Imaginary part of a complex value	—	AIMAG	1	Complex	Real
Conjugate of a complex value	—	CONJG	1	Complex	Complex
Square root	SQRT	— DSQRT CSQRT	1	Real Double Complex	Real Double Complex
Exponential	EXP	— DEXP CEXP	1	Real Double Complex	Real Double Complex
Natural logarithm	LOG	ALOG DLOG CLOG	1	Real Double Complex	Real Double Complex
Common logarithm	LOG10	ALOG10 DLOG10	1	Real Double	Real Double
Sine (in radians)	SIN	— DSIN CSIN	1	Real Double Complex	Real Double Complex
Cosine (in radians)	COS	— DCOS CCOS	1	Real Double Complex	Real Double Complex
Tangent (in radians)	TAN	— DTAN	1	Real Double	Real Double
Arcsine	ASIN	— DASIN	1	Real Double	Real Double

Description of function	Generic name**	Specific name	Number of arguments	Type of argument	Type of function
Arccosine	ACOS	—	1	Real	Real
		DACOS		Double	Double
Arctangent	ATAN	—	1	Real	Real
		DATAN		Double	Double
Arctangent (a_1/a_2)	ATAN2	—	2	Real	Real
		DATAN2		Double	Double
Hyperbolic sine	SINH	—	1	Real	Real
		DSINH		Double	Double
Hyperbolic cosine	COSH	—	1	Real	Real
		DCOSH		Double	Double
Hyperbolic tangent	TANH	TANH	1	Real	Real
		DTANH		Double	Double
Lexically greater than or equal to	—	LGE	2	Character	Logical
Lexically greater than	—	LGT	2	Character	Logical
Lexically less than or equal to	—	LLE	2	Character	Logical
Lexically less than	—	LLT	2	Character	Logical

**Except for conversions of type, any generic function name, when used, will return a value having the same type as its argument.

A Summary of FORTRAN 77

Statement	Nonexecutable statements Description	Sections
BLOCK DATA	Block data subprogram heading. `BLOCK DATA ONE`	7.6
CHARACTER	Type statement used to specify character variable. `CHARACTER*4 NAME,SYMB*1,TITLE(5)`	2.4, 8.2.2
Comment statements	Clarifying comments included with source code. C or * must appear in the first column. `C       PROGRAM COMMENTS...`	1.4.2, 1.5

COMMON	Specifies variables within different program units that share a named or unnamed common storage area.
	`COMMON RED,ALPHA`
	`COMMON/NAME/ARC,PERIM`
		7.6

COMPLEX	Type statement used to specify complex variable. `COMPLEX A,VORTEX(2,3)`	2.4, 8.5.3
DATA	Initializes specified program variables during program compilation. `DATA ARC,X(2),J/2.3,1.0,2.0,8/`	2.7, 6.2.1, 6.3.1, 8.2.3, 8.3.2, 8.4.3, 8.5.6
DIMENSION	Specifies dimensions of arrays. `DIMENSION INDEX(30),A(2,3)`	6.2, 6.3, 6.4, 7.4
DOUBLE PRECISION	Type statement used to specify double-precision variable. `DOUBLE PRECISION ROW,ARC(3)`	2.4, 8.4.2
ENTRY	Specifies alternative entry point in a subprogram unit. `ENTRY SUB1(X,Y,N)`	10.5.1
EQUIVALENCE	Specifies variables within the same program unit that share a common storage area. `EQUIVALENCE(X,Y,ARC),(B(1),R,S)`	10.2.2
EXTERNAL	Specifies the arguments of subprograms that are externally defined subprogram names. `EXTERNAL NAME,SUB,FUNC`	7.7
FORMAT	Statement used to specify field makeup for formatted input/output operations. `10 FORMAT(1X,E14.7,2X,I4,2F6.3)`	3.4
FUNCTION	Function subprogram heading. `FUNCTION AXIS(R,X,M)`	7.3
IMPLICIT	Used to specify rules for typing variables. `IMPLICIT REAL(I-J,M),INTEGER(A-C)`	10.2.1

Statement	Nonexecutable statements Description	Sections
INTEGER	Type statement used to specify integer variable. `INTEGER INDEX,AXIS,VEL(3)`	2.4, 6.2
INTRINSIC	Specifies the arguments of subprograms that are intrinsic function names. `INTRINSIC COS,EXP`	7.7
LOGICAL	Type statement used to specify logical variable. `LOGICAL LA,ABLE(2)`	2.4, 8.3.1
PARAMETER	Defines symbolic constant. `PARAMETER (PI=3.1415927)`	2.7, 8.2.3, 8.3.2, 8.4.3, 8.5.6
PROGRAM	Program heading. `PROGRAM NAME`	1.4.2
REAL	Type statement used to specify real variable. `REAL Y,INDEX(2,2)`	2.4, 6.2, 6.3, 6.4, 7.4

SAVE	Specifies that local variables in a subprogram are
	to be saved.
	`SAVE`
	`SAVE FLOW,INDEX,PRESS`
		7.3

Statement function	Defines a one-statement-long function. `VEL(VX,VY)=(VX*VX+VY*VY)**.5`	7.2
SUBROUTINE	Subroutine heading. `SUBROUTINE METER(X,Y)`	7.5

Statement	Executable statements Description	Sections
Arithmetic IF	Selects one of three possible branches to follow depending on whether the result of an arithmetic expression is positive, zero, or negative. `IF(Y-X+2.) 23,24,33`	4.6
ASSIGN	Assigns address label for GO TO statement. See assigned GO TO. `ASSIGN 15 TO INDEX`	4.6
Assigned GO TO	Transfers program control to a specified statement whose address is given by the value assigned to an integer variable. See ASSIGN. `GO TO INDEX(3,15,17)`	4.6

Assignment Statement	Variable name=expression
	`X=3.2*Y+Z**3`
	`Y=44.2`
	`I=I+1`
		2.5

Statement	Executable statements Description	Sections
BACKSPACE	Rewinds file to the last record processed. `BACKSPACE(UNIT=3)`	9.5
CALL	Statement used to call out a subroutine. `CALL SUB1(X,Y,A,N)`	7.5
CLOSE	Statement used to close a file. `CLOSE(UNIT=3)`	9.3
Computed GO TO	Transfers program control to one of several statements depending on the computed value of a given integer expression. `GO TO (3,15,17) N=K-1`	4.6
CONTINUE	Last statement of a DO loop. `3     CONTINUE`	5.2

DO	First statement of a DO loop.
	`DO 9 I=1,7,2`
	`DO 3 J=1,N`
	`DO 7 X=0.0,3.0,.1`
		5.2, 5.4

ELSE	Permitted in an IF-block structure. It provides alternative responses within the block if the leading IF-THEN logical expression is false. `ELSE`	4.4.2
ELSE IF	Permitted in an IF-block structure. It presents an alternative IF structure for processing if the leading IF-THEN logical expression is false. `ELSE IF (Y.EQ.SIN(Z)) THEN`	4.3
END	Required last statement for any program or subprogram unit. The statement stops compilation of the program. `END`	1.4.2
END IF	Last statement of an IF structure. `END IF`	4.4
ENDFILE	Places end-of-file special record at the end of a sequential file. `ENDFILE(UNIT=3)`	9.5
GO TO	Transfers program control unconditionally to a specific statement. `GO TO 7`	4.5, 4.6
IF statement	Performs a specific executable statement if a given logical expression is true. `IF(X.GT.Y-3.2) PRINT *,X`	4.3
IF-THEN	Block IF statement. The block is performed if the given logical expression is true. `IF (X.GT.Y-3.2) THEN`	4.4

INQUIRE	Used to inquire about file properties or current status of a specified unit number.
	`INQUIRE(FILE='DATA',OPENED=LOP)`
	`INQUIRE(UNIT=3,EXIST=LEX)`
		9.4

Statement	Executable statements Description	Sections

OPEN	Statement used to open an external auxiliary file.
	`OPEN(UNIT=3,FILE='DATA',`
	`    STATUS='OLD')`
		2.8, 9.3

PAUSE	Allows program execution to be temporarily halted. `PAUSE 'PROGRAM INTERRUPTION'`	10.4

PRINT	Statement used to output data to the terminal.
	
	`PRINT *,'THE ANSWER IS = ',TIME`
	`PRINT 10,S,R,Z,K`
		3.3, 3.4, 6.2.1,
		6.3.1, 8.2.5, 8.3.4,
		8.4.5, 8.5.6

READ	Statement used to read data from the terminal or from a specified unit number.
	`READ *,A,J,(C(I),I=1,3)`
	`READ(3,10) A,J,(C(I),I=1,3)`
	`READ(3,13,END=99) BETA`
		2.8, 3.2, 3.7, 6.2.1,
		6.3.1, 9.6
		8.2.5, 8.3.4,
		8.4.5, 8.5.6

RETURN	Statement used in subprogram unit to return control back to calling program.
	`RETURN`
	`RETURN 2`
		7.3, 7.5, 10.5.2

REWIND	Rewinds auxiliary external file. `REWIND(UNIT=3)`	2.8, 9.5
STOP	Statement used to stop program execution. `STOP`	1.4.2

WRITE	Statement used to output data to the terminal or to a specified unit number.
	`WRITE(6,10) VEL,(C(I),I=1,3)`
	`WRITE(9,11) TIME,DIST`
		2.8, 3.3, 3.4, 6.2.1,
		6.3.1, 8.3.4, 9.6

Glossary

access arm   The arm that positions the read/write head of magnetic disk storage devices.

actual arguments   The arguments or values that will actually be used for the listed dummy arguments of a subprogram.

adjustable dimensions   Dimension specifications of a subprogram's array that are defined by an integer variable. The dimensions are adjusted in the main program by assigning a value to that integer variable.

algorithm   A step-by-step set of rules, or procedures, for the solution of a problem.

alphanumeric data   Data consisting of characters from the FORTRAN 77 character set: letters, special characters, and numbers.

ALU   The arithmetic logic unit.

analog process   A process described by continuously varying physical quantities.

analog-to-digital converter (ADC)   A device that converts an analog signal to digital representation for computer processing.

arithmetic logic unit   The part of the central processing unit that controls and performs all arithmetic and logic operations.

array   A set or list of related data values stored in memory under one common name. The elements of the array have subscripts to distinguish them from one another.

ASCII code   American Standard Code for Information Interchange. Each alphanumeric character entered into a computer is internally represented by one of two standard numeric codes: ASCII or EBCDIC. Code use is computer dependent.

assembler   A program used to convert assembly language to machine language.

assembly language   A programming language that uses symbols and abbreviations to denote the computer process to be performed, for example, add or store.

assignment statement   A statement that assigns a value to a variable name.

assumed-length specifier   A means of specifying an adjustable length for a character-type subprogram formal argument.

auxiliary external memory   Memory area that is located external to the computer. It is used for long-term storage, since it will not be affected when the computer is turned off.

BASIC   A high-level computer language that is widely used on personal computers.

batch processing   A means of collecting and later processing programs as a group in the order they were received.

binary   An adjective used to describe a quantity that can only assume one of two values, for example, 0 or 1, on or off.

binary digit   A binary digit can be either 0 or 1.

bit   The smallest unit for data storage. A binary digit.

blank common block   An unnamed common area of memory locations accessible to both the main program and its subprograms.

branch   To leave main top-down sequential program flow and follow an alternative processing path.

bubble sort   A common algorithm used by programmers to sort quantities in ascending or descending order.

byte   A group of bits operated on as a unit of information. Typically a byte consists of eight bits: the amount of memory required to store a character.

cathode ray tube (CRT)   An electronic tube used to display information. It is used for video terminal monitors.

central processing unit (CPU)   The brain of the computer, consisting of the arithmetic logic unit, the processing unit, and internal memory.

character string   A string or group of characters.

character variable   A variable quantity that can only be assigned a character value from the FORTRAN character set.

COBOL   COmmon Business Oriented Language. A high-level computer language used for business purposes.

collating sequence   The given order of characters on the computer.

comment statements   Program descriptive statements that are not executed.

compilation   The process of using a compiler to create an object program in machine language from a source program.

compiler   A program that converts a source program written in high-level language to machine language.

complex variable   A variable that stores both a real and an imaginary part. It is usually written as $z = x + yi$, where $i \times i = -1$.

compound logical expression   A logical expression formed by combining single logical expressions whose values are true or false.

concatenation   The process of combining character strings.

data file   An auxiliary external file containing data information.

debugging   The process of finding a program's errors in either logic or syntax.

decision structure   See selection structure.

diagnostic message   An error message generated by the system regarding either compilation or execution.

direct-access file   A file whose information can be accessed directly rather than in a sequential manner. Each record of the file is numbered.

discrete process   A process that is not continuous but instead begins and ends at isolated times.

disk drive   A device used to read or write information on disks treated with magnetic coatings. The disks are either hard or soft (floppy).

documentation   Usually used to refer to any descriptive or instructional material relating to a program and its operation.

double-precision variable   A real variable that is assigned twice the storage space in memory to contain its numeric value. Since the mantissa assigned to a double-precision constant is larger, more precision is provided for the number's representation. See single-precision variable.

driver   A dummy program created to test a subroutine.

EBCDIC code   Extended Binary Coded Decimal Interchange Code. Each alphanumeric character entered into a computer is internally represented by one of two standard numeric codes: ASCII or EBCDIC. Code use is computer dependent.

executable statement   A statement that causes an action to be carried out by the computer during program execution.

explicit typing   A means of specifically declaring the type of a given variable for storage purposes. Type statements must be used.

exponential notation   Similar to scientific notation. A notation that represents a real number as a signed fractional part greater than 0 but less than 1 multiplied by a power of 10.

external sort   A means of updating and sorting a file's contents without the necessity of entering the entire file at once into internal machine memory.

FEM   An abbreviation for Finite Element Method. A general technique for solving engineering problems in solid and fluid mechanics.

field descriptors   Parameters that describe the length of each field of a data record and also give information as to the type of data that is found within the field and the location of any decimal point.

file   A named group of records that contains data or a program.

fixed-point variable   A variable that can only assume whole (integer) numeric values.

floating-point variable   A variable that can only assume decimal values, or decimal numbers multiplied by a power of 10. Also known as a real variable.

flowchart   A pictorial representation of a solution algorithm.

flow lines   Lines drawn on a flowchart indicating the direction of program flow.

formal arguments   The arguments of a subprogram. Also known as dummy arguments.

format free   Referring to computer input/output operations. This specification defaults to the computer for selection of the type of format to represent the data.

formatted input or output   Input/output operations controlled by user-defined field descriptors. Provides for carriage control, and vertical as well as horizontal spacing.

FORTRAN 77   An acronym for FORmula TRANslation. A high-level computer language, useful in engineering and science, that originated in 1954 and was most recently revised in 1977.

function subprogram   A separately compiled program unit that may be called out by the main program. It returns only one value to the calling program, and that value is assigned to the function's name. It is called by employing the subprogram's name in the main program along with providing the values of the actual function arguments. See subprogram and subroutine.

hard copy   Referring to computer output printed or plotted on paper.

high-level language   Computer programming languages that employ English words rather than the internal binary code of the machine (machine language).

I/O   Input/output operations.

IF structure   See selection structure.

IF-block structure   Decision structure. This block allows for alternative branches of a program. The actual program path followed is dependent on the logical value of a specified test condition.

imaginary part   In a complex number, the imaginary part consists of a real value multiplied by the imaginary value i.

implicit typing   A means of typing variables by default: the first letter of the variable name is used to specify its type. Variable names beginning with the letters I through N are considered to be of the integer type. Variable names beginning with the letters A through H or O through Z are considered to be real variables.

implied-DO list   An abbreviated means of listing the elements of an array in a READ, WRITE, PRINT, or DATA statement.

index variable name   The name of the control variable used in a DO statement. It is incremented or decremented each time the loop is traversed.

initialize   The process of assigning initial values to a variable when the program, or a loop or program substructure, begins.

integer variable   A fixed-point variable. A variable that can only assume whole number (integer) values.

internal memory   Memory that is built into the computer. Part of the central processing unit.

intrinsic function   An external function available from the FORTRAN library. It is called out by using the function's name. An object-code version of the function will be placed within the program at compilation time.

key field   A field used for identification of data records in a sequential file. Typically used for sorting purposes.

lexical functions   Referring to the intrinsic character functions provided by the FORTRAN library for comparisons of character variables.

LISP   A language used in artificial-intelligence programming.

list-directed format   See format free.

listing   A copy of the source code or program.

literal descriptor   Apostrophes used to enclose a character string for output purposes. The string or literal will be sent directly to the output unit.

literal   A character string.

local file   A file stored in internal memory.

logging on, logging off   The process of connecting with or disconnecting from the computer.

logic errors   Errors in a program caused by logical mistakes or omissions in the program's design.

logical expression   An expression whose value can only be true or false.

logical operator   An operator used to combine logical expressions.

logical variable   A variable that can only assume a true or a false value.

machine language Instructions written in machine (binary) code that require no translation by the computer for their immediate execution.

main program A complete program unit. The main program must be opened before any subprogram unit can be accessed.

mainframe computer A computer capable of handling large amounts of data at great speeds.

mantissa of a number The fractional part of a logarithm. It is commonly used to refer to the decimal portion of any floating-point number.

matrix An array of constants or subscripted variable elements that may be operated on using prescribed mathematical rules.

microcomputer A small, inexpensive computer consisting of a microprocessor, a keyboard, internal storage, a built-in disk drive, and a CRT. See PC.

microprocessor An entire central processing unit chemically etched on a single silicon chip.

minicomputer A medium-sized computer.

mixed-mode operation Arithmetic operations involving operands with different types. Mixed-mode operations should be avoided, as they can lead to calculation error.

modem An acronym for MOdulator/DEModulator. A communication device that enables the computer to receive and send information via the telephone lines.

named common block A named common area of memory locations accessible to both the main program and its subprograms.

nested block A structured programming block that is completely contained within another block. DO loops, IF structures, and WHILE structures can all be nested.

nonexecutable statement A statement that is not executed by the computer during program execution. Nonexecutable statements are typically specification statements that are processed during compilation time.

object program The translated version of the source program (high-level language) produced by the compiler. The object program is usually written directly in machine code.

overloaded input/output list Usually refers to the labor-saving method of using fewer field descriptors than there are variables on the input/output list to facilitate program preparation. Some field descriptors will apply to more than one variable.

parallel processing A means of processing program steps simultaneously instead of in a serial manner. As a direct result of parallel operations, program execution time can be reduced dramatically. Though parallel processing is still in the development stage, it has already been established as the design goal for the next generation of computers.

PC Personal computer.

program, program unit   A set of instructions written to solve a problem.

program bugs   Any logical or syntactical errors present in a computer program.

PROLOG   A language used in artificial-intelligence programming.

pseudocode   A code that allows program instructions to be symbolically written. The code is used in developing the problem's algorithm.

random-access file   See direct-access file.

real variable   See floating-point variable.

record   The unit of information that is entered or written in a single input/output operation. A record typically contains several data fields. A collection of records forms a data file.

relational operator   An operator used to make comparisons between variables. The result of the comparison is a logical value, either true or false.

repeated edit descriptor   A means of having edit descriptors control successive fields of a record. The edit descriptor is prefixed with an integer coefficient.

repetition structure   See WHILE structure.

roundoff error   An error caused by insufficient memory word size for the resultant fractional part of an operation involving real variables. The overflow (the least-significant digit) is discarded, resulting in lost precision. See double-precision variable.

running product   A product that is increased every pass through a DO loop or WHILE structure.

running sum   A sum that is increased every pass through a DO loop or WHILE structure.

selection structure   One of the main building blocks of structured programs. It provides for the possibility of program branching based on the logical value of a specified expression or variable. Several variant types are available in FORTRAN. Each one begins, or is entered, via a block IF statement. The IF structure terminates with an END IF statement.

serial processing   A means of executing a program's statements in a sequential order. See parallel processing.

short-list technique   A means of having an array's name substitute for all the subscripted elements in an input/output list.

single precision   A term used to describe the precision carried by a real (floating-point) value. See double-precision variable.

slash descriptor   An edit descriptor that signals that the processing of a new record should begin.

software   A term used to denote computer programs.

software packages   A term used to describe large, commercially available programs that are leased or bought.

source program   A computer program written in a high-level language.

statement function   A function that is defined in one assignment statement. The statement function must appear before the first executable statement of the program. It is compiled along with the main program.

string   A group of consecutive characters. See substring.

structured programs   Programs built from the basic building blocks of the elementary program structures: DO loop, IF-block, and WHILE structure. Program flow is maintained in one direction, from top to bottom. Programs that are structured are easier to read and work with because of their organization.

stub   A dummy subprogram written to allow testing of the calling program. See driver.

subprogram   Program units that are compiled and stored separately from the main program. They are portable and can be used with other programs. They are not complete programs but must be called from the main program to function. On the other hand, they may call out other subprograms. Subprograms are joined to the main program through a process called linking.

subroutine   A powerful form of subprogram that may pass several values back and forth to the calling program. It is activated via a CALL statement, which specifies the subroutine's name along with the actual values of the dummy arguments. See function subprogram.

subscript   The integer index identifying an array element.

subscripted variable   An array element.

substring   A consecutively ordered subset of a string. See string.

syntax   Used here to mean the grammatical rules of a computer language.

system commands   Commands that can be executed by the operating system of the computer, for example, LIST, RUN, and COPY.

tape drive   An auxiliary external storage device that uses magnetic tape. Alternatively, the drive for such a device.

temporary storage   See internal memory.

text editor   A word processor. A program that allows for creation and correction of text.

time sharing   A means of data processing whereby the user can interact directly with the computer from a local terminal. The user simultaneously shares the computer with other users who may also be processing their programs on-line but at different terminals.

top-down code   An approach to writing program code whereby the main program is coded (and tested) first, using subprogram stubs as a substitute for the still-to-be-developed subprograms. After this coding has been completed, the next level of subprograms in the calling hierarchy is written, and so on, until the program is finished.

top-down design   An approach to program design whereby the primary areas to be solved for are defined first. Next, the supporting submodules or subprograms are developed in descending order. When the smallest modules become assignment statements, the program is ready to be coded.

trailer line A coded record entry appearing at the end of a data file to signal to the program that the file is empty.

truncate To cut off; to discard. This term usually refers either to the dropping of the least-significant bit that will not fit the storage allocated for the variable, or to the discarding of terms in a series approximation of a function. In the latter case, we speak of truncation error. See roundoff error.

truth table A table created to help the user understand all the possible logical states a logical expression may assume.

turnaround time A term used with batch processing. The amount of time needed to process a program from the moment the job is submitted until it is returned to the user.

type statement A statement that specifies the type of a variable: REAL, INTEGER, CHARACTER, LOGICAL, COMPLEX, OR DOUBLE PRECISION.

unformatted file A data file that is written in machine code. A standard formatted file is written using alphanumeric characters.

updating program A program whose purpose is to manage the updating and sorting of an external data file.

user friendly A term describing a program that is easy to use, especially by the uninitiated. Good documentation and clear visual program-generated cues to the user so that he or she may make appropriate responses are essential ingredients in making a program user friendly.

WHILE structure This structure allows for repetitive processing of a block of program instructions while a specified test condition remains true. The test condition must be initialized before entry into the block, it must be tested every pass through the block, and it must turn false at some time so that an infinite processing loop is not formed. Also known as repetition structure. See structured programs.

Answers to Selected Review Questions

CHAPTER 1

2. The arithmetic logic unit, a part of the CPU, performs the arithmetic and logical operations for the computer.

4. Programs that are written for the computer.

6. Binary (machine) code. Instructions written in machine language can be executed by the computer without further translation.

8. Logic errors are logical errors or omissions in the program algorithm; syntax errors are errors in language usage, that is, in spelling, punctuation, or grammar.

10. Cathode ray tube

12. In batch processing, programs are collected and assembled to be performed as a group at a later time. In time sharing, programs are performed on-line, providing interactive dialog with the computer, and direct and immediate control of computer operations.

14. A local file resides in internal machine memory and will be lost when a new working file is entered or the user logs off. A permanent file is stored in an auxiliary external memory device and will not be lost by logging off.

16. Syntax, as used here, refers to the spelling and grammatical rules of a programming language.

18. A comment line or statement is denoted by a "C" or "*" appearing in column 1.

20. An END statement.

CHAPTER 2

2. An executable statement requires an action to be performed by the computer during execution time. A nonexecutable statement is typically a statement that is processed during program compilation and requires no further computer action during the program's execution.

4. Integer, real, double precision, and complex.

6. Implicit typing is a means of typing variables by default. The first letter of each variable name is used to determine its type. If the first letter of the variable name is any letter between *I* and *N*, the variable is assumed to be of the integer type. Otherwise, the variable is assumed to be real.

8. Truncation occurs when the least-significant digit (bit) of a resultant from an arithmetic operation is discarded as a result of insufficient storage allocation for the fractional part of the number. Truncation also occurs when the entire fractional part of a resultant is thrown aside because the result is stored as an integer variable.

10. Radians.

12. Calculation errors can result from the unintentional loss of a resultant's fractional part (truncation). This occurs when a real value is assigned to an integer variable name, or when the result is from an arithmetic operation involving integer values.

14. d, e, f, g, i, j, k, l

16. **(a)** 11.76 **(b)** 7 **(c)** .525 **(d)** 0 **(e)** 6.6 **(f)** 42.875 **(g)** 4.6 **(h)** 26.59797331

18. **(a)** 403 **(b)** 3.2E4 **(c)** 84.0 **(d)** 84.0 **(e)** 186624.0

CHAPTER 3

1. **(b)** -56   **(d)** ****   **(f)** -108.1   **(h)** *****   **(j)** _.93426E+03
    (l) ****   **(n)** _.8E-03   **(p)** .0008   **(r)** _.0008

2. Values are assumed to appear on the line printer.
 (a) __-210____56_____****__934.26
 (c) ___TRIAL___56_____X = __.93426E+03
 (e) 210___THE RESULTS ARE = __.934E+03_____.00080

3. **(a)** __-210____56_____-108.1
 (c) -108.1
      7.632E4.008E-01
 (e) __-.1081E+03_____.934259E03

4. **(a)** (I4, 2(2X,F6.3))   **(c)** (2E8.0)   **(e)** (E8.0,T20,E9.6)

CHAPTER 4

2. IF(T.LT.TIME) Z=ALOG(T)

4. IF(T.GT.5.0.AND.T.LT.7.5) PRINT *,X

6. IF (Q.LT.0.0) THEN
        PRINT *,'THE RADICAL IS COMPLEX'
        STOP
    END IF

8. IF (X.GE.0.) THEN
        BMOM=7500.*X
    END IF
    IF (X.LT.15.) THEN
        PRINT *,BMOM
    END IF
    IF(X.GT.30.) STOP

10. IF (KCASE.EQ.0.OR.KCASE.EQ.4) THEN
        STRESS=P/AREA
    END IF

12. IF (T.GT.4.0) THEN
        IF (RHO.LE.ZETA) THEN
            PRINT *,EPSI
        END IF
    END IF

14. The variable PRESS will always equal ALLOW regardless of the results of the logical expression.

16. IF (T.GE.0.0.AND.T.LE.10.0) THEN
        F=3.0*T
    ELSE IF (T.GT.10.0.AND.T.LT.15.0) THEN
        F=30.0
    ELSE
        F=0.0
    END IF

18. IF (I.LE.7) THEN

CHAPTER 5

1. (a) 10   (c) 11

2. (a) 21   (b) 1

3. 3.4

4. (a) 0.0   (c) 5,

5. (a) The last statement in a DO loop cannot be a STOP statement.
   (c) The index cannot be changed inside the loop.

6. (a)
```
      SUM=0.0
      DO 7 I=1,11
          READ *.X
          SUM=SUM+X
    7 CONTINUE
```

7.
```
      DO 14 I=1,100
          READ(9,5,END=11) R
          SUM=SUM+R
          K=K+1
   14 CONTINUE
   11 AV=SUM/K
```

8.
```
      DO 14 K=1,100
          READ(9,2,END=5) R
          IF (R.GT.RMAX) THEN
                RMAX=R
          END IF
   14 CONTINUE
```

9.
```
      DO 5 I=1,40,2
          I1=I**2
          IF((I1-ISQ).GT.40) GO TO 11
          ISQ=I1
          NSUM=NSUM+ISQ
    5 CONTINUE
```

11. (a)
```
      NFACT=1
      DO 7 I=2,9
          NFACT=NFACT*I
    7 CONTINUE
```

```
(c)    SUM=0.0
       DO 7 I=1,9,2
           SUM=SUM+1.0/(I*(I+2))
     7 CONTINUE
```

CHAPTER 6

1. (a) 2   **(c)** .199   **(e)** 2   6
     4   .389   4   8
     6   .565
     8   .717
    10   .841
          .932

2. (a) 13.5   **(b)** 13.5   **(c)** 13.5   **(d)** 13.5   10.4   12.7
     10.4   24.6   10.4   24.6   6.2   13.8
     12.7   14.3   12.7   14.3   7.8   −9.4

3. (a) 13.5   **(b)** 13.5   **(c)** 13.5   24.6   14.3
     24.6   10.4   10.4   6.2   7.8
     14.3   12.7   12.7   13.8   −9.4

4. (a) 3.20000 9.40000 6.80000   **(c)** 3.20000
                                      9.40000
                                      6.80000

5. (a) rows   **(c)** columns

6. (a)
```
DO 7 I=0,10
    A(I)=0.0
```

7. (a)
```
DO 7 I=0,10
    B(I+1)=A(I)
```

8. (a)
```
DO 5 I=-5,5
    IF(A(I).EQ.VAL) J=J+1
```

10.
```
   AMAX=0.0
   DO 8 I=1,10
       DO 7 J=1,10
           IF(ABS(A(I,J)).GT.AMAX) AMAX=A(I,J)
     7     CONTINUE
   8 CONTINUE
```

12.   DO 9 K=1,J
           IF (A(I,K).LT.A(I,K+1)) THEN
               T=A(I,K)
               A(I,K)=A(I,K+1)
               A(I,K+1)=T

14.   PROGRAM SORT
           REAL A(20)
           READ *,A
           DO 1 K=19,1,-1
               DO 2 J=1,K
                   IF (A(I).LT.A(I+1)) THEN
                       T=A(I)
                       A(I)=A(I+1)
                       A(I+1)=T
                   END IF
       2   CONTINUE
       1   CONTINUE
           .
           .
           .

16.   PROGRAM TRANS
           REAL A(4,4)
           READ *,((A(I,J),J=1,4),I=1,4)
           DO 4 I=1,4
               DO 5 J=1,4
                   T=A(I,J)
                   A(I,J)=A(J,I)
                   A(J,I)=T
       5   CONTINUE
       4   CONTINUE
           .
           .
           .

CHAPTER 7

1. **(a)** V(R,H)=3.14*R*R*H/3.0
        (c) DB(V2,V1)=20.0*ALOG10(V2/V1)

2. **(b)** Y=P2(P1,T2,T1)
        (c) Y=DB(V2,.5*V2)

3. **(a)** -2.6
        (c) -4.269231

4. (a) ‑0.25000   0.50000

6.
```
      SUBROUTINE TRANS(A,B,N,M)
      REAL A(N,M),B(M,N)
      DO 7 I=1,M
          DO 8 J=1,N
              B(I,J)=A(J,I)
    8         CONTINUE
    7 CONTINUE
      RETURN
      END
```

CHAPTER 8

1. An apostrophe can be represented within a string by a double apostrophe.

3. A character constant's string length is the number of characters appearing between the apostrophes, including blanks. Double apostrophes are counted as one character.

5. If the string is too long, the extra characters are truncated on the right.

7. If I is missing, it is assumed to have the value 1; if J is missing, it is assumed to be the last position of the declared string length.

9. No.

11. When comparisons are made between numbers and letters or special symbols, or letters and special symbols, no standard collating sequence exists, and the results become computer dependent.

13. If the character variable length is larger than the string, the string will be left-justified when entered, and blanks will be added to the right of the string. If the string is larger than the character variable's length, it will be truncated on the right.

15. If w is less than the length of the character variable, blanks are added to the right of the data string to fill the word, that is, the character variable is left-justified. If w is greater than the character variable's length, the word is truncated on the left-hand side, that is, only the rightmost positions are kept.

17. In evaluating a complex logical expression, all parenthetical expressions are treated first. Arithmetic expressions are evaluated within the parentheses, and then the truth value of relational expressions is determined. Logical operators are applied last, from left to right, and in the following order:

.NOT.
.AND.
.OR.
.EQV. and .NEQV.

19. List-directed output is displayed simply as a "T" or an "F."

21. On formatted input, the first nonblank character entered must be a period or a "T" or an "F." Periods are optional and are ignored. Any trailing letters appearing in the field are also disregarded.

23. If a single-precision constant is assigned to a double-precision variable, only twenty-four bits of significant (mantissa) information are present, and the second part of the storage word (the extra thirty-two bits) is filled automatically with zeros. A double-precision constant, on the other hand, has full fifty-six-bit representation to begin with (that is, its last thirty-two bits are nonzero) and therefore can represent a decimal value more accurately.

25. When a complex value is assigned to a real or an integer variable, the imaginary part of the number is discarded.

26. (a) WIND   (c) _ENERGY____

27. (a) WIND'S   (b) WIND

28. (a) FIFT   (b) TWO_

29. (a) FALSE   (c) FALSE

30. (a) WAVE POWER   (b) WAVE LENGTH

31. (a) DYNAMICS   (b) DYNAM

32. (a) 21   (c) 5

33. (a) FALSE   (c) TRUE

34. (a) __.1612747D+03   (b) ___T

35. (a) 1.01.0   **(c)** 3.0 +13.0

36. (a) 5 - i2   **(c)** 30 - i10   **(e)** 5   **(g)** 5e$^{i \cdot 9723}$

CHAPTER 9

2. To access a record in a sequential file, all previous records must first be opened, in consecutive order. A direct-access file permits addressing an individual record directly.

4. An integer unit code number is a code number for the specific input/output device that is to access a referenced file.

6. Formatted files are written using alphanumeric characters; unformatted files are written using the machine's internal binary code.

8. Both parameters provide for program recovery from a file-control error. The IOSTAT parameter contains the system error code identifying the type of failure that occurred. The ERR parameter contains the statement number of the next executable statement in the event of error.

10. Each record of a direct-access file must have the same record length, *LREC*. For formatted files this record length is measured by the number of characters present (per record). When a direct-access file is opened, the record-length parameter must be given. It must be large enough to accommodate the data fields specified by a referring FORMAT statement.

12. Since records in a sequential file are not numbered, a record field itself is used instead for identification purposes. This field is called a key field.

CHAPTER 10

1. (a) __.1235E-01   **(c)** __1.235____   **(e)** __123.5____
   (g) __.1235E+05

2. (a) __.123450E-01   **(c)** __1.23450____   **(e)** __123.450____
   (g) __12345.0____

3. (a) _____.004   **(c)** _____.414   **(e)** ____41.356

4. (a) __.004E+02   **(c)** __.414E+00   **(e)** _41.36E-02

5. (a) __.004136D+02   (c) __.413562D+00   (e) _41.35618D-02

6. 3, plus

7. (a) 413.5   (c) 4.13   (e) .4135

8. (a) X=4.135E−10 K=300
   (c) X=.4135 K=300

Index

Adams